The Afterlife of Race

The Afterlife of Race

An Informed Philosophical Search

LIONEL K. McPHERSON

OXFORD
UNIVERSITY PRESS

OXFORD
UNIVERSITY PRESS

Oxford University Press is a department of the University of Oxford. It furthers
the University's objective of excellence in research, scholarship, and education
by publishing worldwide. Oxford is a registered trade mark of Oxford University
Press in the UK and certain other countries.

Published in the United States of America by Oxford University Press
198 Madison Avenue, New York, NY 10016, United States of America.

CIP data is on file at the Library of Congress

ISBN 978–0–19–762684–9

DOI: 10.1093/oso/9780197626849.001.0001

Printed by Sheridan Books, Inc., United States of America

To my parents, Aretha C. and Lionel B. McPherson, for so much

Contents

PART III. NON-EXCLUSIONARY BLACK (AMERICAN) SOLIDARITY

Acknowledgments

scholars, scientists, journalists, and writers whose work I was fortunate to read

students in my "race"-related philosophy courses at Tufts through the years

editor Peter Ohlin of Oxford University Press, for unwavering interest and faith

longtime colleagues, present and former, in the Department of Philosophy at Tufts University: Jody Azzouni; Nancy Bauer; Avner Baz; Norman Daniels; David Denby; Daniel Dennett; Brian Epstein; Patrick Forber; Ray Jackendoff; Erin Kelly; Monica Kim; Kathrin Koslicki; Jeffrey McConnell; Dilip Ninan; Christiana Olfert; Mark Richard; Susan Russinoff; George Smith; Sigrun Svavarsdottir; Stephen White

special contributors or supporters: Lucy Allais and the University of the Witswatersrand; Lydia Amir; Albert Atkin, Adam Hochman, and Macquarie University; Avner Baz; Gabriel Benaim; Jinann Bitar; Benjamin Brent; Hannah Carrillo; Mario De Caro; Elizabeth Branch Dyson; Saskia Eschenbacher; John Heil; Evan Linn; Mara Marin; Lionel B. McPherson; Lucius Outlaw Jr.; Irene Stapleford; Caroline Streeter; Fabienne Toback; Jeremy Toback; Jonathan Wilson and the Center for the Humanities at Tufts; Nathan Witkin; Naomi Zack; each of the peer reviewers

longtime fellow members of the Cambridge Race and Philosophy Reading Group: Lawrence Blum; Jorge Garcia; Sally Haslanger; Adam Hosein; José Mendoza; Ifeanyi Menkiti; Megan Mitchell; Tommie Shelby

cherished interlocutors: James Bland; Michael Klein; Renu Nahata; Orlando Patterson; Tommie Shelby; George Smith

philosophical mentor always, Richard Moran

life partner and philosophical companion, Erin Kelly

our extraordinary daughters, Iman and Rayha

Reader's guide

In every section of the book, a sentence in bold print indicates the section's main point.

***

An abridged early draft of Part I was published as "Deflating 'Race'" in the *Journal of the American Philosophical Association* 1, no. 4 (2015).

Some passages in Part III borrow from "Black American Social Identity and Its Blackness," published in *The Oxford Handbook of Philosophy and Race*, ed. Naomi Zack (OUP, 2017).

PART I

RACIAL FAITHS, ABSURD PURPOSES

"When *I* use a word," Humpty Dumpty said in a rather scornful tone, "it means just what I choose it to mean—neither more nor less."
"The question is," said Alice, "whether you *can* make words mean so many different things."
"The question is," said Humpty Dumpty, "which is to be master—that's all."
—Lewis Carroll, *Through the Looking-Glass* (1871)

The fear that in some way or other a social equality between the races shall be enforced by law or brought about by political measures really has no foundation except in the imagination of those who fear such a result.
—William Howard Taft, President-elect of the United States "The South and the National Government" (1908)

Accurate scholarship and free, dedicated artists would reveal a singularly important thing: that racism . . . was and is a monumental fraud. Racism was never, ever the issue. Profit and money always was. . . . The very serious function of racism . . . is distraction.
—Toni Morrison, "A Humanist View" (1975)*

1

A Socratic Device

We draw strong connections between "race," whatever the word is supposed to mean, and skin color. In this first part of the book, I probe how the idea of human races has gone from a flimsy notion about physical differences between people by continent, to a weighty notion about possible mind/brain differences between those people, to a slippery notion about some *X factor* for grouping people by (sub)continental ancestry profile. No theory or school drives my discussion. Special training is not needed for entry. Race inquiry only gives the appearance of deep dispute, which I try to uncomplicate through analysis and evidence.

My philosophical method steers away from high abstraction by enlisting support across disciplines to gauge the race concept's shapeshifting persistence. The record shows that we lack a stable, informative conception: different speakers make "race" mean different things, with no end in sight. The unruliness prompts me to take a nonpartisan stance toward "race" theory. Trying to win control over the word is worse than wasteful, despite the standard assumption that there does or should exist some distinctive thing we mean. I do not claim, though, that "race" must have no meaning(s).

Part I presents a comprehensive argument for a more productive mode of continental color consciousness. The United States is my primary, not my exclusive, concern. Social plus political lineage proves to be more important than any real thing race could be. This is subconscious knowledge the Western race concept has obscured—mainly in service of moral *pseudo-rationalizations* for Western slavery and colonialism as well as their active legacies of social inequality.[1]

Color-conscious propaganda in the United States shied away from crass bigotry after the 1960s. "Civil rights" progress spurred White America to strive for public "colorblindness."[2] Good citizens now seem motivated more by jealous selfishness than anti-black animus; liberals treat disparities in well-being by free or slave lineage as regrettable, if impractical to half repair.[3] "Race," as Americans evasively say, remains a factor. The race concept's

The Afterlife of Race. Lionel K. McPherson, Oxford University Press. © Lionel K. McPherson 2024.
DOI: 10.1093/oso/9780197626849.003.0001

lifeblood is speculation and counter-speculation about important natural differences between reputed major races.

In early days of Western race intrigue, ideology and rhetoric became hard to disentangle: thus *the Western race ideology-rhetoric complex*, the materials supplier for color-conscious propaganda. There was never anything other than absurdity in the thought that "racial" difference (revealed by skin color) could be relevant to moral justification or excuse for slavery. Try telling a story to rationalize owning other human beings and their future children as property, with attendant horrors, from generation to generation, that does not sound unhinged.[4] Do try.

Self-liberated abolitionist and philosopher of freedom Frederick Douglass succinctly dismissed the moral possibility of inherited slavery. In a speech known by his question, "What, to the American slave, is your 4th of July?," he used national Independence Day as a metaphor for (un)freedom itself:

> But, I submit, where all is plain there is nothing to be argued. . . . On what branch of the subject do the people of this country need light? Must I undertake to prove that the slave is a man? That point is conceded already. . . . The slaveholders themselves acknowledge it in the enactment of laws for their government. They acknowledge it when they punish disobedience on the part of the slave. . . . What is this but the acknowledgment that the slave is a moral, intellectual and responsible being. . . .
>
>   . . . Would you have me argue that man is entitled to liberty? That he is the rightful owner of his own body? You have already declared it. Must I argue the wrongfulness of slavery? . . . To do so, would be to make myself ridiculous, and to offer an insult to your understanding.[5]

Philosopher Immanuel Kant anticipated Douglass's point about "the enactment of laws" to guide "a moral, intellectual and responsible being." Slave codes could communicate to the enslaved themselves—and in America forbade education along with everything else conducive to the slave's humanity.[6]

Application of law to the enslaved was proof by itself of law's recognition that they had rational power to exercise self-governance. There lies Kant's plain intellectual dishonesty as an apologist for slavery, in real time, under the guise that "black" Africans are mentally disabled.[7] Generally, the idea of continental human races sponsored extreme pretexts for injustice against non-European peoples. Enslaved Africa-identified Americans were born into legal nonpersonhood and visceral terror under those pretexts for two hundred years.[8] Civil war, not moral reason, brought that to an end, ushering in one hundred years of violent segregation.

The organizing thesis of Part I is this: *Western commitments to exploiting non-European peoples and resources principally explain race intrigue, not the other way around.* Of Western cultural psychology, anthropologist Franz Boas wrote, "The superior power that the European owes to his inventions and that enables him to subject and exploit foreign peoples . . . gives emphasis to the feeling of superiority."[9] Such feelings per se will be a minor theme in my philosophical narrative. Their translation into race sentiment, on my interpretation, expresses attachment to social values and priorities that are already in effect. Western race ideology is a peculiar tribal branch of imperialist political ideology, normalized by a denialist cultural psychology. In America, a settler Anglo-Saxon "white" identity eventually allowed its incorporation into a pan-European "white" identity, in absolute contrast with African-origin blackness of American slave lineage.[10]

By "racism," a term I will use sparingly, I mean personal or institutional attitudes and behaviors supportive or expressive of color-conscious social dominance. Racism in the final analysis answers to political power and influence, not to kinder attitudes or greater human understanding. Although American antiblackness is especially discouraging, grassroots activism does offer glimmers of hope.[11] The moral arc of the United States could bend at last from dreary progress toward corrective justice.

In a 1908 address delivered in New York to "the South," US president-elect William H. Taft assured his nation that "equality of opportunity before the law," White America's law, would continue to foster inequality "between the black and white races."[12] The so-called problem of race was politics of American slavery, segregation, and nonrepair.[13] Race hype, with an implacable science, has provided ideological-rhetorical structure and cover for these politics.[14] In American context, "race" and cognates generally perform as euphemisms that signal social hierarchy revolving around (or in comparison with) "black" social disadvantage.[15] "Racial" segregation, for example, immediately brings to mind Black American exclusion from equal society or isolation in urban ghettos, while White self-segregation seems normal and reasonable to insiders.[16]

"It is sufficient to say, as I think we all now realize, that the institution of slavery was a bad thing," Taft told his people—as if their realization took serious effort requiring no further thought of centuries-long atrocity. He offered exculpatory empathy to White America: "Had the Northerners been [as] interested in slaves, they would have viewed the institution exactly as the

Southerners viewed it and would have fought to defend it because [it was] as sacred as the institution of private property itself."[17] The "bad" Anglo-American institution of inherited slavery belonged to the "sacred" Anglo-American institution of "private property," ruled by tribal Anglo-American law. All manner of protest from the enslaved property did not count: those people were outside the president-elect's moral community.[18] Taft personified the White establishment—moving from chief of the executive branch of the United States, to a professorship at Yale Law School, to appointment in 1921 as chief of the Supreme Court. He is a beacon for the post-1960s "conservative" movement: "an extraordinary man by any standard," said Justice Antonin Scalia in 1988, and "one of the greatest Chief Justices."[19]

The Supreme Court had set forth the doublespeak doctrine of "separate but equal" in *Plessy v. Ferguson* (1896). Yes, the 14th Amendment gave citizenship rights to Black Americans,

> but, in the nature of things, it could not have been intended to abolish distinctions based upon color, or to enforce social, as distinguished from political, equality.[20]

Barely interested in details of racial nature, White America's elite court made explicit that segregation after the abolition of slavery would conserve a social hierarchy "based upon color", in the name of the Constitution's framers.[21] Formal equality under White law did not happen for Black America until the mid-1960s, a hundred years after the Civil War—courtesy of the legislative and executive branches of federal government, not the judicial. Yet legal chicanery has remained the coin of the country's colored realm. Taft would be proud.

Social and political demarcation by color fueled race ideology's broader function, which is to instill distraction from unconscionable social practices and arrangements. The details of racial distraction are forever fungible. Novelist and critic Toni Morrison sums up the pseudo-rationalization climate: "Somebody says you have no language and so you spend twenty years proving that you do. Somebody says your head isn't shaped properly so you have scientists working on the fact that it is. Somebody says that you have no art so you dredge that up. . . . None of that is necessary. There will always be one more thing."[22] Some racial thing has ultimately been irrelevant to slavery, segregation, and a nation's steadfast refusal to make amends for morally inconceivable gross injustice.

The race concept might do nothing special other than sustain the Western race ideology-rhetoric complex, whether for reactionary or liberal purposes. Race inquiry was more unified about its motivations when surrounded by naked commitment to social dominance.[23] Nevertheless, countries that combined colonialism and slavery—as the United States did before turning to "legal" segregation, then mass incarceration—would be among the worst social environments in the world to study natural differences of racial mind.[24] Sincere observers will grasp as much, without race intrigue.

The word "*varna*" is Sanskrit for "color." In the 1830s monograph *Caste*, William Lloyd Garrison's American Anti-Slavery Society compared the local color-conscious hierarchy to the Hindu *varna* system:

> Their degradation is generally inhuman. [The Sudras or lowest caste] are compelled to work for the Brahmins, being considered as created solely for their use. They are not allowed to collect property, "because such a spectacle would give pain to the Brahmins." . . . How striking the resemblance between this and *American slavery*! Who does not recognise the same feeling and principle which creates the barrier between the whites and blacks in this country?
>
> . . . Surely this spirit of caste, lurking among our free institutions, like the devil in paradise, is the offspring of slavery, and lives in no country apart from its parent abominations.[25]

Visible African ancestry was the telltale sign of "blacks" under American slavery. Western race ideology was invoked by "whites" after the fact, to deflect from the evils of an exploitation system they consciously built through terror, law, and complicity. As abolitionist William Goodell reported, "Slaves Can Possess Nothing."[26]

By 2019, social progress enabled a nonwhite future Vice President, Kamala Harris, to confront Black American voters with a liberal axiom: "So I'm not gonna sit here and say I'm gonna do something that's only gonna benefit Black people. No."[27] Descendants of American slavery would be due nothing special and tangible. Ever, regardless.

A century earlier, social and literary critic H. L. Mencken explicitly defended the American caste hierarchy:

> The distinction I have described is the product, not so much of varying environment as of inborn differences. [I]t might be possible to make an

appreciable improvement in the stock of the American negro, for example, but I must maintain that this enterprise would be a ridiculous waste of energy, for there is a high-caste white stock ready to hand, and it is inconceivable that the negro stock, however carefully it might be nurtured, could ever even remotely approach it. The educated negro of today is a failure . . . because he is a negro. His brain is not fitted for the higher forms of mental effort. . . . He is, in brief, a low-caste man, to the manner born.[28]

For Mencken, African "inherited marks" would indicate a people naturally beneath "the superior white race." Americans of European stock were to biologize their anti-black casteism: "Castes are not made by man, but by nature," he declared.[29] "Racial" caste rhetoric suited both inherited slavery and enforced segregation. Since Mencken was an atheist ("Religion is fundamentally opposed to everything I hold in veneration . . . above all, love of the truth"),[30] his advocacy of white caste supremacy in the United States cannot be blamed on an anti-black deity.

Black was the figurative color of enslaved Americans. The physical appearance or inner biology of individuals of that known heritage was of no legal consequence.[31] What the US government described by census as "color"— then "color or race" (1900), and at present "race" alone—foundationally tracks Europe-identified freedom and Africa-identified slavery.[32] For most of American history, until the 1950s, there was little confusion about this. Anglo-America's legal model for reproducing a slave caste (1662) preceded the idea of continental races (1684).[33] The doctrine that established inherited slavery, *partus sequitur ventrem*, was modified from English livestock law to apply to the "condition of the mother" as either free or enslaved.[34] Commitment to American slavery, or to its accommodation, guaranteed "colored" unfreedom.

Color caste ideologies strongly recall the "myth of the metals" in Plato's *Republic*. "When you hear it, you'll realize that I have every reason to hesitate," Socrates said. "Speak, and don't be afraid," Glaucon urged. "I'll tell it, then," Socrates replied, "though I don't know where I'll get the audacity or even what words I'll use" for an old "Phoenician story which describes something that has happened in many places":

We'll say to them in telling our story, "[T]he god who made you mixed some gold into those who are adequately equipped to rule, because they are most valuable. He put silver in those who are auxiliaries ['guardians'] and iron and bronze in the farmers and other craftsmen. . . . So the first and most important command from the god to the rulers is that there is nothing that they must guard better or watch more carefully than the mixture of metals in the souls of the next generation. If an offspring of theirs should be found to have a mixture of iron or bronze, they must . . . give him the rank appropriate to his nature and drive him out to join the craftsmen and farmers. . . ." So, do you have any device that will make our citizens believe this story?

"I can't see any way to make them believe it themselves," Glaucon answered, "but perhaps there is one in the case of their sons and later generations and all the other people who come after them."[35] The myth would be governed by hypodescent, the custom of assigning children of parents of different inherited social rank the status of the parent of lower rank. This would create an enduring coincidence between social standing and, among citizens in ancient Athens, the "noble falsehood" that there exist heritable metallic divisions of human beings.

The Western race concept became the color-conscious "device" that conveyed a slavery variation of the Socratic metals myth. But there has been a shadow tradition that directly decodes race as caste—without extravagant myth, theory, or science. A "wayfaring fugitive from American slavery or American caste" is how New York slavery-born physician James McCune Smith described fellow abolitionist Frederick Douglass sheltering in the British Isles.[36] Swedish social economist Gunnar Myrdal, in his famed 1944 study *An American Dilemma*, used "the term 'caste' to denote the social status difference between Negroes and whites in America"—a difference that was "fixed and rigid."[37] Narrative journalist Isabel Wilkerson brought the shadow tradition to mainstream awareness with her 2020 book *Caste: The Origins of Our Discontents*; the title echoes historian, sociologist, and activist W. E. B. Du Bois's 1904 speech "Caste in America: That Is the Root of the Trouble."[38] At America's color-conscious heart, antiblackness (aka anti-black "racism") is casteism that originates with American slavery.

In India, polymath scholar and statesman B. R. Ambedkar emphasized race in his landmark 1936 speech "Annihilation of Caste": "Some have dug a biological trench in defense of the Caste System. It is said that the object of Caste was to preserve purity of race and purity of blood."[39] Ambedkar envisioned political solidarity between Black Americans and the Dalit ("untouchables") undercaste of his birth. He sought and found an ally in Du

Bois.[40] The Hindu caste system was a racial system, Ambedkar argued. Caste ideology and race ideology had become symbiotic.

Buddhist philosophy scholars G. P. Malalasekera and K. N. Jayatilleke "stress the close analogy between the inequalities created by the caste system" and Western race systems. "The resemblance is particularly striking," they observe, circa 1958, "when it comes to the behaviour of those who claim superiority on the strength of membership of a privileged caste, the colour of their skin, or even the type of their hair."[41] Philosopher Akeel Bilgrami describes the "social psychology" of caste as consisting of "an exclusionary attitude": each subordinate *varna* group "was to be excluded" from a higher group's "way of life," using "the most brutal physical and psychological violence."[42] So too the color-conscious race device is viciously ideological in the United States. The word "race" has named a weird public rationale for gross injustice that includes its perpetual nonrepair.

"Geoancestry" is a term I have coined to displace "race" and redirect attention toward the core of Western color consciousness. In a slogan, continental ancestry in combination with social-political lineage matters, not race. Philosopher and activist Cornel West, author of *Race Matters*, would agree: his frame of reference is Black Americans as a homegrown national people.[43] They are, to be clear, Africa-identified descendants of American slavery that was foundational to an emerging nation.

The concept of geoancestry reflects Western practices of attaching importance to visible continental ancestry. I will argue that this color consciousness has been boosted by but is not hostage to the modern idea of race as continental groups of people. Race theorists doubt that debate over the reality of race has reached diminishing returns.[44] As an analyst of the afterlife of race, I prefer to accept a stalemate than join an unwinnable contest. The geoancestry concept is a demystifying alternative to the race concept.[45]

Capital-"B" Black Americans are my working example. They distinctively exist, regardless of the nature of something that might be called "race." The typical Black American is estimated to have around 25 percent non-African ancestry.[46] Around thirty million Africa-identified Americans identify as a (nonimmigrant) national people. A 2019 public opinion survey attests to their reality: "More than eight-in-ten black adults say the legacy of slavery affects the position of black people in America today, including 59% who say [the legacy] affects it a great deal."[47] Descendants of American slavery constitute around 80 percent of "black" raced Americans.[48]

Africa is the most human genetically diverse continent.[49] A basic empirical-conceptual question is this: In which respects might Africans or blacks/Negroes be distinctly alike "racially," apart from the looks or biology of skin color? A 2020 genetics study found that for persons in the Americas who have greater than 5 percent "sub-Saharan African ancestry," the average component was highest in the British Caribbean (~76 percent) and the United States (~71 percent). The broad "sub-Saharan" label is misleading:

> There are four local ancestries corresponding to regions in Atlantic Africa: Nigerian (Nigeria), Senegambian (Gambia, Guinea, Guinea-Bissau, Senegal), Coastal West African (Sierra Leone, Ghana, Côte d'Ivoire, Liberia), and Congolese (Angola, Democratic Republic of the Congo). [T]he majority of individuals within the United States (93%) . . . tend to have ancestry from all four of these Atlantic African populations.[50]

Simply put: "blacks" in the New World are of substantially less than full African descent, and their African ancestors come from a specific western chunk of a vast continent.

What kinds of generalizations could be made about "black people" based on an Atlantic Africa subgroup? How could Black Americans—despite also having substantial European ancestry—belong to a single black/Negro race? Could physical appearance of any African ancestry amount, or the genetics of greater than 5 percent "sub-Saharan African ancestry" (the study's threshold), suffice for membership? Why, anyway, care about a continental "race" thing?

The idea of continental human races stays searching for clarity amid a haze of preconceptions. Globally and often locally, we don't know what we mean in using "race." Does it merely refer to continental ancestry and physical appearance? Certain genes traced to certain geographic regions? Color-conscious historical and social processes? Social identities shaped by having enough associated continental ancestry? (Sub)continent-wide biological or cultural family relations? Stories told by certain European peoples to distinguish themselves from an "of color" world? Some additional or other things?

How to determine the best methods for answering such questions is itself a challenge. Race inquiry will have to seek conceptual clarification and guidance. Philosophical metaphysics, however, reaches for abstraction that bears a questionable relation to "race" in social practice. Theorizing the existence of "race" groups—when we will not agree even on the type of natural or social phenomenon the "race" thing is supposed to be—seems to me a lost and superfluous cause.

At least we should be wary of race science's fixation on categories from the eighteenth century. Grouping people by alleged "Caucasoid," "Negroid," or "Mongoloid" descent type (the terms remain in technical use) precedes the formation of biological science.[51] When modern biology did arrive, it borrowed the old way the groupings were made—through casual judgments about physical appearance as a manifestation of relevant continental (or subcontinental) ancestry. Race science has offered thin and circular findings to support further "natural race" research. Meanwhile, optimistic "social race" partisans suggest that educated people hardly believe in natural races anymore.[52] Either way, race is supposed to be real, in different senses of reality.

This is an occasion for Humpty Dumpty's insight that a word's meaning will have to depend on who, if anyone, gains control over it. Under conditions of inexhaustible disagreement, we may lose faith that trying to capture the meaning of "race" isn't a wild-goose chase. In turn, we might disengage from the abstract question of whether races exist: different speakers can mean different things, and some of those things may be real in some sense. With consensus never nearby, partisans could abandon their overheated word and go their separate semantic ways, each cadre stating concisely what they mean by "race" and giving their thing a new name. We have no master theory, principle, or practice to govern the word nor even to provide nontrivial ground rules for engagement and dispute about the things named by it.

Constant battle over a best use for "race" serves slippery purposes after the decline of faith in essential differences between continental groupings of people. Nonetheless, my nonpartisan approach to the word may seem misguided to quite a few readers. Why shouldn't science keep trying to uncover the facts about race? Am I pushing a public culture of "colorblindness"? What about righting racial injustice, or studying race for biomedical research? Hasn't "race" become part of everyday language?

While the race concept is past its dubious prime, race intrigue continues to exert influence in public life and scholarly work. I agree with theorists who argue that race is more plausibly conceived as some social, not natural, thing. Yet I advise adopting a less divisive concept that can do any positive work the race concept can do and more, by doing less, with lower risk of mass confusion. My analysis of race inquiry draws on historical, social scientific, natural scientific, and journalistic evidence. No stable, informative understanding of race seems feasible.

Race theorists, natural and social, underappreciate that the modern idea of race involves (sub)continental ancestry plus an elusive X factor: this formula for misdirection became the conceptual innovation. Race cannot be seen simply as physical features that indicate various thresholds of continental ancestry.[53] That would leave little oxygen for "race" mystery. Nor can belief in natural races be explained away as a natural psychology of tribal behavior. No anxious debate would then arise over the reality of race as compared, say, to the psychology of a dogged faith that some color-conscious people have in their own natural superiority.

Continental (skin) color became a divine and a scientistic sign of innate mental differences. **The race concept's main source of intrigue concerns whether nature deeply divides people by continent, as Western race ideology claimed on behalf of the "white" European.** Race ideology, not any continental "race" thing itself, has the enduring power of an illusion.

2

Grappling with the Race Concept

The modern idea of race is strangely difficult to describe. My attempt tries for fairness: "race" would refer to one of a few discrete human groups that formed when geographic collections of people became biologically distinct, with each group forming on separate landmasses later regarded as continents. Skin color and other physical features that indicate continental ancestry would be the permanent clue for suspecting that these groups might naturally differ in mentality—a suspicion that might seem unpromising to pursue for three hundred years and counting.

The discipline of geography was central to race. Geographer Martin Lewis and historian Kären Wigen observe that "the sevenfold continental system of American elementary school geography did not emerge in final form" until the mid-twentieth century, without new knowledge.[1] Antarctica and six humanly populated regions—Africa, Asia, Australia/Oceania, Europe, North America, and South America—became "the continents," each being "one of Earth's seven main divisions of land" (per National Geographic).[2] Western geographic convention has judged the Caucasus and Ural Mountains a natural barrier dividing the (super)continent of Eurasia into Europe and Asia. There is no geological basis for this continental division. But the modern idea of race was committed to an image of geographically separated and isolated peoples as distinct biological groups.[3] This requires ignoring, for example, that Silk Road trade routes (c. 130 BCE–1453 CE) linked Europe, Asia, and Africa.

Political scientist Walker Connor emphasizes that "'the cradle of Western civilization' actually consisted of the intertwined civilizations of Asia and Africa, as well as Europe."[4] Their connectedness aligns with the seventeenth-century perspective of English historian Peter Heylyn: "A Continent is a great quantity of Land, not separated by any Sea from the rest of the World; as the whole Continent of *Europe, Asia, Africa*."[5] The geographic unity of these regions was suggested by ancient Greek historian Herodotus: "For my part I cannot conceive why three names . . . should ever have been given to a tract which is in reality one, nor why the Egyptian Nile and the Colchian Phasis

The Afterlife of Race. Lionel K. McPherson, Oxford University Press. © Lionel K. McPherson 2024.
DOI: 10.1093/oso/9780197626849.003.0002

[river] . . . should have been fixed upon for the boundary lines."[6] Radical geo-graphic and cultural separation of European, Asian, and African pre-modern peoples is a myth.

Some race theorists have responded to nonbelief in natural races by trying to remove nature from the race concept: they argue that race is socially constructed and not a biological thing. This conceptual reform mission is more ambitious than redefining the idea of meat, for example. Some vegans claim that "meat" no longer refers only to animal flesh but also to plant-based substitutes intended to mimic the taste and texture of animal flesh. Omnivores reply that vegan "meat" isn't really meat. A core understanding of meat as animal flesh is not really in dispute; at issue is using the word "meat" to include plant-based substitutes said to be meat-like.[7] By contrast, there is endless debate about whether and how race is real in the first place. We have no core understanding of race—except in the narrow sense that human beings descended in full or part from respective continents tend to look dif-ferent in characteristic ways.

Suspicions that we might be unequally divided by race—in the broad sense that people would naturally and importantly differ by continent—do not presume that each race must come close to purity. Western conventional wisdom has it that non-European racial types are better able to incorporate persons who have ancestors from more than one continent (or raced sub-continent). Barack Obama, with a father of non-northern African descent and a mother of European descent, is widely recognized as "the first black president" of the United States and could not be "white": "the black race" X factor is supposed to work that way.[8] (I introduce "non-northern" to replace "sub-Saharan," for reasons that become clear in discussion of ancient Egypt.) Despite public confidence about such judgments, a shared understanding of race eludes us. Race calculus gets more arbitrary or involuted with more investigation.

Typically, we follow "know it when we see it" guidelines for ascribing a color or race identity, also known as "racial identity," to individuals. This customary method can be adequate for everyday purposes. To modify US Supreme Court Justice Potter Stewart's statement on the nature of pornog-raphy: "I shall not today attempt further to define the kinds of [human being] I understand to be embraced within that shorthand description ['race'], and perhaps I could never succeed in intelligibly doing so. But I know it when I see it."[9] The trouble comes from heavy confusion about what the racial *it* truly is and why. Purposes for race classification have gone far beyond conveying visible continental ancestry.

We are more or less where we were in the 1940s, when anthropologist Ashley Montagu flagged the unhelpfulness of semantic gamesmanship: "In biology a race is defined as a subdivision of a species which inherits physical characteristics distinguishing it from other populations of the species. In this sense there are a number of human races. But this is not the sense in which most of the older and many of the modern physical anthropologists, race classifiers, and racists have used the term."[10] Montagu was ahead of his time and ours. The latitude for putting "race" to varied uses has enabled equivocation, nonsense, and dishonesty.

Consider recurrent fascination with whether Beethoven was black. A liberal-minded commentator claims to have the answer: "He surely wasn't, but some insist otherwise.... He was simply swarthy." The verdict comes with scant guidance for knowing an individual's race. Two musicologists, we are told, refuted the rumor of the composer being "genealogically African and thus black."[11] But how much African ancestry might suffice for being black and not merely sort of black-looking? Rather than going down that road, the commentator would have us preempt "the myth of black Beethoven." We are to reject an "either/or [social] construction of race along a strict black/white binary": if the composer was not strictly white, he was "mixed-race or interracial," not black.[12] Of course, the rumor wasn't that Beethoven had nearly full African ancestry. Much less has been required for being racially black: witness lighter-skinned Americans of mixed African and European descent—Dorothy Dandridge, Malcolm X, Stephen Curry, and countless others—whose categorical (if not social) blackness routinely goes unchallenged.[13]

Raced confusion has a habit of intruding. I was in the comments of a philosophy blog post when mention of the profession's complacent whiteness triggered one professor to defend possible "IQ differences between races." In his opinion, "all that's necessary [is] for there to be genetic differences, on average, between the populations of people we categorize in different races, whatever the basis of those categorizations." He cited "hair and skin color variations" and "variation in sickle cell anemia incidence" as evidence of racial differences.[14] The alleged evidence, however, implies circular criteria for distinguishing "racial" populations: who belongs to which race would already have to be known in order to study natural differences between natural races. Persons can indeed be grouped by some or other natural properties, and the groupings can then be compared and measured. But credible science usually proceeds from a hypothesis, which in this case would concern the criteria for racial groupings to begin with and the research rationales.[15]

More is needed to sustain race inquiry than "white people have some European ancestry and look white," "black people have some African ancestry and look black," etc.[16] Otherwise, race would merely be continental ancestry that manifests as certain types of physical appearance. Researchers might later discover that those physical types correlate with certain genetic phenomena, which by itself does not invite wondering whether some reputed major race is smarter than another.[17] If credible science hopes to designate and compare natural races, customary "know it when we see it" guidelines could not possibly do.

Africa-identified persons are an obvious test case. Are "blacks" of the racially "native" non-northern African variety to be grouped with whatever-the-basis blacks/Negroes of the American variety, ignoring that the latter also have substantial European ancestry? (By substantial, I'll roughly mean any component, at a minimum, equivalent to one great-grandparent.) Must the African component be visible for persons to be racially black? How about persons in Latin America who are not categorized there as black despite having visible African ancestry? Is there a biological reality to being black, apart from the genetics of physical features variously seen to "look black" in different societies?

I am not questioning whether there are any characteristic natural differences between people categorized by something called "race." I doubt that stereotypes about brain differences between purported racial peoples would be confirmed.[18] Still, I do not assume that race science is only in the business of lending support to rhetoric about racial superiority, a tradition that also treated different settled peoples of Europe as different races.[19] Science generally hasn't prided itself on going down littered rabbit holes.

At an early stage of writing this book, I presented a lecture in Rome. The university audience struggled to process the notion of important differences in natural intelligence by race, and I was met by disbelief that the topic is a live issue.[20] In the United States, I tried to explain, conjecture persists that "blacks" are naturally less intelligent and more violent than "whites." Philosopher Sebastiano Maffettone politely clarified that inquiry into racial mind seems weird to educated Italians, who learn under the influence of German philosopher G. W. F. Hegel: they believe that history and culture are responsible for nonphysical differences between geographic descent groups. I will return to that Hegelian irony.

In the rest of Part I, I set forth the development of racial thinking, then address race ideology, race science, race as a social construct, and race skepticism. Philosophers, natural scientists, and social scientists have been arguing for centuries about the reality and importance of race. The debate is fueled by refusal to believe that it cannot be won. **We collectively do not know what "race" thing we are talking about, and neither the public at large nor specialists will agree on what *it* is supposed to be.** A more helpful mode of continental color consciousness is overdue.

Please note my own usage of continental color labels: "black" will simply mean Africa-identified (given non-northern African ancestry); "white" will simply mean Europe-identified (given European ancestry); "yellow" will simply mean Asia-identified (given Asian ancestry).[21] My hope is to discourage the presumption that individuals grouped under a certain label share some distinctive race quality, natural or social, that importantly makes them members of a same people.

3

Some "Race" Things

My approach to continental color consciousness might spark unease. As long as individuals are identified through labels such as "black," "white," or "yellow," aren't we committed to the modern idea of race? No, we are not. If nothing else, continental color labels can signal visible African, European, or Asian ancestry.[1] We also can refer to social phenomena such as "racial" discrimination and "racial" injustice without assuming that races must exist somehow. Intelligible use of the adjective does not require a stable underlying meaning for "race"—though the etymological dependence may recommend displacing "racial" as well.

Continental color labels are not globally confined to standard application. Many Australian Aboriginals, for example, affirm a "global black" identity (inspired by Black American boxing icon Jack Johnson, who won against an Anglo-Canadian world boxing champion in Sydney in 1908) as a people marked not by racial negritude but by color-conscious oppression they associate with Black American liberatory resistance.[2] Australian Aboriginal "black" identity is a nonracial, aesthetic-political mode of blackness articulated in the settler colonial context of "White Australia" policy (1901–1973) and ideology.[3] (White/European Australia today has the moral decency to publicly acknowledge by custom its war of conquest.)[4] Color labels for social groups can be put to alternative use.

The standard global sense of "black" social identity refers to Africa-identified persons. In the United States, "Black" Americans could more readily be recognized as a distinctive national people if widespread faith in natural race did not persist. Subjugating this Africa-identified people—through force of violence and law—has been a pillar of White American nationhood. Antebellum "agriculturists" proudly harnessed "the cheapest labor in the world."[5] The moral of such history escaped future First Lady Hillary Clinton: "I was told that using prison labor at the [Arkansas] governor's mansion was a long-standing tradition, which kept down costs"; she "agreed to abide by tradition" and, with future President Bill Clinton, "became friendly with a [few] African-American men."[6] The Anglo-American tradition of

The Afterlife of Race. Lionel K. McPherson, Oxford University Press. © Lionel K. McPherson 2024.
DOI: 10.1093/oso/9780197626849.003.0003

extracting value from descendants of American slavery also encompasses human experimentation, college sports, and municipal funding schemes, among many other aspects of public life.[7]

Virtually all Africa-identified persons in the United States before the Civil War had roots in American slavery. There were virtually no African or Caribbean immigrants prior to the twentieth century. By the 1960s, "the foreign share of all U.S. blacks" was around 1 percent. Laws severely restricting "black immigration" were eased in the 1980s, despite the country's unceasing hostility toward homegrown Africa-identified people.[8] Descendants of American slavery—not "the black race" or "black people" in general—are the fundamental reference of American antiblackness.

Sociologist Ira Reid wryly summed up, in 1938, the US method for ascribing racial blackness: "'Africans, Black' is the cover-all term employed by the Department of Labor's Bureau of Immigration and Naturalization to cloak with racial identification all persons of Negro extraction. . . . ['Black'] is a conjure-word that metamorphoses persons who, prior to embarking for the United States, may have been known as 'coloured,' 'mulatto,' or 'black,' some having nationality identification without benefit of race."[9] The US policy was less a statement of any science than a political reconstruction of blackness—which has clouded the justice claims of homegrown "blacks."

US immigration policy turned more welcoming to foreign "blacks" after the Black (American) rights movement of the 1950s and 1960s.[10] This has fostered a general public sentiment that "black identity" in America is now "elusive."[11] The feeling is one-sided: recognizing that Black Americans are a non-native, nonimmigrant group is not about "dividing" Africa-identified persons, leaving those of immigrant background "in limbo," or dictating who "counts as black."[12] Such charges presuppose the primacy of alleged racial sameness across different national peoples. In the historical sense, Black Americans are descendants of American slavery; this feature of our social identity is not subject to negotiation with people who do not share the social-political lineage.[13] National origin is also pertinent to understanding which "blacks" are owed reparations by the US government for intergenerational gross injustice tied to inherited slavery and enforced segregation.

The formal practice of not distinguishing different Africa-identified peoples is a deliberate policy choice in the United States. Informal distinctions are made and have a tendency to disfavor Black Americans.[14] Immigrant "blacks" of Haitian, Nigerian, Ethiopian, etc. heritage have social freedom to affirm with pride their national background as a distinctive feature of their social identity. (A recent headline, for example: "For Some Children of Immigrants, 'African American' Doesn't Fit Their Unique, Black Experiences in the US.")[15] This contrasts with the civic habit of discouraging a national consciousness among Black Americans in their homeland as descendants of American slavery.[16] Frederick Douglass, Ida B. Wells, W. E. B. Du Bois, Richard Wright, Billie Holiday, John Coltrane, Aretha Franklin, Toni Morrison, and so on are of Black American social-political lineage—not of nationless, generic racial stock.[17] No one is simply American or "just black."

A published anecdote about Ghanaian diplomat and former United Nations secretary-general Kofi Annan illustrates Black American specificity. As a college student in Minnesota, he joined a road trip through the American heartland in 1961: "Kofi needed a haircut. The guy said, 'We don't cut niggers' hair,' and he said, 'I'm not a nigger, I'm an African,' and the guy said, 'That's O.K., come on, siddown'—and cut his hair."[18] Per American race ideology-rhetoric, all niggers are "black" but not all "blacks" are niggers. Visitors to the United States soon learn that the "niggers" were specifically Black Americans, an Africa-identified people comprising descendants of American slavery.[19] Despite public pretense to the contrary, the N-word is passed down in White America as social reaffirmation of the bottom caste status of homegrown "Slaves"/"black" people.

Labeling persons by color or continent hides an odd racial condition. Unmoored from its modern inception, "race" has been in search of an intelligible idea to govern. Believers keep moving the goalposts with every attempt to win the meaning game; nonbelievers compare the idea of race to occult "witches" and the fire element "phlogiston." No other word attracts more controversy about what type of earthly thing we are trying to refer to. The conceptual fault line lies between continental groupings that supposedly constitute natural "race" kinds and social groupings constituted in uncertain terms of continental ancestry and physical appearance.

Disagreement has splintered into competing efforts to impose a best biological or social account of the "major races" phenomenon. Consensus about a conceptual basis is nowhere on the horizon. Various meanings for "race" may be made to fit some things or others that are naturally or socially real about continental groupings of people. **This recommends a stance I call *deflationary "race" pluralism*: races might intelligibly be said to exist in various senses, or not, depending on what different speakers believe the "race" thing is.** The semantically messy situation is not surprising. On arrival, the modern idea of race had to discount continental mixtures produced wholesale via European transatlantic slavery.[20] That was in addition to seventeenth-century unawareness of major human migrations from Africa to India and Europe many thousands of years earlier.[21]

Different notions of race may be inferred from different designs that yield various continental groupings. This suggests a moderate version of Alice's paradox of meaning: words may be made to mean many different things, hence meaning no particular thing that is stable and informative. Humpty Dumpty's self-willed solution admits too many dissenting possibilities for "race." Reminiscent of the famously simple plan for getting the United States out of an unwinnable war in Vietnam, partisans could declare victory and go home—satisfied enough that "no potential enemy is in a position to establish its authority" over what "race" means.[22]

A deflationary "race" pluralist like myself thus abstains from claiming that races do not exist. While less tolerant theorists would eliminate the "race" thing from our empirical-conceptual understanding of reality, I prefer to avoid metaphysical drama.[23] No understanding of race is needed for knowing that peoples from Europe, Africa, and Asia tend to look different in characteristic ways and have different national and global histories. The geoancestry concept captures such information with greater clarity and less baggage; there is no elusive X factor to contend with. Terminologists might insist that "race" can be made to mean geoancestry: hypothetically, wide agreement could form around the old word's now-specified meaning. Then there is reality: of various candidate meanings, none—including what I mean by "geoancestry"—seems bound to gain a consensus following under the "race" banner and thereby become master of the word.

Most believers today might reject the notion that every member of each racial group has some quality distinctive to their group, a race essence. This has not dispelled the illusory promise of natural kinds; race science refurbishes support for roughly the same eighteenth-century continental groupings of people.[24] There remains a global abundance of awkward cases due to indeterminate racial geography (e.g., "the Middle East"), anomalous racial territory (e.g., northern Africa), or rampant racial hybridity (e.g., Afro-European "blacks" in the New World). Race classifiers grasped such pitfalls early: the elastic, borderless terms "Caucasoid," "Negroid," and "Mongoloid" came to replace (or sometimes mix with) the straight continental designations "European," "African," and "Asian."

Here is an excerpt from the 2008 "Human Races" entry in a global publisher's *Encyclopedia of Genetics, Genomics, Proteomics and*

Informatics: "The main human races are Caucasoid, Mongoloids (including Chinese, Japanese, Koreans, and American Indians, etc.), and Negroid. Khoisanoids or Capoids (Bushmen and Hottentots) and Pacific races (Australian aborigines, Polynesians, Melanesians, and Indonesians) may also be distinguished." An assurance follows: "There is no genetic incompatibility among the various human races and there is no well-founded scientific evidence that interracial marriage would lead to the disruption of co-adapted gene blocks resulting in biological or mental deterioration in the offspring."[25] The case for racial eugenics is evidently not closed. Nor is there much concern over how to properly determine race membership.

Customary vagueness is on full display when persons who have mixed continental ancestry are racially grouped by the component deemed most relevant at the local time. Take the case of Sally Hemings, Thomas Jefferson's slave and young concubine through whom he fathered a number of children.[26] The quick answer that those children were racially mixed, not racially black—because they had substantial non-African ancestry—would apply to typical Black Americans. That mixed Afro-European solution conflicts with local thought and practice: visible evidence of lineage in the slave caste has been the regular if not "the proper way to define a Negro" in the United States.[27] Inherited slavery and enforced segregation did not require a more exacting account of black racial being.

Science has never led the way in designating natural races. As we will see, race science enters the fray after the race concept has been deployed for political purposes. One such purpose was to legally govern whether the Jefferson-Hemings children, for instance, could be enslaved by their own father; they could and they were. Their fate was a byproduct of the American Founders' commitment to protect the enslavement of Africa-identified persons for an indefinite future. My point here is to illustrate the social background against which race inquirers have loose rein to attach their own sensibilities to "race." The word is too messy to be dominated by a master meaning.

In effect, raced confusion helps enable ideological-rhetorical pretexts that suffuse gross injustice with mysteries of nature and fate. Motivation for race inquiry was easy to discern before the political foundation splintered. On the old story, skin color and other physical features indicate continental ancestry that is the source of a race quality purportedly relevant to which human beings have claims to equal humanity. That ideology is not openly championed in mainstream circles anymore. The "race" thing, with various replacement stories and never an emergent consensus, is only harder to pin down. Contemporary race inquiry has been in a slow-motion stalemate over what we're supposed to be talking about.

Geoancestral thinking offers a feasible alternative. There is already movement in this direction. For instance, my physician advised early screening for a disease that "men who have African ancestry" are at higher risk of getting. His phrase was clearer than the notion that a person who "looks black" and identifies as Black American is racially black. I don't know how to think I might naturally or socially be a member of "the black [sub-Saharan African] race" or any other continental race. I also don't know what others think they mean by "race," beyond some facts of (sub)continental ancestry. My physician easily communicated his message through direct reference to my visible African ancestry. In doing so, he did not imply any racial statement concerning European ancestry that I, like Black Americans typically, also have.

Color-conscious reality is far less about science, metaphysics, or civility than about social group relations of power and vulnerability.[28] This would remain the case if the race concept were to fade away after its dismal struggle for credibility. Two hundred years of academic investigation and theory have yet to deliver a stable, informative idea of race as continental groups. Modern racial thinking is ultimately about the intersection of history, subjugation, exploitation, and nonrepair—mainly surrounding European transatlantic slavery and colonialism, with their enduring national and global legacies. "Racial" beliefs have been a distraction from gross injustice.

4

(Sub)continent-Wide Human Types?

François Bernier, a French physician and traveler writing in the late seventeenth century, has been described as the first to use the word "race" ("*espèces ou races d'hommes*") in "its modern sense to refer to discrete human groups organized on the basis of skin color and other physical attributes."[1] He did not propose that these groups could be characterized by differences of mind. While he had much to say about cosmopolitan diversity in female beauty, his classification concept did not imply the notion of a natural hierarchy of the world's people as grouped by continental ancestry and appearance.[2]

Races as Bernier saw them in 1684 were slightly more than skin deep: "Geographers up to this time have only divided the earth according to its different countries or regions. [My travels] have given me the idea of dividing it in a different way. [T]here are four or five species or races of men in particular whose [exterior] difference is so remarkable that it may be properly made use of as the foundation for a new division of the earth." He suggested no crucial application for that division. "Africans," like other "species or races," were marked by skin color: "The blackness which is peculiar to them [is] not caused by the sun. . . . The cause must be sought for in the peculiar texture of their bodies, or in the seed, or in the blood."[3] Bernier's classification of people by "race," as he termed it, represented little more than continental descent indicated by skin color and other physical features.

The idea of race became less innocuous a century later, at the height of the Enlightenment. In the 1770s, German philosopher Immanuel Kant added some rigor to "race" thought: "Negroes and whites are clearly not different species of human beings [but] they do comprise two different races. This is because each of them perpetuate themselves in all regions of the earth [regardless of environment]." His "Hun" and "Hindu" types rounded out a tentative classification of four races. Africa's native inhabitants were the Negroes, except for "Moors (Mauritanians from Africa)," who were whites. "The reason for assuming that Negroes and whites are the base races is self-evident," Kant asserted, despite believing early that there is a single line of

The Afterlife of Race. Lionel K. McPherson, Oxford University Press. © Lionel K. McPherson 2024.
DOI: 10.1093/oso/9780197626849.003.0004

human descent.[4] He could have noticed that his notion of many "varieties" under four general "races" under two "base races" was hopelessly unstable.

Kant instead pursued truths that would reveal more than the cause of certain patterns of physical difference: "The Negro [is] well-suited to his climate, namely, strong, fleshy, and agile. However, because he is so amply supplied by his motherland, he is also lazy, indolent, and dawdling." Thus began the tradition of biologizing anti-black/African stereotypes. "My opinions in these matters are only preliminary," Kant allowed, "and I offer them only for the purpose of stimulating further investigation."[5] His previous opinions were cruder: "The Negroes of Africa have by nature no feeling that rises above the ridiculous. . . . So essential is the difference between [black and white] human kinds, and it seems to be just as great with regard to the capacities of mind as it is with respect to color."[6] Eventually, he decided that slavery violates a cosmopolitan right covering all human beings.[7]

Thomas Jefferson emphasized skin itself in his 1785 apologia for chattel slavery: "I advance it therefore as a suspicion only, that the blacks . . . are inferior to the whites in the endowments of both body and mind. . . . The unfortunate difference of color, and perhaps of faculty, is a powerful obstacle to the emancipation of these people."[8] In our time, the nonprofit corporation that owns his Monticello estate extols "the author of the Declaration of Independence," acknowledges that he "enslaved more than 600 people over the course of his life" (some of his own children and their mother included), and embraces moral euphemism: "To Jefferson, it was anti-democratic and contrary to the principles of the American Revolution for the federal government to enact abolition or for only a few planters to free their slaves."[9]

In other words, "the blacks" in America were not to have an "inalienable" right to freedom. They could be turned into personal property by "the whites" through violence and law, to be possessed and used in any manner without restraint. America's Founders were determined to protect the property interests of slaveholders as a central feature of American democracy and a key to the nation's wealth.[10] Jefferson lent a veneer of caution to the civic doctrine that some human beings may justly be kidnapped into or born into slavery . . . because of some elusive factor indicated by an "unfortunate difference" of skin color. Even he was too judicious to speculate further.

When the Civil War dawned seventy-five years later, Northern proslavery activist John H. Van Evrie threw rhetorical caution to the wind: "God has made the negro an inferior being, not in most cases, but all cases. . . . The same Almighty Creator has also made all white men equal." Van Evrie's epigraph for *Negroes and Negro "Slavery"* (1861), his tome on the "great 'Anti-Slavery' delusion that originated with European monarchists," borrows the "suspicion" passage from Jefferson's *Notes on the State of Virginia*.[11] Post-slavery Jeffersonianism was enacted by Woodrow Wilson, who as president of the United States screened D. W. Griffith's racist film classic *The Birth of a Nation* (1915) at the White House.[12] In 2022, the Senate Republican leader, who hails from a southern slaveholding family, let slip a public distinction between "African American voters" and "Americans."[13]

The story of Anglo-America's antiblackness is morally uncomplicated and savage. Africa-identified descendants of American slavery—a conspicuous out-group that could be mined for absolute exploitable value as manual and sexual labor—were chained by the Founders outside the reach of minimal human decency. There are no sober conflicts of principle, conscience, or science to morally contextualize in that intergenerational atrocity. Historians Sven Beckert and Seth Rockman sum up the business climate: "Antebellum America privileged 'the rule of law,' an ideological support for the status quo, property rights, and the legitimacy of federal and state legislation upholding slavery."[14] Anglo-American capitalism, slavery, and a public-private culture of terror were inextricable.

Tradition guided President Abraham Lincoln in signing the "District of Columbia Compensated Emancipation Act" of 1862: "The law provided for immediate emancipation, compensation to loyal Unionist masters of up to $300 for each freed slave, voluntary colonization of former slaves to colonies outside the United States, and payments of up to $100 to each person choosing emigration. Over the next nine months the federal government granted almost $1 million for the freedom of approximately 3,100 former slaves."[15] To translate: Washington, DC, slaveholders were paid money for their loss of property. The human property never received compensation and neither have their Black American descendants.

Lincoln stated his core "racial" values during his 1858 Illinois campaign for the US senate:

> I will say then that I am not, nor ever have been, in favor of bringing about in any way the social and political equality of the white and black races—that I am not nor ever have been in favor of making voters or jurors of negroes, nor of qualifying them to hold office, nor to intermarry with white people; and I will say in addition to this that there is a physical difference between the white and black races which I believe will forever forbid the two races

living together on terms of social and political equality. And inasmuch as they cannot so live, while they do remain together there must be the position of superior and inferior, and I as much as any other man am in favor of having the superior position assigned to the white race.[16]

The American problem of race was slavery and its inevitable aftermath. White America was "a house divided against itself" by law protecting human property ownership. After abolition, Black America would still comprise a bottom caste—though Lincoln added, "I do not perceive that because the white man is to have the superior position the negro should be denied every thing."[17] Freedom from enslavement was going to be better than nothing besides noblesse oblige.

Fast-forward to the twenty-first century. Harvard University cognitive psychologist Steven Pinker touts the "dangerous idea" that "groups of people may differ genetically in their average talents and temperaments." He highlights a biologist who rejects the liberal wisdom that "race does not exist" and a social scientist who infers that "average racial differences in intelligence are intractable and partly genetic."[18] To Pinker, resistance to these possibilities is obstinate egalitarian ideology. Critics are "unwilling to grasp" that innate group differences "pertain to the average or variance of a statistical distribution," not to individuals: "Large swaths of the intellectual landscape have been reengineered to try to rule out these hypotheses a priori (race does not exist . . .)," at a time when "genetics and genomics will soon enable us to test hypotheses about group differences rigorously."[19] Science could finally demonstrate that "blacks" are naturally less intelligent than "whites." Confirming the biological reality of race would be the first step, though Pinker mentions nothing about how genetics and genomics might accomplish that.

The history of modern science and medicine is strewn with racial quackery. In the 1850s, American physician Samuel Cartwright hypothesized two "maladies of the negro race." *Drapetomania*—the term combined Greek words for "a runaway slave" and "mad or crazy"—was the disease of "absconding from service." The afflicted were "sulky and dissatisfied without cause." But "with proper medical advice, strictly followed, this troublesome practice that many negroes have of running away, can be almost entirely prevented." Popular opinion near free-state borders favored "whipping them out of it," whereas medicine prescribed compassion: "If treated kindly [and] not overworked or exposed too much to the weather, they are very easily governed. . . . When all this is done, if any one or more of them . . . raise their heads to a level with their master or overseer, humanity and their own good require that they should be punished until they fall into that submissive state which it was intended for them to occupy."[20]

A second malady required less violent remedy. *Dysaesthesia aetheopica* (aka "rascality") caused "a partial insensibility of the skin, and so great a hebetude of the intellectual faculties, as to be like a person half asleep" (acute laziness). From their "careless movements," the afflicted were "apt to do much mischief, which appears as if intentional, but is mostly owing to the stupidness of mind and insensibility of the nerves induced by the disease." The condition was "easily curable, if treated on sound physiological principles": after the blood's "atmospherization by exercise . . . the negro seems to be awakened to a new existence, and to look grateful and thankful to the white man whose compulsory power . . . has restored his sensation." Hence "he is no longer the *bipedum nequissimus*, or arrant rascal, he was supposed to be, but a good negro."[21] Medical diagnosis of the enslaved Negro's noncompliant tendencies was taken seriously enough, as propaganda, to waste antebellum time and energy debunking.

Race science advocates are still telling us there might be natural differences, expressed as functional differences, between people assigned to different race categories. That is not news regarding certain "talents." We have understood that average differences in physique can help explain average performance differences in short-distance running, long-distance swimming, and so forth. Reportedly, for instance, Polynesians are predisposed to be "heavier, have more muscular limbs and smaller proportions of body fat and can produce 'greater force in explosive movements' than [persons] of other ethnicities" (who play rugby).[22] The fact that heritable group differences of body type can affect group performance in different physical endeavors is uncontroversial.

Efforts to study possible racial brain differences, by comparison, are detached from any robust explanation for corresponding performance differences on tests said to measure natural intelligence.[23] There is also never much explanation about who would truly belong to which natural race. Self-identification by customary "race or ancestry" labels has become the norm for demographic queries.[24] Individuals typically identify as "white," "black," "Asian," etc. because they are socially seen and labeled as such where they're from—not because anyone could know their natural race by underspecified criteria.

Pinker's complaint about "a priori" resistance to research into IQ differences between races is embarrassing.[25] Racial mind/brain has been researched since the late eighteenth century, by every available (alleged) technique.[26] Why continue to fixate on the prospect that a social group dehumanized and marginalized for hundreds of years is less intelligent by nature?[27] In any event, how would science rebut the view that major races do not exist? We already understand that certain physical features roughly indicate continental ancestry, for which there are genetic correlations. If race amounted to differences of narrow consequence, there wouldn't be strident debate. Race science has welcomed the open question of whether more is at stake.

Paleontologist Stephen Jay Gould suggests a leading motivation for the search for something more to race. "Biological determinism," as he skeptically describes it, "holds that shared behavioral norms, and the social and economic differences between human groups—primarily races, classes, and sexes—arise from inherited, inborn distinctions and that society, in this sense, is an accurate reflection of biology."[28] Intervening to offset racial biology could be inefficient or futile. Swiss-born American naturalist Louis Agassiz, who founded Harvard's Museum of Comparative Zoology in 1859, pleaded for scientific objectivity in assigning Negroes to a lowest human subspecies that threatened European natural superiority.[29] As science would have it, he should have been stressed about Neanderthals, not Negroes: genomics establishes that his racial interbreeding nightmare was misplaced.[30] (I will elaborate in Part I, Section 9, "Renewed Race Science.")

A century before Agassiz, Scottish philosopher David Hume articulated anti-black/African contempt through the method of armchair empiricism, in a footnote: "I am apt to suspect the negroes, and in general all the other species of men (for there are four or five different kinds) to be naturally inferior to the whites. There never was a civiliz'd nation of any other complexion than white, nor even any individual eminent either in action or speculation."[31] He later revised the passage to specify only "the negroes to be naturally inferior to the whites."[32] These days, defenders of the philosopher's honor resort to claims like "Hume's discussion of human species must be recognized as scholarly in its motivation" and "of Hume's attitudes toward Blacks it must be noted that he opposed slavery."[33]

Amplifying Hume's race sentiment, Kant wrote of "a Negro" who dared object to a white man's reproach: "There might be something here worth considering, except for the fact that this scoundrel was completely black from head to foot, a distinct proof that what he said was stupid."[34] The anecdote

cannot be swept aside as a statement of underinformed or irrational prejudice. **Africa-identified skin color became racial color and prima facie evidence of disabled racial mentality, invoked to contrive doubt about an
otherwise universal human capacity for reason, autonomy, and freedom.**
Kant professed that "black" Africans were fit for enslavement by "white"
Europeans.[35] Near the end of his life, he reconciled disrespect for the Negro
mind with strong criticism of slavery.

Needless to say, Kant did not believe his racist caricature: it serves the function of distracting from, and obscuring in a haze of color-conscious absurdity,
the obvious fact that "Negroes of Africa" reasonably objected to their enslavement.[36] Europeans also encountered moral and social resistance from their
colonized subjects everywhere.[37] The color-conscious dishonesty of a great
philosopher is not driven by unfortunate inner biases or complex systems of
thought: racist nonsense is driven by brute politics of advantage for peoples
whose lives and well-being he prioritizes beyond the limits of conscience.[38]

Hume was only more explicit about connecting blackness to pseudo-
rationalizations for categorical wrongs practiced by "the whites" of "civilized"
nations against outsiders:

> Not to mention our colonies, there are NEGROE slaves dispersed all over
> EUROPE, of which none ever discovered any symptoms of ingenuity.... In
> JAMAICA, indeed, they talk of one negroe as a man of parts and learning;
> but 'tis likely he is admired for very slender accomplishments, like a parrot,
> who speaks a few words plainly.[39]

Unbridled loyalty to European empire's exploitation of Africa-identified peoples would explain such patently false and unbelievable claims. Hume's "parrot"
was, in fact, Francis Williams (c. 1702–1770), the black Jamaican scholar
and writer whose "extraordinariness was indisputable," writes a twenty-first-
century historian.[40] While a great philosopher's anti-black animus might be
of some biographical interest, pertinent worldly politics will shed brighter
light on his conception of morality and value.[41] In brief, canonical Western
philosophers dissembled in ideological defense of active European supremacy.

Echoes of Hume and Kant can be heard in the US Supreme Court's *Dred
Scott v. Sandford* decision (1857), which excluded from citizenship all Black
Americans, enslaved or free. Chief Justice Roger Taney wrote the majority
opinion that laid out a constitutionally valid rationale for the American caste
hierarchy:

The question is simply this: Can a negro, whose ancestors were imported into this country, and sold as slaves, become a member of the political community formed and brought into existence by the Constitution of the United States, and as such become entitled to all the rights, and privileges, and immunities, guarantied by that instrument to the citizen?

. . .

[T]he legislation and histories of the times, and the language used in the Declaration of Independence, show that neither the class of persons who had been imported as slaves, nor their descendants, whether they had become free or not, were then acknowledged as a part of the people, nor intended to be included.

. . .

They had for more than a century before been regarded as beings of an inferior order . . . ; and so far inferior, that they had no rights which the white man was bound to respect; and that the negro might justly and lawfully be reduced to slavery. . . . This opinion was at that time fixed and universal in the civilized portion of the white race. [M]en in every grade and position in society daily and habitually acted upon it in their private pursuits, as well as in matters of public concern.[42]

The *Dred Scott* Court accurately interpreted the nation's founding documents.[43] A "fixed" white/European "opinion" sufficed to legalize atrocity.

By whichever theory of American legal interpretation, constitutional reverence has been more important than correcting gross injustice.[44] Nazi Germany was impressed, as legal scholar James Q. Whitman examines in *Hitler's American Model*.[45] Historian Ira Katznelson highlights a disturbing reality: "during the 1933–45 period of the Third Reich, roughly half of the Democratic Party's members in Congress represented Jim Crow states, and neither major party sought to curtail the race laws so admired by German lawyers and judges."[46] American apartheid did not come to a formal end until the late 1960s. Segregation scholar Richard Rothstein describes the impact of the Fair Housing Act of 1968: "It was not primarily discrimination (although this still contributed) that kept African Americans out of most white suburbs after the law was passed. It was primarily unaffordability. . . . The advantage that the FHA and VA loans gave the white lower-middle class in the 1940s and '50s has become permanent."[47]

The Constitution's 14th Amendment (1868) had granted citizenship to "all persons born or naturalized in the United States." In 1896, the Supreme

Court responded with the *Plessy v. Ferguson* decision that approved "equal, but separate, accommodations for the white and colored races"—a legal fiction for purposes of continuing to subjugate the former "Slaves" caste.[48] In *Dred Scott* forty years earlier, the Court had explained why the Declaration of Independence, despite words that "would seem to embrace the whole human family," was not an exercise in hypocrisy:

> Yet the men who framed this declaration were great men . . . incapable of asserting principles inconsistent with those on which they were acting. They perfectly understood the meaning of the language they used, and how it would be understood by others. . . . [N]o one misunderstood them. The unhappy black race were separated from the white by indelible marks . . . and were never thought of or spoken of except as property.[49]

To repeat, that decision specifically targeted "the class of persons who had been imported as slaves, [and] their descendants." Legal reaffirmation of white/European caste supremacy did justice to the American Founders' integrity and conviction. "No one misunderstood them," the Supreme Court said.

In a "civilized" country, people of European nature could proudly own or exclude Africa-identified human beings. *Dred Scott* licensed "white" entitlement to "daily and habitually" violate "black" adults and children; "private pursuits" was a grotesque euphemism for extralegal brutality and coercive sex.[50] There were no other privacy issues of legal note concerning human property.[51] White America's elite court protected, without limit, tribal tyranny and perversion. The institution of inherited slavery was gilded by pseudo-rationalizations to feel tolerable enough to the pan-European collective, who were "forever" to occupy "the superior position," as Lincoln put it.[52] Perpetual nonrepair of the "unhappy black" caste is an extension of that malign political tradition.

Belief in a biological reality of race is surely not the culprit. What if racial differences amounted only to characteristics of narrow consequence? Skin pigmentation (amounts and kinds of melanin), sickle cell disease (a rare genetic blood disorder), cystic fibrosis (a rarer genetic lung disorder), aldehyde dehydrogenase 2 deficiency (aka "Asian flush syndrome") . . . ?[53] Nothing of profound interest would depend on race. Natural differences of racial mind, however, could be of broad consequence.

Conjecture about racial mind is what drove the scientistic ideology known as scientific racialism: the human species would fall into a few distinct biological groups, as subspecies/races, based on distinctive heritable brain traits tied to geographic descent. In turn, race was claimed to be relevant to the merits of slavery, colonialism, segregation, etc. This has normalized color-conscious oppression that Westerners can culturally bracket as perhaps not utterly wrong in hindsight. Otherwise, apologetics for European supremacy icons ("But the idea that they should be judged entirely by today's, justifiably higher standards is too harsh"[54]) would sound absurd, whether regarding the common sense and experience of enslaved persons themselves or Kant's own "categorical imperative" variation of the Golden Rule.

The race concept enables all sorts of intrigue for intellectual distraction. Explicit or implicit "racism as an idea" has followed suit, as Toni Morrison elaborates: "It really is the red flag that the toreador dances before the head of a bull. Its purpose is only to distract; to keep the bull's mind away from his power and his energy; to keep his mind focused on anything but his own business."[55] There was nothing remotely serious for Kant, Hume, Jefferson, etc. to think about how some continental "race" thing could be relevant to morally justifying inherited slavery or settler colonialism. Nor were great Western minds seriously trying yet incredibly failing at race inquiry.[56] Getting to any truth about the nature of human beings was an optional purpose.

Late in the twentieth century, the American Association of Physical Anthropologists tried to skirt the issue of racialism. "Pure races," the AAPA's 1996 "Statement on Biological Aspects of Race" declared, "do not exist in the human species today, nor is there any evidence that they have ever existed in the past."[57] That left slight room for the possibility there had been pure races. In 2019, the AAPA withdrew that possibility: "Race does not provide an accurate representation of human biological variation. . . . Humans are not divided biologically into distinct continental types or racial genetic clusters."[58] Physical anthropology used to be the discipline most dedicated to espousing the existence of major races.

5

Framing a Racialist Placeholder

Racialists seem to imagine there is some natural quality, simple or compound, that itself determines an individual's race. Without such a quality, how to know the true membership of any reputed major race is murky: persons who do not have all of a group's defining traits could not be objectively identifiable as members. Contemporary race science is quiet or evasive about this puzzle of racial being. On a standard account of natural race, blacks/Negroes need not have full, nearly full, or mostly non-northern African ancestry. By virtue of what could these persons be racially black—unless there are heritable traits they distinctively have in common?

I return to Kant, often considered the greatest modern Western philosopher. He was more than a man of his time: the idea of continental human races led him to a groundbreaking theory of essential difference. If Africans and Europeans had roughly the same human potential, African ways of life would not be radically different from those of Europeans.[1] Culture reflects racial nature, erasing a blurry distinction between biological and cultural racism.[2] By that line of thinking, "black" skin would mark a group of humans whose gross natural mental inferiority could be inferred from their cultural inferiority to a group of humans with "white" skin.

Kant's racialism was repudiated by his former student, German philosopher Johann Gottfried von Herder: "Some have [called] four or five divisions among humans, which were originally constructed according to regions or even to colors, *races*; I see no reason for this name.... [T]here are neither four nor five races, nor are there exclusive varieties on earth.... [Race] belongs less to the systematic history of nature than to the physical-geographic history of humanity."[3] Herder was not confused in referring to a "race" thing while questioning "race" thought. Rather, his point was that race did not objectively amount to much other than physical features characteristic of continental ancestry. The idea of human races was redundant and misleading. Montagu would make a similar observation 150 years later.

Some prominent eighteenth-century scientists also took no interest in broad notions of race. Johann Friedrich Blumenbach, German founder of

The Afterlife of Race. Lionel K. McPherson, Oxford University Press. © Lionel K. McPherson 2024.
DOI: 10.1093/oso/9780197626849.003.0005

physical anthropology, shunned "race" altogether. In 1775, he distinguished four "varieties" of people by continent:

- Europe, plus "the Russian Far East."
- Africa, in its entirety.
- Asia, plus Australian territories.
- The Americas, minus Newfoundland.[4]

By 1795, with scant explanation, he revised the categories to "five principal varieties" and gave them technical names:

- *"Caucasian,"* for peoples of Europe plus "Northern Africa."
- *"Ethiopian"/"Negro,"* for peoples of non-northern Africa.
- *"Mongolian,"* for peoples of Asia.
- *"American,"* for peoples of the Americas.
- *"Malay"* (a new category), for peoples of Australia, New Zealand, and the Pacific Islands.

These were "arbitrary kinds of divisions" because "innumerable varieties of mankind run into one another by insensible degrees."[5] Blumenbach was not conceiving of natural races. He found "no single character so peculiar and so universal" among "black" Africans that isn't seen "everywhere in other varieties of men [and] that many Negroes are seen to be without."[6] His classification scheme was nevertheless expropriated by racialists and lives on in race science.[7]

Although Blumenbach declined to theorize natural mental differences between his varieties of people, he did believe falsely that "white [was] the primitive colour of mankind." He also proposed an aesthetic standard that awarded "first place to the Caucasian" (from the Caucasus Mountains region) for beauty.[8] Apart from that, he defended a narrow notion of what others called "race." Literary scholar Peter Kitson notes that Blumenbach was "a firm opponent of slavery and the slave trade" and that his *Contributions to Natural History* (1806) "contains a sustained plea for the full humanity and equality of the negro and enumerates many instances of the scientific and literary excellence of the achievements of black people."[9]

Scientific antiracialism failed to convince Hegel. In principle, he affirmed social equality: human beings are "implicitly rational" is how he cast "the possibility of equal justice for all men and the futility of a rigid distinction between races which have rights and those which have none."[10] Yet "the mental and spiritual characteristics" of Africa-identified peoples called into question their claim to humanity:

> Negroes are to be regarded as a race of children. . . . They are sold, and let themselves be sold, without any reflection on the rights or wrongs. . . . [T]heir mentality is quite dormant, remaining sunk within itself and making no progress, and thus corresponding to the compact, differenceless mass of the African continent. . . .
>
> It is in the Caucasian race that mind first attains to absolute unity with itself. Here for the first time mind enters into complete opposition to the life of Nature . . . achieves *self*-determination, *self*-development, and in doing so creates world-history.[11]

Hegel would have us believe that the natural environment caused separate paths of human potential. There would be natural mental differences between race groups, not simply differences of skin, history, and culture.[12] His fantastical biologizing of race gave rhetorical cover for Europeans to pursue global exploitation of non-European peoples through sheer violence.[13] Most miserably, Africans suffered from "natural difference" obscurely "connected with" their continent.[14] Hegel was wrong on the geography and the biology: Africa is vast, not "compact," and has greater human genetic diversity than the other continents.[15] He preferred to theorize that Negroes, on their own, did not have rights as full persons.

When English naturalist Charles Darwin joined the fray in the 1870s, Western consensus held that continental races of people exist. He agreed that "the various races" are "very distinct; chiefly in their emotional, but partly in their intellectual faculties." He disagreed that races are distinct natural kinds: who belongs to which race was "much influenced [by] the mere color of the skin and hair"; there was also "the greatest possible" difference of opinion about whether human beings "should be classed as a single species or race, or as two . . . five . . . twenty-two . . . or as sixty-three."[16] Race classification could not be a straightforward representation of nature. Darwin deduced that races, whatever their designations, had "similar" natural "mental powers."[17] Different racial mentalities would mainly result from differences in cultural background.

While Darwin did not rule out natural mental differences between race groups, he drew a broader conclusion: "The variability of all the characteristic differences between the races . . . indicates that these differences cannot be of much importance; [else] they would long ago have been either fixed and preserved, or eliminated."[18] There might be mental differences above the minimum level of fitness for any group's survival. But even when lapsing into racist rhetoric, Darwin steered in a cultural direction that contrasted "civilised races" and "savage races."[19] He believed that natural races do not exist, so neither would differences of racial mind as a powerful force of nature.

Indeed, Darwin's theory of evolution by natural selection is a poor fit for conjecture about racial mind. The fact that nature selects for darker skin closer to the equator seems understandable before learning the details. (Dark skin protects against UV radiation that destroys folate/B vitamin, and light skin benefits vitamin D production.)[20] Far less understandable is why evolution would select for differences in natural intelligence by continent or any other sizable, environmentally diverse area. Lack of effort to study natural selection pressures on racial mentality is consistent with prejudices that predate the modern idea of race as continental groups. Blackness has remained shrouded in a pre-Enlightenment worldview replete with curses and monsters of color, a theme I revisit in "Renewed Race Science" (Part I, Section 9).

Racialism never went away, as a mainstream entry from 2014 demonstrates: "Analysis of genomes from around the world establishes that there is a biological basis for race. . . . [W]ith mixed race populations, such as African Americans, geneticists can now [analyze] an individual's genome, and assign each segment to an African or European ancestor, an exercise that would be impossible if race did not have some basis in biological reality."[21] We are to bring our intuitions about what, besides the genetics of continental ancestry, the racial X factor is. Thus a former *New York Times* science writer would have us believe in major races because "mixed race" individuals can be, as we already knew, biologically traced to ancestors from different geographic regions.

After bluffing that genetic confirmation of "mixed" continental descent is specially relevant, the science writer questions whether racial differences of mind "solely" reflect culture: "If that's so, why is it apparently so hard for tribal societies like Iraq or Afghanistan to change their culture and operate like modern states? [A] genetic system, based on the hormone oxytocin, seems to modulate the degree of in-group trust, and this is one way that natural selection could ratchet the degree of tribal behavior." Likewise, skin color might have evolved in tandem with different "selected brain genes" because of "different challenges on each continent," though those genes are "not yet understood." We are told, "The three principal races are Africans (those who live south of the Sahara), East Asians (Chinese, Japanese, and Koreans), and Caucasians (Europeans and the peoples of the Near East and the Indian subcontinent)."[22] Being of Afro-European descent, Black Americans would not be true "Africans." Iraqis, as racial natives of "the Near East," would be "Caucasians."

What most needs explaining, of course, is whether humans evolved into a few distinct groups with significantly different mental natures. As Kant and Hegel realized, that possibility would have to depend on the natural environment, not a preexisting race quality. This secular inference led to floundering opinion about the effects of heat and humidity on heritable mind: "Climate, therefore, has a fixed and absolute control over the existence of the negro. God has adapted him, both in his physical and mental structure, to the tropics."[23] Modern humans would be critically different by racial nature of "the Caucasian," "the Mongolian," and "the Negro," in ways informally apparent yet elusive to verify or even offer a passable theory for. Ultraviolet radiation intensities, allergens, or anything else might coincidentally set off certain biochemical reactions that variably affect heritable mental talents and temperaments in human populations by continent.

Race science has grown reticent about hypothesizing natural selection pressures on racial mentality. Moreover, the science is silent on when distinct human races formed on their respective (sub)continents: major races are just there, separated and isolated, in some long-ago time, marked by physical characteristics. Racialists today, rather than fret about what would cause racial brain differences, tout data and anecdata that might indirectly corroborate old suspicions. The never-ending promise of new methods to prove the reality of race has been an inexhaustible source of hope.

In sum, the tradition of searching into racial brain differences began in pre-Darwinian thought. Kant's protoscience loaded the modern idea of race with such conjecture, beating out better Western approaches to color consciousness.[24] From Prussia, he imagined that Europeans are by nature mentally superior to "the Negroes of Africa." Skin color was superficial: some deeply important quality would have to determine the membership and human potential of each continental race group. The spirit of that essentialist mode of racialism is not defunct—otherwise, the stakes in natural race inquiry would be too low.

My main takeaway from the rise of the race concept is that we should not overestimate the power of reason over ideology. There is a popular liberal misconception that false beliefs about race have been leaders in defense of color-conscious injustice—*as if inherited slavery, imperial colonialism, enforced segregation, etc. could ever seek sober rationalization through racial mind/brain stories.*[25] I am not denying that race ideology-rhetoric has influence as propaganda.[26] Nor am I denying that there can be value in pushback against performatives of race science inquiry. Rather, I am cautioning against stubborn battle over the sideshow that is the idea of continental human races.

Reasonable thinkers, including slavery abolitionists, plainly understood in the eighteenth and nineteenth centuries why "major races" speculation was conceptually and morally ill-conceived.[27] Kant, Jefferson, Hegel, Mill, Voltaire, and many others instead chose intellectual dishonesty and tribal chauvinism.[28] There is nothing in an Enlightenment world to excuse or mitigate Western dehumanization of non-European peoples (please see the footnote).[29] Nor can Western thought leaders be charitably graded on a relativistic curve; human catastrophes under cover of whiteness were undeniable and outrageous, in real time.[30] Philosophers, scientists, and statesmen of the European "race" built race ideology for the sake of obfuscating the obvious.

We are still waiting for an account of racial brain hardware, where the waiting is to give us the impression we exist as natural race kinds. **The "race" placeholder lent a guise of seriousness to stories about innate mental differences between Europeans vis-à-vis other (sub)continent-wide populations—in the midst of Western commitments to acquiring natural resources, human labor, and larger markets for untold profit, which called for extreme subjugation and exploitation of non-European peoples.** If race inquirers would come to terms with this, we could all move on to the living politics.[31]

There are no race concept puzzles that demand theoretical or empirical solutions. We can study any of the many impacts of race ideology-rhetoric without concern over a metaphysics or science of race, and without missing anything.

6

Racism versus Racialism

A Marginal Difference

Had scientific antiracialism prevailed, "race" might have fallen into popular disrepute along with "Mongolian" for "yellow" Asian. In the United States, "Caucasian" would not be a polite partial synonym for "white" European. Researchers could stop assuming that persons of a sort (e.g., Black Americans) known to be of substantial mixed continental descent belong to a single race group (e.g., black/Negro).[1] Debate over the reality of race could fade into philosophic and scientific twilight.

When populations are genetically differentiable and in certain respects look different—as northern and southern Europeans can in relation to the Alps and Pyrenees mountains—this is not (now) typically viewed as marking a racial difference between them.[2] Population geneticist Luigi Cavalli-Sforza and colleagues conclude that the project of race classification is biologically dubious: "It may be objected that the racial stereotypes have a consistency that allows even the layman to classify individuals. However, the major stereotypes, all based on skin color, hair color and form, and facial traits, reflect superficial differences that are not [racially] confirmed by deeper analysis."[3] Studying racial biology would be anticlimactic if it ended in an account of differences of narrow consequence.

My philosophical-historical survey of the modern idea of race illustrates why broad notions of racial nature are the ones that arouse interest. Ultimately, the reason is uncomplicated. **Led for hundreds of years by some of the Western world's greatest figures, the main business of the race concept has been to sponsor *ideological-rhetorical pretexts* for subjugation, exploitation, and nonrepair of non-European descent peoples.** The content of race ideology-rhetoric is incomparably less consequential than this overarching function it serves or tolerates.[4] An elusive racial factor can cast indefeasible doubt on a people's worthiness for equal respect and social justice.

Getting drawn down dim roads of race intrigue is therefore to be avoided. The idea of race possesses Wizard of Oz-like capacity for misdirection.[5]

The Afterlife of Race. Lionel K. McPherson, Oxford University Press. © Lionel K. McPherson 2024.
DOI: 10.1093/oso/9780197626849.003.0006

Endless angst surrounding a continental "race" thing distracts from a society's dominant priorities. American law, for example, entangles itself in feeble discursive constructs such as "colorblindness," "affirmative action," and "diversity."[6] Even liberal-minded responses are geared to grind very slowly toward social progress, under auspices of a "supreme law of the land" that has a reliable record of thwarting social equality for Black Americans.[7]

Contemporary White America's color consciousness is laid bare by sociologist Lawrence Bobo and colleagues:

> In post–World War II U.S. society, the racial attitudes of white Americans involve a shift. . . . [W]e witnessed the virtual disappearance of overt bigotry, of demands for strict segregation . . . and of adherence to the belief that blacks are the categorical intellectual inferiors of whites. [That], however, has not resulted in its opposite: a thoroughly antiracist popular ideology . . . embracing [a] democratic vision of the common humanity, worth, dignity, and place in the polity for blacks alongside whites. Instead, the institutionalized racial inequalities created by the long era of slavery followed by Jim Crow racism are now popularly accepted and condoned under a modern free market or laissez-faire racist ideology.[8]

Since that time in the 1990s and starting in the 1960s, social inequality has hardly improved for Black America overall. The enormous wealth disparity between Whites versus Blacks has grown.[9] In Part II, I explore American color caste dynamics in detail.

Consider for now the January 2021 Capitol riot by supporters of President Donald J. Trump after his loss in the 2020 US election—the latest in centuries of violent American whiteness.[10] Capitol rioters made their point: the US government would stand down for unruly people of a certain hue to occupy the congressional chambers, where inherited slavery and enforced segregation were governed from 1800 through 1965, in concert with every president and the Supreme Court. "In the early decades of America's history as an independent country," the *Washington Post* reports, "more than half of all congressmen voting on the laws forming the country's framework were enslavers."[11]

White America swerved toward "race neutral" doctrine after 1965. Paradoxically, this gambit insists on applying the idea of race to descendants of American slavery.[12] The "black or African American" category under *race* on the 2020 US census form expressly includes Africa-identified persons of immigrant lineage (e.g., Haitian, Nigerian, Ethiopian)—as if a global "Negro" thing is most relevant in America. (The race category does not specify a general category for Asia-identified persons; they get boxes for specific national origin identities.) *Slaves* had been the main US census category for Africa-identified persons from 1790 to 1860. A "race" question did not appear on the census until 1900. (See the Census Bureau's website to compare actual forms.)[13] The government simply wanted to track visible or reputed American slave lineage—because of the comprehensive ramifications of inherited slavery regarding the social distribution of wealth, opportunity, and entitlement.

Less than two weeks after the Capitol riot, on Martin Luther King Jr. Day, President Trump released his "1776 Commission" report. Its authors rejected the view that America's Founders "were hypocrites who preached equality even as they codified it in the Constitution and held slaves."[14] Trump's hyper-originalist reading of American history is not false. Black Americans were legally denied everything, for reasons of another people's economic, political, and psychosocial expediency: the "Slaves" caste was consciously disappeared from the Founders' collective vision of humanity. There was no hypocrisy: moral and legal pronouncements on equality were only for persons who belonged in the *Free White* census category (1790–1840). White American liberty and capitalism were ungovernable by reason or conscience.[15]

Some Capitol rioters might previously have visited the official "History of the U.S. Capitol Building" webpage, which discusses design and construction details.[16] Unmentioned are the persons who mostly built the building. In a separate "Art" section, a brief essay begins: "On Tuesday, February 28, 2012, Congress unveiled a marker to commemorate the important role played by laborers, including enslaved African Americans." We learn: "These slaves, as well as other laborers, quarried the stone used for the floors, walls and columns of the Capitol . . . and became skilled in brick making and laying."[17] The US government is again trolling descendants of American slavery. (In 2022, the *Washington Post* uncovered "more than 1,700 people who served in the U.S. Congress in the 18th, 19th and even 20th centuries [who] owned human beings at some point.")[18]

Enslaved Americans forced to labor at the Capitol could not benefit from skills they acquired. Worse, using those skills would further enrich slaveholders and White America's economy, strengthening the institution of inherited slavery. Steadfast White opposition to redressing legacies of that vicious circle led Martin Luther King Jr. to declare shortly before his assassination in 1968: "We are coming to Washington to get our check."[19] King was asserting Black America's reparations claim against the federal government. The country's moral redemption would depend on political commitment to social equality for descendants of the "Slaves" caste.[20]

Americans who oppose King's vision of worldly justice will reject Black social equality as a fair measure of "racial" progress.[21] Believing that innate differences between race groups could explain social fate, however, does not entail feeling that the groups have unequal moral status and should be treated accordingly.[22] We can draw a tentative distinction between racialism and racism.[23] Mere racialists will deny that natural differences between the groups could be morally justified grounds for social discrimination. Latter-day racists will support measures that neuter antidiscrimination protections and forestall corrective justice possibilities.[24] *Racism* is more activist than *racialism*, its fraternal twin.

What if there were differences between race groups in mathematical ability, with many members of one group having somewhat more natural talent than the majority of another? Such attributes would be irrelevant to treating the less talented group as a subordinate caste. Thus racists add attributes like innate laziness and hypersexuality to the equation: there would be natural differences in predisposition that necessitate social control and quarantine. Inferior racial persons would threaten the safety and biological integrity of superior races. Earnest racists are not simple bigots, though the word "racist" has acquired so much baggage they now prefer other names (e.g., "racial nativists").[25]

Knowing when racialism crosses over into racism is, in my estimation, of marginal importance. Sins of commission and omission usually speak for themselves. Social psychologists Jim Sidanius and Felicia Pratto describe a "principle-implementation gap" facing Black Americans: there is "the apparent contradiction between White Americans' expressed support for the principle of [social] equality and their consistent opposition to the implementation of any concrete policies that might actually promote [social] equality in practice."[26] Opposition can take the form of reactionary obstructionism or liberal incrementalism. But we usually can sidestep the question of when conduct definitely counts as racism and instead use a less fraught description, say, "quasi-racism." Of chief public concern are patterns of conduct and outcome, not sincere beliefs or inner feelings about race.

A constant theme underlies American race intrigue: "blacks" might well be mentally inferior to "whites." For as long as that ideological-rhetorical theme has work to do—normalizing White power and Black social disadvantage—fervent interest in or around racial mentality will endure. No matter how many incoherent, shoddy, or ridiculous claims about natural races have come and gone, more have been on the way. Two hundred years of conventional wisdom about racial mind/brain differences between Caucasians and the Negro can be willed into circumstantial evidence for believing there is something there—as many intellectuals, politicians, slaveholders, and segregationists of European descent said they suspected—perhaps explaining a lot. The prospect of racialism, rather than meeting "a priori" resistance Steven Pinker complains about, has received ceaseless attention under the banner of knowledge.

British historian and statesman James Bryce summed up, in the early twentieth century, the academic landscape of a self-perpetuating debate: "No branches of historical inquiry have suffered more from fanciful speculation than those which relate to the origin and attributes of the races of mankind. . . . Hypotheses are tempting, because though it may be impossible to verify them, it [is] almost equally impossible to refute them."[27] Ritually questioning the natural intelligence of Africa-identified peoples is less about finding truth than stoking doubt. Interest in racial mentality provides a public rationale for gentler versions of biological determinism that might explain away entrenched social inequality.[28] In the United States, a dominant White American caste has forever sought to make peace with "racial" injustices and disparities that otherwise would be inconceivable and heartbreaking.

Try to imagine a wealthy country whose establishment institutions have fostered undying speculation about the mental and moral inferiority of Ashkenazi Jews compared to European gentiles.[29] For centuries, the "Free Gentile" caste subjugated a Jewish "Slaves" caste through means of terror, law, and complicity.[30] Descendants of Jewish slave laborers—still nearly wealthless after gaining their "civil rights" in the 1960s, and dealing with disastrous legacies of social destabilization and restricted access to education—are stuck in a rearguard fight for social equality.[31] Gentile liberals, in the spirit of kindly and respectful moderation, support "diversity, equity, and inclusion" ("DEI") protocols that would leave intact a stark social hierarchy by ethnoreligious identity. Nevertheless, scientific truth might require further investigating whether Jewish nature could appreciably explain the Jews' perpetual bottom caste status in that anti-Jewish society.[32]

7

How Deep the Egypt Worry Goes

There is an overlapping history of anti-black ideology voiced by the Judeo-Christian and Western academic traditions.[1] In Hebrew mythology, Noah cursed the descendants of his son Ham—who included the ancient Egyptians—to be black in color and natural slaves. The Bible does not tell of major races or an analogue. Regardless, biblical "Hamites" were turned into racial "blacks."

Anthropologist Edith Sanders examines why Christians revised a "curse of Ham" myth in the sixteenth century: "Identification of the Hamite with the Negro, a view which persisted throughout the eighteenth century, served as a rationale for slavery."[2] Color-conscious geography bolstered the impression of natural unity. The moniker "negro," from Spanish or Portuguese for "black," applied since the 1400s to any "member of a dark-skinned group of peoples originally native to sub-Saharan Africa," as the *Oxford English Dictionary* puts it.[3]

Napoleon Bonaparte invaded Egypt in 1798, bringing scholars who determined that ancient Egyptian civilization (c. 3100 BCE–30 BCE) was a primary source of European culture. This posed a problem for the likes of Hegel, who had opined that Africa is a "differenceless mass" inhabited by mentally "dormant" Negroes.[4] In his reformulated opinion, the continent "must be divided into three parts": "south of the desert of Sahara" was "Africa proper"; "to the north of the desert" was "European Africa"; and "the river region of the Nile" was linked with "Asia."[5] Civilization in Africa could be explained if there were natively not-black regions. "Napoleon's expedition to Egypt," Sanders argues, "became the historical catalyst that provided the Western world with the impetus to turn the Hamite into a Caucasian."[6] A geographic reproductive barrier between "sub-Saharan" Africa and "North Africa" would serve that cause.

Some Western thinkers did grasp ancient reality and its modern reverberations. French historian and traveler Comte de Volney drew immediate lessons from the remnants of early civilization he saw in Egypt a decade before Napoleon: "How are we astonished . . . when we reflect that to the

The Afterlife of Race. Lionel K. McPherson, Oxford University Press. © Lionel K. McPherson 2024.
DOI: 10.1093/oso/9780197626849.003.0007

race of negroes, at present our slaves, and the objects of our extreme contempt, we owe our arts, sciences, and even the very use of speech."[7] Volney was confident about that people's racial being: "The ancient Egyptians were real negroes, of the same species with all the natives of Africa; and [though] mixing for so many ages with the Greeks and Romans . . . they still retain strong marks of their original conformation."[8] Whatever he meant by "real negroes," they looked like they had substantial ancestry from non-northern Africa—an imprecise territory below the middle of the Sahara desert. His racial observations went ignored.

An image emerged of the Sahara as a barren, virtually uninhabitable zone separating a single (sub)continent-wide black African race located south of the zone from a not-Negro race north of the zone. This is a fuzzy geographic construct. The southern Sahara includes much of the territory of Mali, Niger, Chad ("black" countries) along with Mauritania and (pre-secession) Sudan, whose populations consist of (black) African and (not-black) Asian Arab peoples.[9] Yet the image persists of "black Africa" as a physically and racially distinct "sub-Saharan" place.[10] There are current regional experts convinced that of the eleven Nile River Basin countries, "two, Sudan and Egypt, are located north of the Sahara Desert"—which is at odds with a good map.[11]

Sudan's name comes from the eighth-century Muslim description "Bilad al-Sudan," that is, "Land of the Blacks." The country had been variously located until 2011: South Sudan seceded and was moved to the "sub-Saharan" place; the reduced Sudan was moved to the north.[12] As for Egypt, 90 percent of its land area is in the Sahara; the preferred names for that territory are the "Western Desert" and the "Eastern Desert." A nineteenth-century distinction between "African Egypt" and "Asiatic Egypt" helped later locate the country in "the Middle East."[13] The modern idea of race was able to reconcile fantasies of European natural superiority with a new reverence for ancient Egypt. The "Asiatic" ancient Egyptians were on a Western trajectory to becoming "Caucasian."

White anxiety over African influence on ancient Greece, the cradle of Western civilization, was a side effect of Napoleon's Egypt campaign. American archaeological Egyptologist James Breasted offered this fable in the 1920s: "The teeming black world of Africa [is] separated from the Great White Race by an impassable desert barrier, the Sahara. . . . Isolated [and] unfitted by ages of tropical life for any effective intrusion among the White Race, the negro and negroid peoples remained without any influence on the development of early civilization."[14] English historian Hugh Trevor-Roper expressed a more restrained version of that dogma in the 1960s: "Perhaps, in the future, there will be some African history to teach. But at present there is none: there is only the history of the Europeans in Africa."[15]

Actual history tells a different story. An African people of "black" skin color, the ancient Nubians, migrated from the Sahara region toward the Nile River valley around 5000 BCE. In what is today southern Egypt and northern Sudan, they developed a civilization around 3500 BCE. Trade between the Nubians and the Egyptians dates to 3100 BCE, the beginning of Egypt's civilization. After periods of mutual conquest mostly dominated by the Egyptians, the Nubians abandoned Egypt around 670 BCE.[16] Archaeologists put together the historical outline by the 1990s. General awareness of non-northern African influence on ancient Egyptian civilization has been suppressed through dispute about the natural race of the Egyptians.

The impassable Sahara is an old myth. Camel trade caravans crossed the desert throughout the Middle Ages, as Europeans had known since the four-teenth century.[17] We now know that transformation of the humid "green Sahara" to arid desert was only complete around 5,500 years ago.[18] This is consistent with genomics research that shows modest "gene flow" from non-northern Africa through the northern region and into Europe as early as 11,000 years ago.[19] In short, populations north of the lower Sahara had black-ness during the time of ancient Egyptian civilization. Visual clues obviously triggered white anxiety.

Written description of the Egyptians as "*melanchroes*" and "*oulotriches*" by ancient Greek historian Herodotus also causes concern. The issue is whether to translate as "black-skinned" and "woolly-haired" or "dark-skinned" and "curly-haired." English historian George Rawlinson chose the former, adding a footnote in his classic 1858 translation: "Herodotus also alludes [elsewhere] to the black colour of the Egyptians; but not only do the paintings pointedly distinguish the Egyptians from the blacks of Africa . . . the mummies prove that the Egyptians were *neither black nor woolly-haired*, and the formation of the head at once decides that they are of Asiatic, and not of African, origin."[20]

We can be confident that Rawlinson's translation is closer than not-black alternatives. He chose "black-skinned" and "woolly-haired" while at pains to deny that the ancient Egyptians were black to any visible degree. Recent translators get rid of the tension by inserting racial ambiguity. Herodotus himself did not have a race problem, and he strongly implied visible blackness—writing two lines later that he could not determine whether the

Egyptians learned circumcision from "the Ethiopians" (of Nubia) or vice versa. The Nubians and the Egyptians were intimately involved.

Anthropology contrived and pursued the "Hamitic hypothesis." The purpose was to preempt questions about "sub-Saharan" (code for "black") influence on the ancient Egyptians. As Sanders summarizes the tale, "everything of value ever found in Africa was brought there by the Hamites, allegedly a branch of the Caucasian race."[21] English ethnologist Charles Seligman made the definitive statement in *Races of Africa* (last reprinted in 1979): "The history of Africa south of the Sahara is no more than the story of the permeation through the ages [of] the Negro and Bushman aborigines by Hamitic blood and culture." To be clear, "the civilizations of Africa are the civilizations of the Hamites," and these Hamites included "the ancient and modern Egyptians"— who were "Caucasians, i.e., belong to the same great branch of mankind as almost all Europeans."[22]

Inventing a European source of ancient Egyptian civilization knew no limits. Henry Morton Stanley, the Welsh explorer and a colonial agent in the Congo for Belgium's barbaric King Leopold II, set the stage for twentieth-century race fabulism.[23] In the 1870s, he reported seeing "light-complexioned, regular-featured people" near the mountains of Uganda's Congo border. These were "the white people of Gambaragara," presumed ancestors of the ancient Egyptians—a European tribe long "lost" in non-northern Africa.[24] The scholarly version of Stanley's yarn became the Hamitic hypothesis and enjoyed academic gravitas into the 1960s.[25]

Although the Caucasian Hamite went out of style, its purpose succeeded. The ancient Egyptians are not widely thought of as black or even as black and not-black mixed. Certain peoples located in "North Africa" or "the Middle East," along with "the Indian subcontinent," qualify under the "Caucasian" umbrella category that chiefly defines whiteness and tacitly confers off-whiteness. Whereas a "European" race category could sound ridiculous if it included migrant tribes who resettled thousands of years ago in Africa and Asia, the "Caucasian" maneuver has proved effective. Off-whiteness, or at least not-blackness, suffices for the ancient Egyptians of Western design: race intrigue did not require true-whiteness.

A 2017 study focusing on three ancient Egyptian "mummy genomes" takes for granted a racial divide that equates black/Negro Africa with the sub-Saharan construct: "Absolute estimates of ['sub-Saharan'] African ancestry . . . in the three ancient individuals range from 6 to 15%, and in the modern samples from 14 to 21%. . . . [This] suggests that African gene flow in modern Egyptians occurred indeed predominantly within the last 2,000 years." In plain English, the three ancient Egyptians were less black than modern Egyptians. The researchers bury their lead: "Ancient Egyptians are more closely related to all modern and ancient European populations that we tested."[26] A caveat downsizes the results:

> We note that all our genetic data were obtained from a single site in Middle Egypt and may not be representative for all of ancient Egypt. It is possible that populations in the south of Egypt were more closely related to those of Nubia and had a higher sub-Saharan genetic component. . . . Throughout Pharaonic history there was intense interaction between Egypt and Nubia . . . and there is compelling evidence for ethnic complexity within households with Egyptian men marrying Nubian women and vice versa.[27]

That understatement about the ancient Egyptians' "ethnic complexity" did not dissuade journalists from conveying the racial message: "But there was one persistent hole in ancient Egyptian identity: their chromosomes"; "Researchers in future want to determine exactly when sub-Saharan African genes seeped into the Egyptian genome and why"; et cetera.[28]

To repeat, archaeologists knew by the 1990s that the Nubians and the Egyptians were intertwined through trade and conquest from the start of ancient Egypt's civilization in 3100 BCE. The mummy researchers prefer highlighting phenotype results based on a single ancient individual's indication of "lighter skin pigmentation" characteristic of Neolithic Anatolia (aka "Asian Turkey").[29] The implicit generalization is that ancient Egyptians were European-adjacent despite their African locale and black admixture.

Color-conscious controversy also follows ancient Egyptians who apparently had darker features. Unease with customary guidelines for ascribing blackness has met the tomb painting of "Queen Nefertari playing senet," for example, and the "Head from monumental red granite statue of Amenhotep III." In the tradition of imagining a geographic reproductive barrier between the Egyptians and the Nubians, archaeologist Kathryn Bard reaches for reasons to suspend judgment: "The shading of skin tones in Egyptian tomb paintings . . . may not be a certain criterion for distinguishing race"; "Black-painted skin could be symbolic of something of which we are unaware four thousand years later"; and "Far from suggesting that the king had black skin, the two guardian figures of Tutankhamen may appear black simply because resin was applied to the skin areas."[30] Bard withholds similar caution about ancient Egyptians depicted with lighter features.

The principle of explanatory simplicity (Occam's razor) tells us that many ancient Egyptians depicted with darker features had substantial ancestry from "black Africa." Whether visible African ancestry could ever be "a certain criterion" for race is beside the point, as is the truism that ancient references to the Egyptians' color "by no means indicate that persons so described were Ethiopians, that is, blacks or Negroes in the modern usage of such terms."[31] We could dampen controversy by heeding archaeologist Stuart Tyson Smith's guidance: "Any characterization of the race of the ancient Egyptians depends on modern cultural definitions, not scientific study. Thus, by [familiar] standards it is reasonable to characterize the Egyptians as 'black,' while acknowledging the scientific evidence for the physical diversity of Africans."[32] Or, as I suggest, we could merely say that visible African ancestry was not uncommon among the ancient Egyptians. We can avoid the distraction of empty debate about which reputed major race they are assigned to or near.

An absurd leap of faith goes into believing that a continental African people were uniformly not of "sub-Saharan" descent despite quite black-looking members among them and close relations with ancient Nubia over 2400 years. Denial about ancient Egypt's visible blackness has fueled evasion of customary guidelines otherwise thought to define black racial being. *The Egypt worry—that "black Africans" evidently influenced ancient Greek civilization and thereby Western culture—challenges a tenet of European supremacy.* Hierarchical racialism is needed for factoring race into the development of a civilization. Without racialism, there is little motive to insist that the ancient Egyptians were closer by racial nature to "white" Europeans than to "yellow" Asians and were "neither black nor woolly-haired" at all.

The Western race scheme makes a point of linking certain non-European peoples to pre-ancient European ancestors. By legend, a certain major race might be naturally more capable of creating a great ancient civilization. By dogma, such people could not have been black/Negro to any meaningful degree.[33] Hence, after 1798, Western tradition moved ancient Egyptians decisively out of "black" ("sub-Saharan") Africa and into "North Africa" or "the Middle East." Western race intrigue can be entertained with less anxiety when ancient Egypt is segregated from blackness and made adjacent to whiteness in "the Caucasian race."

Through a Western device that groups "white" natives of Europe along with continental resettlers of European origin (now gently mixed) in the same major race group, the ancient Egyptians were ascribed Caucasian racial being for civilizational evidence, in Africa, of the European's natural superiority. This racial branching approach to cultural appropriation calls for racialism, which provides an X-factor throughline. The "Caucasian" umbrella category, meanwhile, conceals its hierarchical division of "white" natives of Europe above (off-white) continental resettlers in Africa and Asia. Western race ideology-rhetoric is shameless.

8

From Racial Theology to "Race" Optimism

Some segments of American Christianity stayed with the tradition of applying race myth to divine agency. "The institution of Slavery is full of mercy," a respected Baptist pastor in Virginia declared: "Under the gospel it has brought within the range of gospel influence, millions of Ham's descendant's among ourselves, who, but for this institution would have sunk down to eternal ruin."[1] The Southern Baptist Convention (SBC) formed in 1845, when Baptists in the South split from those in the North to become a proslavery denomination. White America's anti-black theology was not confined to Southern Baptist practice. Pseudo-rationalizations for inherited slavery generally borrowed from Christian missionary discourse about "civilizing the savage."[2] Personal, institutional, and cultural depravity was made righteous.

Frederick Douglass had no patience for blasphemous distraction:

> They strip the love of God of its beauty, and leave the throne of religion a huge, horrible, repulsive form. It is a religion for oppressors, tyrants, man-stealers, and *thugs*. . . .
>
> The American church is guilty, when viewed in connection with what it is doing to uphold slavery; but it is superlatively guilty when viewed in connection with its ability to abolish slavery.
>
> The sin of which it is guilty is one of omission as well as commission.[3]

American racial theology turned a Christian god of love into an anti-black racist to be worshipped as the almighty.

The Catholic Church went from accepting to tolerating slavery. Jesuits "owned" hundreds of Black Americans, and bishops opposed abolitionism during the Civil War. Contemporary Catholic scholars have sketched the backstory: "Nineteenth-century Catholic views on race and slavery grounded racist arguments in Biblical passages . . . the writings of various Church fathers, including . . . Augustine [and] Aquinas, and a number of papal statements. . . . Many American Catholics supported proslavery

The Afterlife of Race. Lionel K. McPherson, Oxford University Press. © Lionel K. McPherson 2024.
DOI: 10.1093/oso/9780197626849.003.0008

positions with the claim that Jesus frequently used masters and slaves in his parables without condemning the institution of slavery itself."[4] The church's anti-black posture continued well into the twentieth century. No pope to date has apologized to Black Americans.[5] American Catholic institutions have done little toward redress other than (Jesuit) Georgetown University's condescending and miserly measures, announced in 2016, to acknowledge enslaved Americans who built campus buildings and were sold as property to help finance school operations.[6]

The Church of Latter-day Saints (Mormons) cited the "mark of Cain" and the "curse of Ham" in prohibiting "Negroes of African descent" from priesthood and full membership until 1978.[7] Official rejection of anti-black theology waited until 2013: "Today, the Church disavows the theories advanced in the past that black skin is a sign of divine disfavor or curse, or that it reflects unrighteous actions in a premortal life; that mixed-marriages are a sin; or that blacks [are] inferior in any way." Blame partly shifted to American history: "The Church was established in 1830, an era of great racial division," though "founder Joseph Smith openly opposed slavery" shortly before his death in 1844.[8] Brigham Young then led Mormon pioneers and their human property to Salt Lake Valley, where he successfully lobbied for "legalizing and regulating slavery" in Utah Territory.[9] The Mormons would be a "white and delightsome people"; Negroes would be uniquely inferior because their supernatural ancestors were "neutral" in the "war in Heaven."[10] Color-conscious exclusion in the material world remains an unofficial theme.[11] The LDS Church has done nothing toward redress for Black Americans.

As the twentieth century closed, the Southern Baptist Convention started wrestling with its reactionary raison d'etre. A 1995 resolution was intended to "lament and repudiate historic acts of evil such as slavery."[12] In 2018, the Southern Baptist Theological Seminary released a report detailing "the legacy of this school in the horrifying realities of American slavery, Jim Crow segregation, racism, and even the avowal of white racial supremacy."[13] The seminary acknowledged doctrinal errancy: "[Trustees] argued first that slaveholding was righteous because the inferiority of blacks indicated God's providential will for their enslavement, corroborated by Noah's prophetic cursing of Ham."[14] Early Southern Baptists hoped God would hear their racist circular reasoning for the sake of atrocity.

Contemporary Southern Baptist leadership has opted for musings on "racial reconciliation" and man's "sinful nature" instead of redress.[15] Yet there was the 2017 failure of a resolution renouncing "the roots of White Supremacy within a 'Christian context' . . . based on the so-called 'curse of Ham' theory once prominently taught by the SBC."[16] In the 2016 US presidential election, 80 percent of mostly Southern Baptist "white evangelical" voters cast their ballots for Donald Trump; the numbers slightly dipped in the 2020 election.[17] A revisionist history has been revived: "The Southern Baptist Convention was created 'to provide a general organization for Baptists in the United States and its territories for the promotion of Christian missions at home and abroad and any other [ends] it may deem proper and advisable for the furtherance of the Kingdom of God.' "[18] The less varnished truth is that the Southern Baptist Convention formed in defense of the institution of inherited slavery.

Current race science, too, seeks to distance itself from explicit anti-black ideology. Remarks by Nobel Prize–winning molecular biologist James Watson illustrate how strained these efforts can be. In 2007, he was "inherently gloomy about the prospect of Africa" and its visible descendants when "all our social policies are based on the [assumption] that their intelligence is the same as ours." Watson saw "no firm reason to anticipate that the intellectual capacities of peoples geographically separated in their evolution should prove to have evolved identically"; his evidence was "all the testing" and the anecdotal stories from "anyone who has worked with black employees."[19] To question whether "equal powers of reason are a universal heritage of humanity," he added, is a scientific inquiry and "is not to give in to racism."[20] Watson's 2019 restatement of his racialism was no friendlier to that marginal distinction.[21]

My purpose in charting the continuity of religious and scientistic racial thinking is threefold. First, to show that slavery as the prime source of American antiblackness precedes the modern idea of race as continental groups. Second, to show that race intrigue has not been sustained by serious empirical inquiry. Third, to show that racialism lives on among scientists and other educated people. The third point responds to styles of race theorizing that are optimistic about the meaning of "race" today and the decline of (super)naturalist race ideology.[22] Supposedly, "race" is no longer mired in bad theology and weird science: the word would now refer, through conventional opinion, to social groups or groupings left over from old rhetoric of belief in a (super)natural hierarchy of major races. Racialism would be more or less dead.

Such "race" optimism moves philosopher Paul Taylor to declare a paradigm shift. His beginning of racialism's end was the US Supreme Court case

United States v. Thind (1923). The Court's chief justice was former President Taft, avowed legal and political opponent of Black American social equality.[23] Defendant Bhagat Singh Thind, a man of "high caste Hindu stock," sought naturalized citizenship as a member of "the Caucasian or Aryan race."[24] Taylor claims that the Court decided "to detach the state's official race-thinking from scientific categories"; American "whiteness" would be recognized "by appeal to common sense."[25] But race science never meant for "white" and "Caucasian" to be equivalent. For its part, *Thind* struck down any question whether the color of law was stricter than major race designation: the equation of whiteness and (nearly) full European descent was again reaffirmed.

"The high-class Hindu regards the aboriginal Indian Mongoloid in the same manner as the American regards the negro," defendant Thind argued before the Supreme Court.[26] Nevertheless, off-white Caucasians were not the equals of white ones, which limited the judges' sympathy for comparing light-skinned Asian Indians with White Americans: "[The statute] does not employ the word 'Caucasian' but the words 'white persons,' and these are words of common speech and not of scientific origin. The word 'Caucasian' not only was not employed in the law but was probably wholly unfamiliar to the original framers of the statute in 1790."[27] White America's law barred all Asian Indians of foreign origin from eligibility for naturalized citizenship. The judges mocked "Caucasian" intrigue: "It is at best a conventional term, with an altogether fortuitous origin, which, under scientific manipulation, has come to include far more than the unscientific mind suspects."[28] An originalist ideology of white/European social dominance was left intact. Legally calling on "race" for post-slavery euphemism and distraction would be reserved for the Court's more pressing political goals.[29]

The *Thind* case provides plain illustration for my deflationary analysis of race. An American land of the free would be protected and promoted for a Europe-identified caste, per its in-group ancestry criteria for social membership—expressly excluding an Africa-identified caste from legal rights or level citizenship and reserving the right to summarily exclude others identified by non-European descent.[30] That, as the Supreme Court implied, is what "the original framers" would have cared about under their law—not the latest stories from anthropology, psychology, genetics, metaphysics, etc. concerning a white, black, or other "race" thing. It was slavery's Constitution, not a deeply racial contract.[31] Bhagat Singh Thind's off-white plea to join the white caste was irrelevant to foundational Anglo-American law and politics.

Not until 1950 did *race* appear as a standalone US census category, enshrining the now-standard practice of retroactively superimposing the word on what the government since 1790 had been tracking by "color" of freedom or slavery.[32] The sleight-of-language is meant to add confusion. Contrary to a popular assumption that Taylor echoes, White America's "official race-thinking" was never attached to race science. All along, the closely monitored thing had been birth into the "Free White" caste or the "Slaves"/ "black" caste (with other peoples marked by national or linguistic identity)—purposes for which figurative color categories served adequately. The American public operated on the norm that "white" meant Europe-identified (including Euro-Latino peoples[33]) and "black" meant Africa-identified (via homegrown slave lineage). The Taft Supreme Court, keeping with ideological precedent, simply did not care about an Asian Indian's race, whether by alleged natural or social measure.

Taylor's other example of fading belief in natural races is the 1950 United Nations Educational, Scientific and Cultural Organization (UNESCO) "Statement by Experts on Race Problems."[34] As he interprets it: "This was the first high-profile, international statement of the view that classical racialism is false. . . . [Some experts] reinterpreted race as a social product. . . . Very soon, the average white person's common understanding of race would bear little resemblance to its analogue in 1923."[35] In reality, two hundred years of racialist (super)naturalism had not largely disappeared. Martinique-born psychiatrist and philosopher Frantz Fanon described the impact on him at the time: "I discovered my blackness, my ethnic characteristics; and I was battered down by . . . intellectual deficiency [and] racial defects."[36] Race ideology-rhetoric has carried its scientistic stereotypes into the twenty-first century. For many Africa-identified persons of transatlantic heritage, lived experience of inferiority narratives often rebuts optimistic theory.[37]

Racialism is also alive through strange misinformation about physiology. A 2016 study in the United States found that a sizable number of "white laypeople and medical students and residents hold false beliefs about biological differences between blacks and whites." The list includes these: "Blacks' nerve endings are less sensitive than whites'"; "Blacks' skin is thicker than whites'"; and "Blacks are better at detecting movement than whites." These beliefs "predict racial bias in pain perception and treatment recommendation."[38] Across generations of White Americans, race superstitions are being conveyed below the surface of public culture. Another study found color-conscious bias in "the attribution of supernatural, extrasensory, and magical mental and physical qualities." This "superhumanization" of Africa-identified persons "predicts denial of pain to black versus white targets."[39] Racialist physiology has accustomed the public to American anti-black cruelty, from slavery plantations to present-day hospitals and police departments.[40]

Overlap between racial theology and race science has an unsung pedigree. Religion-and-science scholar Terence Keel argues the following: "Modern scientists construct race and explain the origins of human variation by transferring the creative power of God onto nature, biology, and genetics. [T]he modern scientific study of race is not merely shaped by Christian intellectual history but is engaged in a secular form of theology, a secular creationism."[41] Searching for important differences in people by (sub)continent-wide natural environment or breeding populations, race science has maintained the supernaturalist tradition. Pre-Enlightenment faith in essential social group differences was grafted onto the idea of continental races, enabling the faith to last through religious, protoscientific, and scientific transitions.

Americans are not alone in thinking and acting as if racial blackness is more than skin deep.[42] **The race concept carries old (super)naturalist suspicions that whatever race is, *it* naturally goes beyond shallow or incidental differences.** Popular as well as specialist notions of race are steeped in a hodgepodge of beliefs, feelings, and attitudes about racial nature.

9

Renewed Race Science

Today's believers in natural race shun the old story where every member of each racial grouping has some quality, an essence, that is distinctive to their group. This raises the puzzle of how members of a race group could be truly identified in the absence of heritable traits they distinctively have in common. What is the racial X factor? Most Black Americans have substantial mixed African and non-African ancestry. Most Brazilians who have that mix do not identify as black. Most Africans do not have substantial non-African ancestry, and many identify as black only in some weak sense. Who counts as belonging to a particular race group, why, and by whom varies across place and time. Contemporary race science needs an answer for what would redeem the idea of continental human races.

The answer seems self-evident to German-born American Ernst Mayr, one of the leading evolutionary biologists of the twentieth century. Anyone convinced " 'there are no human races,' " he complains, is "obviously ignorant of modern biology." The word "race" is to get its meaning solely from taxonomy, the scientific practice of classification: "A subspecies is a geographic race that is sufficiently different taxonomically to be worthy of a separate name. [A] geographic race . . . is restricted to a geographic subdivision of the range of a species." How race classifiers could objectively decide on what counts as "sufficiently" different about humans assigned to one group rather than another, Mayr does not explain. He is confident that racial difference—before proof of distinctive "cultural as well as genetic elements"—can be known by physical appearance along with facts of continental descent. The American Academy of Arts and Sciences published that 2002 restatement of Mayr's lifelong quasi-racism.[1]

Biologists Edward Wilson and William Brown Jr. disagree that racial difference only needs genetic confirmation. To summarize their 1953 criticism of Mayr's general "subspecies concept": the notion of a "genetically distinct geographical fraction of the species," using "whatever 'diagnostic' [differences] are chosen to delimit races," is "subjective and arbitrary in taxonomic practice."[2] There are no objective criteria for distinguishing human

The Afterlife of Race. Lionel K. McPherson, Oxford University Press. © Lionel K. McPherson 2024.
DOI: 10.1093/oso/9780197626849.003.0009

races or any other type of subspecies. Which differences are "worthy" enough to mark a subspecies/race will be a judgment call about the significance of certain genetic traits. Wilson and Brown Jr. do not deny that taxonomists are responding to something or other found in nature. But as Blumenbach figured out over two hundred years ago, race classifiers will not have discovered objective major race divisions.

Physical and genetic differences within (sub)continental populations have been discounted as compared to patterns of physical difference by (sub)continent. Luigi Cavalli-Sforza and colleagues describe an inherently flexible scenario: "The classification into races has proved to be a futile exercise for reasons that were already clear to Darwin. [M]odern taxonomists [define] from 3 to 60 or more races. To some extent, this latitude depends on the personal preference of taxonomists, who may choose to be 'lumpers' or 'splitters.'"[3] **Any biology of familiar racial groupings will proceed from some preconceived racial geography, searching for natural traits that persons of any same race distinctively have in common—which may call for (re)adjusting the boundaries or revising the race concept itself in order to get the desired fits.**

Evolutionary biologist Marta Lahr explains how race science hoped to reach its subspecies/race determinations: "The most meaningful criterion for [taxonomy] is phylogeny, whereby organisms are classified into [flowchart] units that reflect evolutionary relationships. Classifications of human races have traditionally by-passed this principle, and classify [groups] in which clear distinctive features may be recognised as a 'race', while other less discrete groups of people are taken as the result of extensive admixture. This line of thought led to the belief that pure races existed, and that these pure races once had pure racial ancestors."[4] In myth, each major race came into being on its own respective (sub)continent.

Georacial boundaries have been determined by social convention, not science. Ancient Greeks launched a xenophobic perspective that filtered through the Middle Ages, settling in the eighteenth century on the Caucasus and Ural Mountains to divide the Eurasian supercontinent; this joined modern racial thinking to create white Europe and yellow Asia.[5] Soon afterward, new Western knowledge of ancient Egypt called for a physical barrier to divide the African continent into a not-black northern part and a black "sub-Saharan" part. Use of these (sub)continental boundaries, or any other, to mark race divisions is not empirically neutral. Race science struggles with unseeing racial geography that has been taught in primary schools and communicated through folk wisdom.

The case of the Indian subcontinent flouts racial geography. Linguist Asya Pereltsvaig and geographer Martin Lewis sum up the issue: "Racial theorists maintained a stark separation between the [white] peoples of Europe [and] the darker-skinned inhabitants of South Asia, yet the philologists argued that Europeans and northern Indians stemmed from the same stock."[6] Only lighter-skinned Indians might be a "Caucasoid" people; the non-geographic term is meant to absorb European tribes that long ago resettled in Africa and Asia. To explicitly exclude the darker-skinned native masses, though, would highlight the absurdity of locating in Asia a distinct off-white people of almost European nature, surrounded for millennia by "Mongoloid" peoples.[7] Implicitly, "the Indian subcontinent" of race science would refer only to off-white (gently mixed Indo-European) inhabitants, not the whole of India with its wide light-skinned to dark-skinned range of inhabitants.[8]

By contrast, racial geography is unequivocal about restricting native blackness to "south of the Sahara." Mayr insists that major race groups were "established before the voyages of European discovery and subsequent rise of a global economy."[9] A (sub)continent-wide European race and "sub-Saharan" African race are supposed to have been nearly pure before then, their racial ancestors conserved by force of radical geographic isolation. Such a story is contradicted by DNA analysis to "reconstruct the last 4000 years of genetic history in African populations": for example, "multiple ethnic groups from The Gambia and Mali all show signs of sharing the same set of ancestors from West Africa, Europe and Asia who mixed around 2000 years ago."[10] Non-northern Africa's radical inaccessibility until the 1400s is a myth.

Racialists will be undeterred. (Sub)continent-wide "white" racial being would have begun forming after the last waves of black out-of-Africa migrants found their way to Europe. Through the passage of pre-ancient millennia, under pressure of the natural environment, blackness was refined out of the migrants' descendants as whiteness spread across the European land. Biological tolerance for "race mixing" would be low for white natives of Europe, as indicated by light skin's sensitivity to race impurity. Race science therefore does not ignore undeniable evidence of non-European admixture. The "Caucasian" umbrella category both includes and downgrades tribes from Europe that continentally resettled at the cost of true-whiteness: they are off-white, a tacit subracial distinction that reifies fantasies of European natural superiority.[11]

The origin myth of white purity is ruined, however, by recent studies of gene flow into Eurasian populations. "Neanderthals and anatomically modern humans overlapped geographically for a period of over 30,000 years," the science shows. More specifically, "The Neanderthal component in the modern human genome is ubiquitous in non-African populations" that include "Europeans" and "East Asians."[12] Natives of Europe had copious relations with Neanderthals. We could wonder whether anatomically modern humans mixed with Neanderthal are a "sufficiently different" racial kind to support marking a major division.

The European voyages of conquest represent a less distant complication.[13] Forced mass movement of persons from Atlantic Africa led to wholesale (coercive) reproductive mixing with Europeans in the New World. Even Mayr's pre-conquest story has to dismiss normal human migration across deserts, mountain ranges, isthmuses, and straits—which is how early migrants must have exited Africa in the first place. Before the race concept took over, scholars ancient (e.g., Herodotus) and early modern (e.g., Heylyn) could plainly view Europe, Asia, and Africa as three regions of one (super)continent.

The doctrine of polygenism hoped to remove doubt about race origins by imagining separate development paths for each race on separate (sub)continents, from the beginnings of human kinds.[14] A polygenist story had gained prominence with Christian scholars in the Middle Ages: "the earth appeared to be a sphere, mostly covered by water, on which four small 'islands' emerged," and "the ocean made any communication difficult, if not impossible," between the human populations that arose. Aided by such geography, the practice of "placing monstrous races in Africa was a commonplace phenomenon" into the fifteenth century.[15] Ham's descendants had become literal monsters of dark color. We are fortunate that modern maps undermine the thesis of major races through geographic isolation.[16] Moreover, DNA analysis now shows that non-northern African "coastal populations experienced an influx of Eurasian haplotypes over the last 7000 years."[17]

Mayr casually cites the overrepresentation of "contenders of African descent" in Olympics sprint finals as evidence that "each human race consists of individuals who, on average and in certain ways, are demonstrably superior to the average individual of another race."[18] Apart from the sports performance fallacy that props up speculation about racial mind/brain (see Part I, Section 4, "(Sub)continent-Wide Human Types?"), notice that these runners are mainly Black American and Afro-Caribbean—diaspora peoples traceable to Atlantic Africa and assigned to the black/Negro category despite typically also having substantial European ancestry. (Non-northern Africans, mainly eastern, are overrepresented among world-class long-distance runners, not sprinters.)[19] Mayr goes on to imply that "people of Jewish descent" comprise a separate (non-continental) race because of their "propensity for Tay-Sachs disease."[20] Such departures from the logic of continental race groups leave us with no serious idea what he means by "race."

Philosophers of race science have grasped that human taxonomy requires conceptual rehabilitation. Major races could be "lineages of reasonably re-productively isolated breeding populations" that are distinctive enough for subspecies or subspecies-like designation—even if all members of a certain race group do not distinctively have in common certain heritable traits.[21] Philip Kitcher's "minimalist" notion amounts to little more than endoga-mous reproductive patterns that continue to generate "three major races," albeit "in highly qualified form" in the New World, if not in Africa.[22] Skin color and other physical features that roughly indicate various thresholds of European, Asian, or African ancestry might be the crux of those groupings. Robin Andreasen's "cladistic" notion construes race "solely in terms of common ancestry," starting from whenever the different reputed major races formed on their respective (sub)continents. This leads her to grant that major race groups "are likely on their way out": race populations that used to be geographically separated and isolated may have experienced "too much gene flow" across them.[23]

European transatlantic slavery does disrupt a static geography of conti-nental human races. But why cling to the notion of racially distinct (sub)con-tinental populations that were nearly intact until five hundred years ago? We would need to believe that pre-ancient collections of anatomically modern humans, each collection on its own large landmass, eventually coalesced into discrete groups that had acquired distinctive biological qualities "worthy" of taxonomic separation. On the basis of physical features alone, descendants of those populations would typically be identifiable as members of a corre-sponding subspecies/race. Only in very recent human history would con-tinental race groups have started to break down, jeopardizing taxonomic redemption.[24]

It is true that after race science has fixed on certain georacial boundaries, investigators may discover some genetic traits that typically distinguish humans within those boundaries from those outside. But racial geography can prove disappointing: a 2017 study found that genetic "variants associ-ated with dark pigmentation in Africans are identical by descent in South Asian and Australo-Melanesian populations."[25] There is a mismatch for the non-Africans between their "black" pigmentation and their racially native habitats; there weren't supposed to be genetically "black-skinned" peoples who aren't blacks/Negroes. Skin color genetics are messier than expected.

To reiterate: the notion of race populations presupposes that descendants of out-of-Africa migrants evolved into distinct subspecies/races of humans. Physical features would have been the only evidence that non-Negroid populations had come into being. In an incredible coincidence, qualities unrelated to physical features might be truer to racial boundaries. Searching into those other qualities was easier to explain when Western conventional wisdom held that the Negro's natural mental talents are grossly inferior to the Caucasian's. Updated biological conceptions of race, Andreasen observes, are "relatively minimalist" in comparison.[26] Renewed race science tries to redeem the race concept by first implying that there are innate differences of narrow consequence.

Geneticist Neil Risch and colleagues cite random genetic traits, disease predispositions, and drug response variation. We are told that the "objective scientific perspective" of population genetics delivers race classification results. A partial list:

- Racial "Africans [are] those with primary ancestry in sub-Saharan Africa," which "includes African Americans and Afro-Caribbeans."
- Racial "Caucasians include those with ancestry in Europe and West Asia, including the Indian subcontinent and Middle East," along with "North Africans typically."
- Racial "'Asians' are those from eastern Asia including China, Indochina, Japan, the Philippines and Siberia."[27]

Old-school and new-school race categories would coincide, supported by a mysterious race calculus. Note the "primary ancestry" criterion for racial "Africans." No continental "blood" quantum criterion is declared for "Caucasians" and "Asians."

In a telling disclosure, Risch and colleagues acknowledge that "there remain individuals of mixed ancestry who will not be easily categorized by any simple system of finite, discrete categories." These individuals are to include "Ethiopians" and many "self-identified Hispanics"—though not subcontinental Indians, "Middle East" peoples, and "North Africans typically," who are truly to be "Caucasians."[28] Set aside the embarrassment of failing for Ethiopians (said to "genetically resemble Caucasians, probably as a result of considerable Caucasian admixture," with no mention of what unadmixed

Caucasian ancestry would be), who in ancient and modern times were seen as an archetypal "black" African people. Also set aside ad hoc concern about "nonwhite" Hispanics who have visible African ancestry. Any racial groupings will be counterintuitive somewhere in the world: by different local or global guidelines, too many or too few persons will count as members of whichever major race.[29]

Revisit the case of Thomas Jefferson's children through his light-skinned Black American slave. Those children, despite having less than 50 percent African ancestry, were counted as Negroes and enslaved by their father. An across-the-board "primary ancestry" criterion for natural race produces this result: persons whose majority or plurality of "mixed ancestry" is "in Europe" (like the Jefferson-Hemings children) are "Caucasian." Risch and colleagues might reject that result by filing such cases under "not easily categorized." But a serious effort at race classification would reject underspecified, uneven guidelines that fit prior color-conscious perceptions. Any continental "blood" quantum for racial "Africans" (e.g., "one drop") will presuppose social convention. That is more honest, at least, than the usual approach in contemporary race science, which is equal silence about quantum criteria for racial groupings.

A 2016 article in the journal *Science* reports that confusion has "worsened with the rise of large-scale genetic surveys that use race as a tool to stratify [research] data." Problems include "haphazard" use of racial and ethnic variables; "failure" to distinguish between racial self-identification and "assigned or assumed racial categories"; and weak reasons for adopting "racial categories relative to the research questions asked." The "use of the race concept in genetics," the authors add, "has vexed natural and social scientists for more than a century."[30] No genetics discovery or revised interpretation (e.g., "genuine biological entity" instead of natural "race" kind[31]) could reveal whatever conceptually shapeshifting thing "race" refers to.

"Know it when we see it" guidelines for who belongs to which race are unworthy of science. There is no useful news in sophisticated confirmation that typically, people who "look white" have some European ancestry, people who "look black" have some African ancestry, etc. Nor is there much explanation for why identifying individuals by reputed major race—rather than by all substantial components of their continental ancestry—is preferable for race biology research. Risch and colleagues merely cite the high rate of self-identification by race in the United States—as could be expected in a color caste society that adopted racial discourse—and downplay pervasive reproductive mixing between Europe-identified males and enslaved Africa-identified females. As is true for other Africa-identified national peoples in New World territories colonized by the British, Black Americans are a people of Afro-European descent.[32]

I have heard the response that fuller georacial inquiry would be too complicated for practical purposes: facts of continental descent need to be edited to fit established single "race or ancestry" categories. Many Americans have mixed continental ancestry, so convenience would recommend asking them to check a single box. Race research programs could struggle if, say, contemporary Black Americans were written in as having 80 percent Atlantic African and 20 percent northwest European ancestry. This fractional estimate would be more accurate than flat identification of Black Americans as racially "African" or "black"—as if they, unlike Risch's Ethiopians, for instance, do not also have "considerable Caucasian admixture."

High-minded fascination with color-conscious human difference by (sub)continent cannot explain endless investment in investigation and artifice. Nor is it credible to suggest that best practices for biological research could be hurt by too much substantial information about an individual's or a group's (sub)continental ancestry profile.[33] Someday, biologists might follow the path of physical anthropologists and decisively renounce the quasi-polygenetic idea of continental human races. Until then, the latest techniques to confirm some preconceived racial geography will bolster belief among the faithful.

10

Racial Metaphysics of Distraction

Pushback against hierarchical notions of race spurred reform of racial thinking. Race is socially real, reformists believe, whatever might be its narrow biology.[1] This now mainstream view that race is a social construct was introduced by W. E. B. Du Bois in "The Conservation of Races" (1897):

> There are differences—subtle, delicate and elusive, though they may be— which have silently but definitely separated men into groups. While these subtle forces have generally followed the natural cleavage of common blood, descent and physical peculiarities, they have at other times swept across and ignored these. At all times, however, they have divided human beings into races, which, while they perhaps transcend scientific definition, nevertheless, are clearly defined to the eye of the Historian and Sociologist.[2]

He proposed that black/Negro "conservation" in the United States was necessary for social equality and human progress. Racial metaphysics fell by the wayside for Du Bois—as I will elaborate by reevaluating his philosophical contribution to thinking about race.

Visible continental ancestry was made important through unbridled Western capitalism that required subjugation and exploitation of non-European peoples. The question now is whether some improved conception of race is needed to help us navigate a color-conscious world. Social constructionists answer yes. Some turn to specialist theories of language and ontology. Philosopher Sally Haslanger, for example, rejects the skeptical view that if natural races do not exist, "the term 'race' doesn't refer" to real groups; she argues that "the best interpretation of our ongoing collective practice using the term 'race' is compatible with races being social kinds."[3] The project of race inquiry has become hazy: whether "race" refers to anything naturally *or* socially real was not the contested issue.

Different natural major races—which would have come into being after dark-skinned African migrants resettled on separate (sub)continents and were refined there by the natural environment over pre-ancient millennia,

The Afterlife of Race. Lionel K. McPherson, Oxford University Press. © Lionel K. McPherson 2024.
DOI: 10.1093/oso/9780197626849.003.0010

until each (sub)continent-wide population acquired its own distinctive not-Negro quality that marked every member—might sound like cryptozoology.[4] Races could more plausibly be said to exist as part of the social world like tribes or citizens. Social constructionists are not bothered if this reframing of the issue changes the subject: they dismiss any broad notion of natural race and emphasize that race is a socially observable phenomenon.

Haslanger argues that "race" refers to certain social groups through a community's use of the word today: her "rational improvisation model . . . invites us to consider the social dynamics, collaboration, and reflective practice required for shared meanings."[5] The race concept, she says, is not "'non-negotiably committed' to a biological basis."[6] Notions of natural race might no longer be part of what a community means by "race" in retaining the word, despite continued angst about what race is. Haslanger's model for meaning(s) invites cherry-picking: a theorist could select from undisputed beliefs about race (e.g., blacks typically have visible African ancestry) and ignore controversial beliefs (e.g., blacks and whites might significantly differ in natural intelligence[7]). This method would locally disqualify meanings for "race" that are disputed within community standards, yielding a local minimum consensus.

In similar fashion, a global minimum consensus could ignore uncertainty and disagreement about ordinary use(s) of "race" (in translation) on the world stage. Races in a global sense would exist insofar as members of the world community act as if there are social groups identifiable by something called "race." A global meaning of the word could emerge that is capable, Haslanger contends, of "doing justice both to the historical collective practice and the worldly facts."[8] When basic dissensus persists—for example, the book *What Is Race?* gathers four philosophers, including Haslanger, to debate their respective "plausible metaphysical views" of race[9]—a theorist may decide the meaning of "race" in the manner she prefers.[10]

Humpty Dumpty returns, wearing fancier sets of philosophical clothes. But opposite the looking-glass world, "race" is stuck in centuries of negotiation, with metaphysical inquiry now witnessing a plethora of self-proclaimed plausible, if incompatible, views. The *What Is Race?* philosophers endorse "race" pluralism: theirs is of a kind that gestures toward respectable stakes (e.g., ameliorative progress, biomedical advancement, cultural reclamation, abstruse insight) in hope that the word is not superfluous and misleading. Such inflationary pluralism seems eager to keep race theory flowing for mutual engagement.

My deflationary kind of "race" pluralism, by contrast, is eager to disengage from metaphysical debate. Philosopher Kwame Anthony Appiah takes a less tolerant stance known as "race" eliminativism. Human races were widely believed to be natural kinds, typically indicated by visible continental ancestry, characterized by significant natural differences of mind. He believes that "there are no races" because those natural kinds never existed.[11] Among his charges: racialist beliefs remain attached to racial groupings; "race" thought is prone to disordered use; and racial identities are a stifling type of social identity. Appiah warns that re-conceptions of race are susceptible to distortion by holdover beliefs about racial nature.[12]

The early Du Bois understood race to be "a vast family of human beings, generally of common blood and language, always of common history, traditions, and impulses." Discernible to him were "eight distinctly differentiated races": four European peoples ("Slavs," "Teutons," "English," and "Romance nations"); "Negroes of Africa and America"; "Semitic people of Western Asia and Northern Africa"; "Hindoos of Central Asia"; and "Mongolians of Eastern Asia."[13] Descendants of American slavery would belong to the same race group as the Maasai of eastern Africa. With "common blood" being complicated through racial admixture and four different language families spoken on the African continent alone, continental and diasporic members of a single Negro race would have to share "history, traditions, and impulses" of a common culture.

Thus the early Du Bois imagined that race carried more than a shallow or incidental biology of geographic descent. "The deeper differences," he asserted, "are spiritual, psychical, differences—undoubtedly based on the physical, but infinitely transcending them."[14] He did not explain how persons who have mixed continental ancestry could belong to a single major race. A physical profile criterion would be too narrow, and visible African ancestry could only indicate Negro being through vague arbitrariness. Moreover, peoples who to outsiders may look like members of the same race (e.g., Koreans and Japanese) frequently make essentialist-type distinctions among themselves.[15] Maybe, for Du Bois, a common (sub)continental culture could represent the "spiritual" thing that binds individuals and peoples to membership in the same major race group, regardless of physical appearance or subgroup antagonisms.[16]

The problem is that members of a "cultural race," even if such a thing is intelligible, would have to have more in common than certain cultural features. Otherwise, persons could culturally opt out of or (in the case of adoption) may never have culturally belonged to their default continental grouping—which violates the logic of major races.[17] The racial X factor would have to be some nonvoluntary spiritual-cultural quality. By 1940, in *Dusk of Dawn*, Du Bois sounded tired of the philosophical pursuit:

> This was the race concept which has dominated my life. . . . It had as I have tried to show all sorts of illogical trends and irreconcilable tendencies. Perhaps it is wrong to speak of it at all as "a concept" rather than as a group of contradictory forces, facts and tendencies. . . . It was for me . . . finally consideration of my connection, physical and spiritual, with Africa and the Negro race in its homeland. All this led to an attempt to rationalize the racial concept and its place in the modern world.[18]

The later Du Bois had come to accept that for descendants of American slavery, deep desire for a homeland—free from habitual disrespect, social marginalization, and traumatic memory—could not be satisfied by (super)naturalist connection to Africa, whose national and tribal inhabitants were mostly indifferent to the American Negro's plight. No race story could accomplish that feat of existential de-estrangement.[19]

Any familiar notion of race, of course, will still include continental ancestry. Appiah contends that "none of them will be much good for explaining social or psychological life, and none of them corresponds to the social groups we call 'races.'"[20] Race ideology in America made persons black/Negro, not always indicated by visible blackness, if they had at least one genealogically traceable African ancestor. The traceable era began around 1619, when "20 and odd Negroes" were forcibly shipped from Atlantic Africa to Jamestown, Virginia.[21] Modern racial thinking (est. 1684) only had to be good enough for White Americans to be complicit in the dehumanizing business of inherited slavery.[22] For everyday purposes of distinguishing a "black" enslaved caste from a "white" free caste, good enough did not require any serious notion of race.

Africa-identified Americans inherited their caste status, by White America's law, for more than three hundred years. Legal historian Warren Billings unpacks the central motivations for that arrangement: "Preoccupation with the roots of modern racism obscures attitudes other than prejudice that allowed Englishmen to find in chattel slavery solutions to their problems with labor and social control."[23] By my further analysis in American context: The major races idea was turned into the conceptual vehicle for racism, principally as an expression of anti-black ideology, pseudo-rationalizing inherited slavery, enforced segregation, and (tacitly) mass incarceration.[24] American color caste has never depended on faulty beliefs and feelings about the nature of race.

The traceable African ancestry norm, later known as the "one-drop rule," was meant to track social-political lineage in American slavery.[25] Race ideology transformed this obvious function into widespread inchoate belief in natural "one drop" blackness.[26] This can prompt asking what a valid standard for racial blackness might be. The easy answer—"blacks" have enough African ancestry to "look black"—is shallow and unstable; individuals who are seen as black in America may count as not-black in Panama, Brazil, or France, for instance.[27] Trying to revise notions of race to keep up with various rationales for ascribing racial identity is a process that has no (informative) universal rules or destination. Appiah is skeptical of our ability to escape racial thinking's tendency to depersonalize those it has othered. The persistence of American slave plantation nostalgia at the atrocity sites themselves, for example, attests to his concern.[28] He would eliminate the race concept from our understanding of reality.

Reformist critics of race skepticism deny that "race" thought is hostage to (super)naturalist belief in race.[29] They object to Appiah's giving short shrift to social reconstruction of the race concept. For Lucius Outlaw, Paul Taylor, and other philosophers oriented toward Du Bois, being black is real as a social phenomenon.[30] The early Du Bois knew that scientific racialism was implausible and introduced a non-hierarchical version of the natural races paradigm. The later Du Bois grew uncertain about natural races without questioning the significance of social identity by color or race. His analysis of the status of Black America, argues Lawrence Bobo, "reflected a foundational concern with prejudice and racial attitudes" in practice.[31] Global race theorizing gave way to local race pragmatism.

In other words, the metaphysics of race was a side issue for Du Bois, a historian and sociologist, as he stated in "The Conservation of Races." Philosophers, starting with Appiah in the 1980s, have inflated Du Bois's side issue into an intricate saga about realism versus anti-realism about race. "Getting the metaphysics right is a philosophical challenge," grants Adam Hochman, who defends the faith that "it can be solved."[32] As he sees things, "we need a theory" of the metaphysical kind to account for continental color consciousness. Hochman's entry pairs anti-realism about race with an explanation of "biological diversity" among other "non-racial" factors that yielded social groups falsely believed to be natural major races, which he contends do not conceptually merit being called "races."[33]

Taylor prefers realism about (social) race. He contends that the word "race" has broken away from its racialist past: "Our Western races are social constructs.... Specifically, they are the probabilistically defined populations that result from the white supremacist determination to link appearance and ancestry to social location and life chances. We no longer actively and intentionally maintain this linkage in the way we used to, but the effects of earlier efforts continue to shape our life chances in ways that disproportionately disadvantage specific populations."[34] Continental color labels used in neighborhoods and schools, social networks, local and global media, etc., steer people to see themselves as members of "race" groups that socially formed through racist histories. (Social) "races *do* exist," Taylor claims.[35] He argues that Appiah's race skepticism neglects patterns of social inequality that have outlasted the natural races paradigm. Race would be a product of social reality, not nature.

Philosopher Lawrence Blum favors a middle ground between race skepticism and social construction, without racial metaphysics. Filtering the race concept through "the language of 'social construction,'" he finds, is "too fraught with confusion."[36] He proposes that by current "sociohistorical consensus," we realize that we are talking about subjective continental groupings, whereas Americans used to believe that "the groups we now call 'whites,' 'blacks,' 'Asians,' and 'Native Americans' were races" of a natural kind.[37] These continental groupings or "racialized" groups, as Blum calls them, are a social phenomenon. We might wonder about the benefit of adding a cautionary suffix to "race" if a reasonable consensus about the word's meaning already existed.

The modern idea of race bears a striking resemblance to the Socratic myth of the metals. To promote a social hierarchy of rulers, soldiers, and laborers, ancient Athens would distinguish citizens by castes said to be heritably endowed with gold, silver, or bronze. Socrates feigned reluctance to retell this "Phoenician story" that would buttress an efficient, cooperative, and anti-democratic society; some "device" would be needed to convince the citizenry to go along with an absurd if "noble falsehood that would, in the best case, persuade even the rulers." Glaucon could suggest nothing, though, to "make our citizens believe this story."[38] Neither he nor Socrates seemed to think the Athenian people's (dis)belief would turn on empirical or metaphysical analysis, visual perceptions, or emotional responses re "metals."[39] Two thousand years later, a slavery variation of the metals myth took hold through the idea of continental races. In White America, for example, "race" became the ideological-rhetorical device through which persons of European descent pseudo-rationalized, normalized, and mystified a caste system grounded in inherited freedom ("white") and inherited slavery ("black").

Blum would rather focus on the mythmaking dynamics of "racialization": he means "the treating of groups as if there were inherent and immutable differences between them; *as if* certain somatic characteristics marked the presence of significant characteristics of mind, emotion, and character."[40] Thinking in terms of "racialized" rather than "race" groups could guard against residual racialism. Highlighting the "sociohistorical consensus" about "racialized" groups might even someday convince natural race partisans that nonbelievers speak the truth.

For social race theorists, though, plain talk of race has become a credible way of describing color-conscious social organization, social identity, and social adjustment.[41] If there is little belief left in natural race, residual racialism could only be a minor factor in current social disadvantage.[42] Talk of racialized groups would appear to preach to the converted or the intransigent. Of more concern, a cautionary shift in vocabulary might destabilize color consciousness that psychologically and politically helps situate members of disfavored race groups.[43]

Whether to talk of race groups or racialized groups is not, at any rate, a pivotal issue. My philosophical intervention sets aside debate about a best terminology or metaphysics for dealing with the race concept and its heavy baggage. Weighty notions of race have been conceptually unserious; ideology is their core function. Inherited slavery obviously could not wait for sensible efforts of moral justification. Race or racialization theorists, drawn into cultures of pseudo-rationalization for entrenched social inequality, vest the race concept with absurd powers of belief and sentiment formation and (im)moral suasion. *What in the natural or social world could possibly be believed or felt about other human beings that might justify, excuse, or mitigate enslaving them upon birth, outside the reach of minimal human decency, for generation after generation, indefinitely into the future?*

My question is unkind to a standard line of denial: "Through law, custom, and popular understanding and behavior, people of African descent [in America] were turned into the racial group 'black.' They were consigned to a subordinate place in society, denied various rights, discriminated against, and subjected to norms meant to keep them separate from whites, all on the basis that they were members of 'the black race.'"[44] Here Blum seems to mistake the pretext for injustice as a major cause of injustice.[45] A capitalist institution of inherited slavery was White America's basis for constructing an Africa-identified slave caste.

The oppression of enslaved Americans and their Africa-identified descendants has not hinged on mere prejudices or possible truths about some "race" thing: violence, law, and greed were essentials. There was and is nothing to feel or learn about the nature of race/racialized groups that might significantly bear on their oppression. There is no enduringly out-of-reach answer to the question of what "race" means. There was no serious age-old question of moral justification for inherited slavery.[46] There is no possible question of moral justification for creating a comprehensively dispossessed people—marked by a stigma of brute violation, grotesque vulnerability, and any other abuse and trauma social groups can inflict on other human beings.[47]

Blame White America's caste society on colonial values dedicated to maximizing wealth. Sociologist Max Weber offered a deadpan analysis in *The Protestant Ethic and the Spirit of Capitalism* (1905):

> It is true that the usefulness of a calling, and thus its favour in the sight of God, is measured primarily in moral terms. . . . But a further, and, above all, in practice the most important, criterion is found in private profitableness. For if that God, whose hand the Puritan sees in all the occurrences of life, shows one of His elect a chance of profit, he must do it with a purpose. Hence the faithful Christian must follow the call by taking advantage of the opportunity.[48]

Anglo-Saxon Protestants assigned God ultimate moral responsibility for carrying out a "white" tribal doctrine of profit above all, with enslaved "black" outsiders sacrificed by tens of millions in honor of a despotic capitalist idol.

Rather than invite digression into questions of individual or collective blameworthiness or what truly counts as racism or racial discrimination,[49] I have focused on the function of race ideology. The contemporary race ideology-rhetoric complex revolves around esoteric natural race inquiry in symbiotic, if indirect, relation to social race inquiry. Race ideology in America more plainly meant that Africa-identified persons comprised a conspicuous out-group, forced through omnipresent terror to labor as a cheap utility producing exceptional returns on investment.[50] After slavery came expediencies of severe exclusion and nonrepair. Only denial would impute White America's vicious priorities to bad information, misunderstanding, or ill-will about some black/Negro quality perhaps had by the enslaved persons and their descendants, a possibility forever crying out for revelation or rebuttal.

Du Bois was undistracted by a post-slavery nation's pretense of guileless ignorance or raw malice about some "race" thing.[51] He argued in *Black Reconstruction* (1935) that White Americans strongly valued a "public and psychological wage" in exchange for their pan-European tribe loyalty.[52] Historian David Roediger describes a class mentality of ambiguous negation: "White workers could, and did, define and accept their class positions by fashioning identities as 'not slaves' and as 'not Blacks.'"[53] The Supreme Court defined a positive special status: "white persons" were the American caste.[54] As I interpret Du Bois's "wage" of American whiteness, it represents caste demands for political, material, and social entitlement—across class lines.[55] Destructive White behaviors happen when those demands are not met.[56]

White America's entitlement looms over progressive dreams of class solidarity, where descendants of American slavery stop thinking that their own social-political lineage matters.[57] Historian George Fredrickson describes an antebellum version of class-first idealism: "Having condemned all forms of subordination of whites to other whites as 'artificial' and unjust, [proslavery activist John H.] Van Evrie then relegated the blacks to abject and perpetual servitude."[58] As Illinois Senator Stephen Douglas had declared: "I am opposed to negro citizenship in any and every form. I believe this Government was made on the white basis. I believe it was made . . . for the benefit of white men and their posterity for ever, and I am in favor of confining citizenship to white men, men of European birth and descent."[59] The famed Lincoln-Douglas debates—on the survival of White America's union and White equality—revolved around inherited slavery and its aftermath.

Social race theorists have overstated a dominant caste's sensitivity to sincere racial belief in governing caste interests or desires. Taylor writes, for example: "The decline of classical racialism was even more evident in the structure of social life. Western nations gave up their colonies, mostly. Racialized schemes of labor exploitation, like sharecropping and contract labor, diminished in significance."[60] That alternate history misconstrues the impetus for such social progress: whether or not a decline in racialist myth played some minor role, economic and political dynamics played the leading role.[61] It was American slavery, not race, embedded in the Constitution.

Race ideology has hardly cared about the rigor of its theological, scientistic, or legal reasoning and never needed to. This can make for strange bedfellows, as Kenyan-raised/US-based writer Mukoma Wa Ngugi confirms: "The end result of the African foreigner privilege . . . is that Africans are becoming buffers between white and black America. . . . Africans experience a patronizing but helpful racism, as opposed to the hostile, threatened and defensive kind that African Americans grow up with. . . . They end up seeing African Americans through a racist lens." Ngugi adds: "Nelson Mandela once said that without African American support, ending apartheid would have taken much longer. But one will not find organisations in African countries that reciprocate. . . . And Africans in the United States tend to stay away from protests."[62] While Black Americans have long linked their freedom struggle to "pan-African" liberation, their awareness of social and political distance from black immigrants has grown.[63] Social-political lineage resists flat blackness in the US and everywhere else.

The apparent paradox of African anti-black racism is preempted when Black Americans are deemed a homeland-less, Afro-European hybrid people unto themselves—which is separable from Western race hype.[64] Indeed, "racial" Africa-identified persons in America were descendants of American slavery: their color caste status rests on this fact.[65] Ascribed membership in "the black race" covers up the specificity of Black American social-political lineage. **Theorizing the existence of continental race groups, whether natural or social, presumes the collective importance of projecting Western "racial" impressions onto the world, flat blackness especially—despite varying cultural responses to skin color and continental ancestry in different geographic, social, and political contexts.**

Capital-"B" Black Americans are known to flatten blackness as well. This appears to stem less from strong belief in natural races than from existential desire for connection with an uncertain place and people our Atlantic African ancestors were stolen from.[66] Black Americans also show a disposition to flatten nonwhiteness in general. I interpret this as an expression of Black empathy for other nonwhite peoples fighting Western domination in their homelands; historical cases include India, Ethiopia, Vietnam, and South Africa.[67] Confusion or mysticism about race does not seem implicated in that cosmopolitan form of solidarity. Nevertheless, Black America's unusual tradition of empathy for non-European national liberationists is baffling to Appiah, who immigrated to the United States as a British-Ghanaian adult.[68]

As Appiah tells it, "Du Bois died in Nkrumah's Ghana," having "never completed the escape from race." Du Bois's early metaphysics are forced upon his cosmopolitanism: "How can something he shares with the whole nonwhite world bind him to only part of it?" Appiah asks.[69] An answer can be found as early as 1915, when Du Bois wrote of international "growing interest" in "a conscious sense of unity among colored races": he cited as a major factor "the common cause of the darker races against the intolerable assumptions and insults of Europeans."[70] Cosmopolitan non-Western solidarity, apart from its political influence, might offer some psychological relief from wearisome Western race ideology-rhetoric.

Race ideology has been more an effect than a cause of subjugation, exploitation, and nonrepair. Toni Morrison distills a prime lesson about the nature of antiblackness: "The very serious function of racism . . . is distraction," particularly from White America's national commitment to unhinged "acquisition of wealth."[71] The post-1960s corollary is White national refusal to reallocate resources and opportunities as justice would demand, at a minimum, for descendants of American slavery.

A *New York Times* headline from 2020 reports: "United States is the richest country in the world, and it has the biggest wealth gap."[72] The article does not disaggregate for the vast disparity in a typical Europe-identified ($188,200) versus Africa-identified ($24,100) family's net worth.[73] Those family wealth numbers are foundational domestic legacies; inherited slavery across three centuries supplied free labor for an enormous cotton industry that was White America's economic engine.[74] Tireless speculation about black/Negro inferiority sustained a culture of race mythology, which normalized successive modes of caste oppression of enslaved Americans and their Africa-identified descendants. The details of major races logic have been inconsequential.

The core function of race ideology-rhetoric is to exert psychological and social influence that lies beyond reason, in service of endless distraction. American attitudes about drug use and police violence, for example, align with studies of racial "superhumanization" bias that reveal default compassion for persons seen as white versus harsh treatment for persons seen as (super)naturally black.[75] If racial thinking were almost cured of race ideology or switching terms from "race" to "racialized" were a nearby remedy, inquiry into the true nature of race would have faded into general disinterest. Racial thinking is still afflicted with centuries-old intrigue, ultimately leading nowhere productive.

On whether races exist, I am sympathetic to race skeptics. Social constructionists turn that question into an argument about philosophical methodology: Can the word "race," previously meant to refer to a natural kind, be reinterpreted to refer squarely to a social kind? They sideline the traditional question of natural differences of racial mind, which remains at the center of ordinary speculation and debate. But I am sympathetic to social constructionists on color consciousness in practice. My caveat is that in a "racial" caste society, the practice is incomparably more motivated by race ideology's overall goals—which principally concern a dominant caste's acquisition and retention of wealth, power, and prestige—than by beliefs or feelings about any "race" thing itself.

Joining epicycles of intrigue, race skeptics and social constructionists seem to assume that high theorizing of "race" in the twenty-first century will produce worldly dividends. This conveys an ahistorical perspective on color caste and what its collapse would require. White America has resisted any approximation of social equality for descendants of American slavery.[76] True or imagined realities of race are tangential to why. Continual variations of earnest race inquiry might be another "symbol of man's capacity for exerting maximum effort to accomplish minimal results," as Rube Goldberg said of his "fantastically complicated" cartoon machines.[77]

The central dispute over the nature of race is about mind/brain, not body and not social ontology. Skeptics and social constructionists agree that natural major races do not exist. This leaves family disagreements over meaning and use of the word "race." Little of worldly value is left to gain on the road of racial metaphysics: what, if any real thing, "race" refers to is an undecidable question until we are told what will be counted as a good answer. Metaphysicians should be even less interested in race science: the question of whether nature-based races exist becomes empirically answerable once believers state what biological X-factor thing they are looking for.[78]

Theorizing whether and how races exist, despite presentation as a common project, is a ruleless contest judged by different schools of thought for variously preferred purposes. This is a recipe for empirical-conceptual impasse. My deflationary "race" pluralism tries to overcome the impasse by readily conceding that different notions might be intelligible by different designs, depending on what different speakers believe race is that would yield their racial groupings.

Instead of always arguing over what the word "race" refers to—as if there were a singular correct or best understanding—we could adopt less loaded terms for whatever different notions are more simply meant. To get started, I propose "geoancestry": the neologism refers to continental groupings recognized through Western practices of attaching importance to visible continental ancestry. This intervention is more productive than forever arguing about "race."

Inquiry into the nature of race is unlike asking whether chairs, tigers, or Greek gods exist. Since we have a good idea what these named things are supposed to be, we are able to refer to them without much controversy. By contrast, the race concept offers no basis for confidence that we mean close to the same thing, locally or globally, other than something in the realm of continental ancestry and skin color, whyever we may care.

11

Enter "Geoancestry"

A critical point emerges when the question of race's reality is deflated relative to who knows how many possible conceptions of the "race" thing. *The widespread (mis)impression that racial thinking is central to social meanings of skin color comes from forgetting that continental color consciousness is rooted in continental ancestry, not in the race concept.* Most human beings fit physical profiles that indicate which continent(s) at least some of their genealogically traceable ancestors are from. Race hype added an elusive X factor, which race science has tried to convert into biological reality.

Above all, the race concept provided a vehicle to pseudo-rationalize inherited slavery. Oppression by way of continental color consciousness was not a new phenomenon. Before the European transatlantic slave trade, Arab enslavers took Islamic license to treat "black" African physical features as proof enough of abject inferiority.[1] The Western conceptual invention of continental race groups came later—and has inspired continuous, if increasingly indirect, inquiry into whether there might be natural truth behind clichéd anti-black/African prejudices.

My deflationary approach to moving on from "race" is uncomplicated in processing color-conscious social realities. Descendants of American slavery, for example, reappropriated the "black" label long used in disrespect by White America; identifying positively as an Africa-identified people challenged mainstream caricatures of the American "Negro" as servile, unintelligent, and brutish. Du Bois had asked, "How does it feel to be a problem?"—making personal the "Negro problem" of being treated as a caste unfit for entry "into the group life of the nation."[2] Original statement of the "Negro element" problem, in a letter to the *New York Times* soon after the Civil War, doubted that "the black population will be diminished one-fourth" through "suffering, … killing by riots, and murders": the anonymous writer lamented that "black people increase about twenty-five per cent, decennially; and you will find that in fifteen years they will amount to six millions of people, and that you have a nation within a nation."[3] Culling the Black

The Afterlife of Race. Lionel K. McPherson, Oxford University Press. © Lionel K. McPherson 2024.
DOI: 10.1093/oso/9780197626849.003.0011

American population was under consideration. Assimilation or integration was out of the question.[4]

Race skeptics may grant that color-conscious identity can become a valid source of affirmation for peoples subjected to color-conscious disrespect. Still, Appiah warns of "racial identity as a species of cultural identity."[5] A stock example is when young Black males are pressured by the criticism "acting white" for speaking standard English, excelling at school, or listening to rock music.[6] In Appiah's opinion, cultural expectations linked to racial identity come with high personal costs and group risks. Maybe Black Americans would be better off with a weaker sense of being black.[7] Color consciousness recedes in the romantic post-racial dream, liberating us from templates for living that may stifle personal expression and race-neutral affinities.

I will set aside Appiah's normative reservations about racial identity for descendants of American slavery. The issue in question is whether continental color consciousness and racial thinking are indistinguishably conjoined. My answer is no. Skeptics, social constructionists, and biological realists share the misimpression that some conception of race is integral to continental color consciousness of political importance.

To bypass irresolvable debate about what the word "race" is supposed to name and whether that thing is real, I have recommended the concept of geoancestry. We can recognize geoancestral groups instead of forever trying to rehabilitate the race concept. Keep in mind that race categories were misappropriated from Blumenbach's 1795 nonracial scheme: "Caucasian," "Negro," and "Mongolian" became his new words for peoples of Europe, Africa, and Asia, after he added a subcontinental exception for (now "Caucasian") "Northern Africa."[8] His initial scheme had proposed "principal varieties" with respect to characteristic physical appearance types by whole continents. That was two centuries before genetics analysis confirmed the implausibility of inner "race" qualities, which Blumenbach did not believe in.

The simple geoancestral membership of Europe-, Asia-, and Africa-identified groups consists of persons who have full corresponding ancestry: their genealogically traceable ancestors come from only one of the continental regions. There is slight difficulty with ascribing geoancestry in this case. Where to place western Asians and northern Africans—whose physical profiles may fail to evoke a white, yellow, or black racial people—does not trigger conceptual stress, which major races logic contrives for itself. For the geoancestry concept, that intrigue is merely a byproduct of nonalignment between geography (e.g., the entire African continent, as per

Blumenbach's initial design), a continental color label (e.g., "black"), and how certain subpopulations are seen (e.g., northern Africans as typically not-black) from a Western encultured gaze.

Many persons who are ascribed membership in a single continental grouping do not have full or nearly full corresponding ancestry. The history of Black Americans as an Afro-European people illustrates that classification by racial geography long ago gave up the pretense of objective rigor. There is neither conceptual nor empirical mystery when a geoancestral grouping does not reference the total facts of a candidate member's continental descent: this incompleteness merely reflects customary guidelines for ascribing racial identity, regardless of the stories that would rationalize racial groupings.

The point of the geoancestry concept is to demystify how Western practices of attaching importance to visible continental ancestry have marked lived experience and shaped social-political lineages. Instead of insisting that everyone use "race" in reference to continental groupings of people, I have coined a new word in order to avoid verbal conflict and mass confusion. "Geoancestry" tracks Europe-, Asia-, and Africa-identified peoples with less fuss and trouble than "race." The geoancestry concept turns away from (super)naturalist inquiry and philosophical meta-inquiry, which the Western race ideology-rhetoric complex sustains.

Geoancestry is unburdened by demands for objectivity. For example, there can be no illusion that continental ancestry alone determines geoancestral identity for persons of substantial mixed continental descent. Ascribed membership in a geoancestral group might diverge from an individual's color-conscious sense of self, where either is supportable by facts of ancestry. Such persons might have a "mixed" (e.g., African and European) geoancestral self-conception they believe is importantly unlike that of a typical member of the ascribed single continental grouping. Part II brings this theme to bear on the notion of "mixed race" identity.

In standard cases, geoancestral identity forms in alignment with Western color consciousness. Black Americans, Afro-Caribbeans, and (non-northern) Africans belong to a geoancestral grouping that comprises Africa-identified peoples. The geoancestry concept is not very revisionist about the basis; it grants the perceptual influence of Western race ideology-rhetoric. Members of different "black" geoancestral subgroups typically have visible African ancestry and see themselves, on some social-political level, as persons of Africa-identified descent despite any other continental ancestry they might have.

As with any Africa-identified national people, there is a distinctive mode of black geoancestral identity for Black Americans: they have substantial African ancestry, normally fit the wide physical profile of "blacks" in America, and identify as a people whose ancestors were enslaved by White America. Africa-identified Americans of immigrant background may well identify with Black Americans, and vice versa, as persons whose visible African ancestry renders them liable to encounter various anti-black stereotypes and biases. Yet the capital-"B" in "Black American" was not meant to be a vague racial honorific of global extension: the "B" specifically designates Africa-identified Americans of homegrown social-political lineage that is neither native nor immigrant.[9]

The binding agent for Black American geoancestral identity is historical memory of American slavery and segregation, confirmed by lived experience and mediated by a tradition of righteous resistance.[10] Specificity of social-political lineage is crucial. White America has imposed on homegrown Africa-identified Americans a stark social hierarchy built through two hundred years of official non-personhood and unfreedom that morphed into partial citizenship after emancipation in 1865, formal equality after Black rights legislation in the 1960s, and little subsequent progress toward Black social equality as a national people.[11] For Africa-identified immigrants, their deeply different relation to the United States anchors their own distinctive social lineages as recent and voluntary "black" peoples in America.[12]

The geoancestry concept would have us focus not on any color-conscious device of distraction and its mythology—be it Socratic metals, Western major races, or Hindu *varna* ("color")—but on gross injustice that the device obfuscates. By comparison, racial thinking flattens historically salient differences between "blacks," for example, Black Americans and immigrant Africa-identified Americans. The distinction speaks to local color-conscious politics and phenomenology.[13] Sociologist Orlando Patterson has shown that existential dishonor accompanied "social death" as a core feature of slavery: "in all slave societies the slave was considered a degraded person; [and] the honor of the master was enhanced by the subjection of his slave."[14] People marked by slave lineage inherited lowest social status in the society that enslaved them.[15] In the United States, descendants of American slavery remain an intensely marginalized, nearly wealthless people.[16] Outside the United States, other Africa-identified peoples may be primary targets of local antiblackness (e.g., post-colonial Suriname immigrants in the Netherlands) while Black Americans enjoy relative favor there.[17]

Racial stigma itself, not American slave lineage, is often and reflexively presented as a leading cause of Black America's caste plight. History instead shows that the American foundation of anti-black maltreatment, systemic and interpersonal, is American slavery. Inchoate notions of race sponsored race ideology-rhetoric that was a smokescreen for turning fellow human beings into heritable property under cover of White America's law. After substituting "caste" for "race," I endorse philosopher Richard Moran's observation on what is required for interpreting history in social practice: "To understand any period of American history is to understand the role of [color caste] in the law, in culture, in politics, in family life and intimate relationships, etc."[18] Coming to terms with the scope and depth of American color caste does not require that races actually exist somehow or are believed to exist.

The geoancestry concept accepts that in the United States and many other societies, persons who have African ancestry and dark skin will still "look black" and, at least in that sense, be considered black and variously treated as such. Does a persistent notion of nonelective blackness demonstrate that color-conscious continental groupings are wedded to the race concept? Not really. Labels such as "black" and "white," besides their use for designating races, can more simply signal visible continental ancestry. Obvious non-European ancestry is the global baseline for distinguishing nonwhiteness from whiteness; obvious African ancestry is the global baseline for blackness. Figurative color labels—few persons literally look white, black, yellow, or red—can be reapplied, without further perplexity, to geoancestral groups.

Some persons feel that no single continental grouping suits them: they would rather identify under a "mixed race" or "other race" category. Philosopher Linda Alcoff has theorized an expansive Latin American model of "mestizo consciousness" that could equally include "racial mixes" of all varieties.[19] In practice, however, a multiracial classification scheme appears to intensify finer-grained status distinctions. "Part of the reason for a multiplicity of descriptions for nonwhite Brazilians, particularly for those whose African descent is visible," political scientist Michael Hanchard finds, is that "such categorizations attempt to avoid the mark of blackness."[20] The distancing from blackness attests to a color-conscious social hierarchy updated from the mulatto/quadroon/octoroon system used in Latin

American slave societies. There is, anyway, no conceptual barrier to "mixed" or "other" geoancestral identities.

Ambiguous color-conscious geography is also not a problem for the geoancestry concept.[21] Are peoples of the South Caucasus countries Georgia, Azerbaijan, and Armenia supposed to be European or Asian natives? Western geographic convention invoked the Caucasus and Ural Mountains to divide the Eurasian supercontinent—which would place South Caucasus peoples in Asia, though Georgians were the "Caucasian" physical archetype. These mountains have stood for a racial barrier between the "white" European "race" separate from all others, including the off-white Caucasian: true-whiteness would have evolved in and spread across Europe over pre-ancient millennia; offshoot white European natives would later migrate, around four thousand years ago, to western Asia, southern Asia, and northern Africa, where they became off-white after slightly but discernibly mixing with the nonwhite natives.[22] Rather than indulge that fanciful scenario, the geoancestry concept ignores tales of racialist being and endogamy.

There arises no need for color-conscious gerrymandering. Where to draw geoancestral boundaries is openly conventional. Global perception roughly yields *Europe-identified peoples* ("whites") and *Africa-identified peoples* ("blacks"), while *Asia-identified peoples* ("Asians") are seen to vary more sharply by region (e.g., South, East, "Middle East").[23] It is a conceptual constraint that members of any geoancestral group must have genealogically traceable ancestry from the respective base continent. Some peoples, including northern Africans, might not fit any single grouping. The geoancestry concept roughly tracks visible continental ancestry, especially for Europe-identified and Africa-identified peoples: this reflects the pervasive influence of Western color consciousness, driven by transatlantic slavery, well before the advent of modern biology.

Today's US Census Bureau willfully labors under mysteries of racialism. Guidelines that "must adhere to the 1997 Office of Management and Budget (OMB) standards on race" define "black or African American" racial being as follows: "A person having origins in any of the Black racial groups of Africa."[24] Traceable origins in non-northern Africa, the place itself, do not suffice. While claiming "to serve as the nation's leading provider of quality data about its people and economy,"[25] the Census Bureau is silent about who "the Black [sic] racial groups of Africa" are and how much of their ancestry is required for a person to be racially black. Collecting color-conscious demographic information without "race" would be too easy for US government purposes.

The Census Bureau has always known what White law explicitly confronted: many Africa-identified persons in America have substantial European ancestry (through antebellum paternity). From 1790 through 1860, all enslaved US inhabitants were officially counted under the category *Slaves*. In 1850, a *color* category introduced "black" to the census. In 1900, a *color or race* category introduced "race" to census forms. As of 1930, demographic distinction between "black" and "mulatto" was canceled by decree that any person "of mixed white and Negro blood should be [marked] as a Negro, no matter how small the percentage of Negro blood."[26] In 1970, "black" returned to the census as "Negro or black." From 1990 to date, census forms specify a *race* (alone) category—but with no standalone category for Asia-identified persons, who get specific boxes for national origin identities.[27] The Census Bureau has compiled and reinterpreted such information under a misleading banner: "Measuring Race and Ethnicity Across the Decades: 1790–2010."[28]

Of course, the US government was not trying to measure a "race" thing when enumerating enslaved Americans. Subsequent "black" and "mulatto" labels, initially under a *Free Inhabitants* category (1850–1860), referred directly back to persons of "Slaves" caste lineage. Race ideology-rhetoric in America was used to mystify a fixed boundary between an Africa-identified "Slaves" caste and a Europe-identified "Free White" caste.[29] This is historical reality, not any type of theory—which can be confirmed by reviewing US census forms.[30]

The association in America between "Slaves" and black/Negro raciality has proven incredibly strong. Demographers, journalists, philosophers, and many others continue to retroactively superimpose a "race" thing onto the "Slaves"/"black" caste.[31] An otherwise helpful article makes this claim, for example: "When the census began in 1790, the racial categories for the household population were 'free white' persons, other 'free persons' by color, and 'slaves.'"[32] To be clear, "color" is not equivalent to "race," and "slaves" is not a race category.

"Color or race" designation for the "Slaves"/"black" caste did not begin until 1900. The 1890 census filled the transition from mere "color" by using a nameless entry: "Whether white, black, mulatto, quadroon, octoroon, Chinese, Japanese, or [American] Indian."[33] Recognition of fractional African ancestry indicates that the government, from 1850 to 1920, was prepared to promote "mixed" subcategories for descendants of American slavery.

These days, the Census Bureau claims that its demographic race categories "reflect a social definition of race recognized in this country and not an attempt to define race biologically."[34] We are to assume that earlier categorizations tried to track—under "pressure [from] race science theories"[35]— some racial X factor. The Census Bureau has entered a new century using an old-school "Caucasian" guideline for contemporary "white" racial being: "A person having origins in any of the original peoples of Europe, the Middle East, or North Africa."[36] Implicitly excluded are persons who might also have visible African ancestry. There is no explanation of "original peoples" by race. Directly referring to traceable origins in Europe, the place itself, does not suffice for the Census Bureau's priorities. American demography in the early twentieth century began officially normalizing racialist distraction from a caste system whose continental color labels marked freedom or slavery.

The notion of "original peoples" has turned out to be double-edged artifice. Archaic humans who lived in Eurasia, the Neanderthals, interbred there with modern humans and remain present in the DNA of non-African peoples.[37] When did Europe's hybrid humans become racially white originals? How was X-factor whiteness imagined to exist and spread across the land to establish a single white race on the European (sub)continent? The Census Bureau's "social definition" mirrors race science that conjures, from unsteady mixtures of geography, physical appearance, and ideological rationale, a distinctive subspecies/race quality had by "original peoples" of Europe (with their genetically traceable Neanderthal ancestry).

The US government is playing pretend about its latter-day "white" race: whiteness would socially extend from Britons, Germans, Italians, Norwegians, and so forth to Saudi Arabians, Iraqis, Iranians, Moroccans ....[38] (On the 2020 census form, Egyptian is listed as an example of "white.") This pretend construction derives from race science, not social practice. Recall the Supreme Court's *Thind* decision (1923), reaffirming the 1790 Naturalization Act, by which only persons of (nearly) full European descent were "white persons."[39] The Census Bureau today prefers to smuggle in race science via dishonest appeal to social whiteness—suppressing the "Caucasian" category's hierarchical distinction between white natives of Europe and off-white resettlers in Africa and Asia. Everyone understands that whiteness has an exclusive ancestry base in Europe and socially excludes Muslim Arab peoples.[40]

Some peoples, such as western Asians, might not be strongly identified with any color-conscious grouping. Ambiguous, unstable, or inexhaustive categories are an open feature of the geoancestry concept. Unlike a racial framework, the geoancestral alternative does not struggle and bluff to apply itself—whether through high theory, scientistic rigor, or bureaucratic directive.

The idea that Black Americans comprise an Africa-identified geoancestral subgroup is compatible with belief among members that they somehow belong to a "black" race. Among Europe-identified Americans, belief could persist that they belong to a "white" race. Nevertheless, reusing figurative color labels for reference to geoancestral identities would not involve a giant semantic leap. Social meanings of skin color were always rooted in visible continental ancestry, not in the race concept, as a marker of social difference.

12

Giving Up the "Race" Ghost

Human beings can be classified on the basis of qualities they have in common with some persons and not others. Criteria for the qualities we sort by—anatomical sex, eye color, nationality, religious heritage, etc.—usually are transparent: we know what we are talking about and looking for. Classification practices become dubious when we sort by features for which criteria are conceptually underspecified, esoteric, or elusive. Such is the fate of race classification. Various phenomena related to continental ancestry have been proposed as the basis for globally sorting people by some "race" thing. The quandary is that regardless of the criteria chosen, classification outcomes will notably depart from previous guidelines for race membership that would support applying those very criteria.

If race is to be strictly a function of continental ancestry, then the sheer number of persons around the world who have mixed ancestry will confound standard classification into major race groups. The American standard had been "one drop" of African ancestry (proxy for American slave lineage) to properly distinguish black from white. By contrast, Western race ideology-rhetoric holds that true-whiteness approximates European purity.

If visible continental ancestry is cited on grounds of common sense, then biological full siblings could be of different races, children could be of a different race from their biological parents of a same race, and no persons would look able to "pass for" being a different race (because they would have to look like the race they are supposed to be). Such scenarios run counter to the major races logic of descent.

If genetic traits are sought in a quest for scientific clarity, then no members of a particular race could lack the distinctive genetic traits—a prospect that breathes life into race essentialism and would kick out presumed members whose genetics do not comply. For example, all blacks/Negroes, and only blacks/Negroes, would have certain genetic traits (the holy grail of scientific racialism).

Nor will a combination platter of race criteria get rid of embarrassments and anomalies that undermine customary guidelines.[1] We are left with

The Afterlife of Race. Lionel K. McPherson, Oxford University Press. © Lionel K. McPherson 2024.
DOI: 10.1093/oso/9780197626849.003.0012

inevitable confusion about the idea of continental human races. Hence fervent believers in natural or social race inquiry trudge on like desert wanderers—theorizing in circles or, eventually exhausted, redrawing their maps so that arrival might seem near.

At this late stage, no new natural or social facts about human beings will be discovered that could best resolve controversy and deliver a master meaning for "race." The controversy turns on which notion of race might ever gain title to the word—not on some prior facts that would settle disagreements over the metaphysics of race, the world of perceived racial being, or the politics of "race" usage. Insistence on race inquiry has become curiouser and curiouser in the twenty-first century. My advice has been to stop following trails to no singular destination in search of some "race" thing(s).

Compare an observation made about Kofi Annan, the Ghanaian diplomat, by a White American journalist friend: "What struck [her] was 'the difference between Africans' and African-Americans' thinking about race.' She said, 'Kofi didn't think about it. It was never a question of race for him,' and she noticed that it was the same for a Nigerian aristocrat he hung around with. 'They didn't think they were in any way involved. They didn't even think of colonialism as a matter of race,' she said." The local race mythology was a phenomenon reflecting American slavery. To recall Annan's verdict: " 'I'm not a nigger, I'm an African.' "[2] Saying the quiet part loud in apartheid America, his exclusion was reversed. Race intrigue proved irrelevant upon further review: "black" Africans could receive service in a Jim Crow "white" space.

The geoancestry concept offers a straighter, cleaner path than "race" for understanding continental color consciousness in local or global contexts. There is nothing deeply mysterious about color-conscious recognition of Africa-identified, Europe-identified, or Asia-identified peoples. Yet and still, designation by social-political lineage will generally be more informative for reasonable purposes—which include demographically distinguishing (Africa-identified) descendants of American slavery from other Africa-identified national peoples, whether in the United States or abroad.

PART II

DECODING "MIXED RACE" AS AMERICAN CASTE

I finally made up my mind that I would neither disclaim the black race nor claim the white race; but that I would . . . let the world take me for what it would; that it was not necessary for me to go about with the label of inferiority.

—the narrator
James Weldon Johnson, *The Autobiography of an Ex-Colored Man* (fiction, 1912)

1

Color-Conscious Identity

I have argued for moving on from the race concept. If natural races are supposed to be real, nothing of broad consequence is at stake unless they are characterized by significant differences of mind/brain. Perpetual suspicion that "blacks" are naturally less intelligent than "whites" mostly explains why race inquiry has generated hundreds of years of ongoing intrigue.[1] If races are said to be groups formed through social construction, this leaves open whether they do not somehow also exist through nature. The target of race inquiry is conceptually hazy and endlessly debated: we collectively do not know what thing we are talking about.

As a reminder on my own usage of continental color labels, "black" will mean Africa-identified (given non-northern African ancestry); "white" will mean Europe-identified (given European ancestry); "yellow" will mean Asia-identified (given Asian ancestry).

Amid European transatlantic slavery and colonialism, modern Western philosophy was an esteemed source of vicious race ideology-rhetoric.[2] The idea of race as continental groups of people came long before modern biology, whose race science presupposes protoscientific categories. Western geographic convention first divided the (super)continent of Eurasia into white Europe and yellow Asia; by the early 1800s, the convention divided Africa into a black non-northern (aka "sub-Saharan") region and not-black "North Africa." Various notions of race have been theorized around these color-conscious boundaries.

Our only shared understanding of race is that it typically involves skin color that roughly indicates continental ancestry in full or part. Customary guidelines for ascribing a color or race identity (aka "racial identity")— "white people have some European ancestry and look white," "black people have some African ancestry and look black," etc.—do not require positing some "race" thing at all. Evidently, the idea of continental human races also involves an X factor, sustained by race inquiry itself.

The racial X factor escapes singular interpretation. Various meanings for "race" may be made to fit some things or other that are naturally or

The Afterlife of Race. Lionel K. McPherson, Oxford University Press. © Lionel K. McPherson 2024.
DOI: 10.1093/oso/9780197626849.003.0013

socially real about groupings of people. To that extent, races might intelligibly be said to exist, or not, depending on what different speakers believe race is. Fundamental confusion and disagreement about the racial factor are not going away. Philosophers of race have conceded as much and kept on philosophizing: for example, "We don't consider our four [very different] views to exhaust all of the plausible metaphysical views one can have about what race is and whether it's real."[3] Meanwhile, race science has clung to centuries-old racial geography and membership norms.[4] Debate about the nature of race has no decidable resolution.

To displace "race," I have proposed the term *geoancestry*. The alternative language clarifies that continental color consciousness is rooted in skin color as sign of geographic descent, not in the race concept. Thus the label "black" gets applied to (non-northern) Africa-identified persons, including Black Americans and many others who have substantial mixed African and European ancestry.[5] By substantial, I have in mind, for reasons of lived genealogy (not biology), any continental component that is equivalent at least to one great-grandparent. De-racing "blacks" would not involve deep difficulty figuring out who could be black—that is, Africa-identified—by geoancestry: they will have visible African ancestry or a parent, grandparent, or great-grandparent who does.

The concept of geoancestry reflects Western practices of attaching importance to visible continental ancestry. Capital-"B" Black Americans comprise an Africa-identified geoancestral subgroup, namely, descendants of American slavery. To grasp this requires no investment in a natural or social reality of "the black race." As I argue in Part I, race intrigue adds superfluous and distracting lines of empirical-conceptual inquiry. Neither some "race" thing nor race ideology is integral to identifying Black Americans as a national people of homegrown slave lineage, 80 percent of Africa-identified persons in the United States circa 2020.[6]

Of course, the geoancestry concept cannot displace the idea of continental human races by academic fiat. Color-conscious social identity will stay tethered to racial thinking for the foreseeable future. In this second part of the book, I use race language merely for reader accessibility. My terms "geoancestry" and "geoancestral" provide ready substitutes.

Continental color consciousness by race is less straightforward than ordinary use might suggest. **Racial identities are not a simple output when persons have continental ancestry that defines a raced group: monoracial identity accommodates mixed continental ancestry.** The connection between racial being and racial identity is mysterious in the statement, for instance, that "the average white person in the United States has 5 percent traceable black genes" and "the average black person [in the United States] has 25 percent traceable white genes," as the *New York Times* imagined in the 1980s—as if genes that indicate continental ancestry would be "race genes."[7] The habit of ascribing monoracial blackness to persons who have substantial mixed continental ancestry is glossed over by race science, which relies on local methods for assigning people to race categories.

Customary racial identifications do come with necessary conditions. It is constitutive of the idea of continental races that black racial identity requires having some genealogically traceable African ancestry, white racial identity requires having some genealogically traceable European ancestry, etc. A person who claimed a black racial identity and had no African ancestry would be confused or mistaken. However, race categories are not defined by a uniform amount of associated continental ancestry. Monoracial blackness typically has been indicated by visible African ancestry, whereas whiteness has required (nearly) full European descent.

In the United States, the default practice of recognizing only monoracial identities has met with growing opposition in recent decades. What if natural major races do not exist or if many people belong to more than one race? Doesn't subjective judgment (e.g., "looks black," "looks white") undermine any pretense of objective race classification? Why, then, should individuals of Afro-European descent accept the racial identity that fits a monoracial label their society would apply to them? Such questions are fair enough.

2

Not Regular Blacks

Many people in the New World have substantial mixed African and European ancestry. Some of these people prefer to think of themselves as neither black nor white. From their perspective, no adequate racial label applies to them if mixed-continental, or "multiracial," identity is unacknowledged in their society. A single-continent, or "monoracial," identity might distort their raced self-conception: they ask why they should espouse racial membership that fails to represent their complete racial self. This gives rise to the *wholeness claim*. Furthermore, a monoracial identity might violate their self-respect: they ask why they should be expected to conceal their true racial being. This gives rise to the *integrity claim*. The wholeness and integrity claims are central to arguments for recognition and affirmation of multiracial identity for people whose ancestry crosses continental boundaries of racial geography.

Feelings of misaligned racial identity are not an abstraction. Consider this from a magazine profile of Eldrick "Tiger" Woods: "Now, in a nod to his mixed heritage, he calls himself Cablinasian and sidesteps African American causes.... And in 2000, when the NAACP asked black athletes to shun South Carolina to force the lowering of the Confederate flag from its capitol dome, Woods declined, telling *Sports Illustrated* magazine: 'I'm a golfer. That's their deal.'"[1] He self-identifies as "Cablinasian"—his neologism for Caucasian, black, American Indian, and Asian. (To deter speculation about his inner motives, I switch to a hypothetical Woods*).

The wholeness and integrity claims could explain why Woods* is unwilling to identify as black. Less than half of his ancestry is African, and he does not see himself as a black person. Not unreasonably, he might believe that being pressured to adopt a black identity depends on unfair monoracial guidelines for racial identification. Who says he has a special obligation to support social and political causes that resonate outside his mixed-race sense of self? If the reply is that Woods* is racially black or that his support as a Black American would serve good purposes, this only begs the question about the nature of his racial self. Critics might insinuate that his nonblack identity is

The Afterlife of Race. Lionel K. McPherson, Oxford University Press. © Lionel K. McPherson 2024.
DOI: 10.1093/oso/9780197626849.003.0014

driven by self-loathing about the African component of his ancestry, opportunism in marketing himself as a post-racial icon, or other dubious motives. Sympathizers might feel that his declaration of a multiracial identity signals social progress.[2]

Multiracial advocacy comes in different varieties. Philosopher Ronald Sundstrom does not view race as primarily a private matter. He renounces any use of *multiracialism—the pursuit of recognition and affirmation for certain people as mixed race*—that would promote a politics of "naive color blindness."[3] His chief worry is that laissez-faire racial identity could become a pretext for eliminating fair consideration of race in law and policy. Overall, Sundstrom's goal is to reconcile multiracialism with a color-conscious vision of social justice. While sensitive to historical patterns of racial bias, he wants multiracial identity to be taken seriously along with monoracial identity.

Gaps between monoracial social reception and multiracial psychology have posed a long-standing challenge. "The criticism that multiracial, or mixed race, is an impossible identity [or] simply a variant of racial passing available to the brown is cruel," Sundstrom objects, "because it dismisses the particular experiences of multiracial persons and precludes any possibility of the existence and legitimacy of multiracial identity."[4] Negative social dynamics that influence the shaping of racial identity can change over time. The "tragic mulatto" from American literature and film—a black-white biracial character suffering from social and existential alienation in a color-conscious netherworld—has grown faint since the mid-twentieth century as an image of persons who identify as racially mixed.[5] Yet this positive shift does not shed doubts about multiracialism.

Western racial culture's global reach has meant that persons of non-European color who look closer to white often experience more favorable social treatment and outcomes. Part of the story involves colorism: also called "skin color stratification," this process "privileges light-skinned people of color over dark in areas such as income, education, housing, and the marriage market" (and in perceptions of intelligence, talent, trustworthiness, etc.).[6] The power of persuasion through racial labeling can be so strong that "black people [*sic*] who identify as multiracial are perceived as more attractive than black people who only identify as black," a recent study finds, "even when controlling [for] skin tone, hair color, and eye color."[7] Unvarnished blackness in twenty-first-century America remains a significant barrier to opportunity.

Racial stigma underlies colorism, with darker skin seen to indicate that a person has relatively more non-European ancestry. As economist Glenn Loury explains, "dishonorable meanings" are ideologically attached to certain "bodily marks," which reproduces "spoiled collective identities."[8] Members of a raced group might well denounce the notion that their "spoiled" identity reflects who they are. Descendants of American slavery, for example, have a group norm that Africa-identified persons are expected to resist anti-black stigma, at least by endorsing a black social identity.[9] (I return to this theme in Part III.) Thus Black Americans tend to be wary of individuals who have obvious African ancestry and claim a nonblack identity.[10] There is the appearance of racial climbing, a self-serving effort to move up from regular blacks, so to speak, in the color caste hierarchy.[11]

The skeptical question, in short, is this: Does multiracial identity convey an ambition to gain social and psychological distance from monoracial groups of lower caste? Psychologist and social scientist John Dollard made the following observation in his 1937 field study of a town in Mississippi: "The Negroes share sufficiently in American society to want to be fully human in the American sense and to this end they prefer to be as light as possible, since the white caste seems to grant some recognition, informally of course, to the lighter colors. Consciousness of color and accurate discrimination between shades is a well-developed Negro caste mark."[12] In the absence of multiracial categories, Black Americans were highly sensitive to degrees of European admixture.

Fast-forward to our time. With helpful candor, Sundstrom describes four objectionable ways that multiraciality functions, especially in relation to blackness:

> First, multiracial identities exist because race theories have constructed nonwhite identities as being inherently inferior and antipathetic. Second, [multiracialism] provides an individual solution to racist oppression that not only fails to question racist social structure but also depends [on] and profits from that structure. Third, [multiracialism] is a vehicle of individual and communal declaration that reinforces [these] types of racism. Fourth, it sets up a multiracial group . . . as a buffer zone [of] brown folks seeking white privilege via the mulatto escape hatch.[13]

Sundstrom suggests that despite the ethical hazards, "responsible multiracial politics" happens more in the United States than in Brazil and South

Africa—a plausible comparative judgment as far as it goes.[14] Studies by research psychologists, however, have suggested that multiracialism plays into the American black-white bipolar social hierarchy.[15] All told, we cannot simply infer that persons who identify as mixed race are seeking a promotion in color-conscious status.

My reference to a bipolar caste system bears emphasizing. Contrary to complaints about a "black-white binary" in the United States,[16] there is an acknowledged spectrum: nonwhite types other than black are recognized between the black and the white poles. Maybe among non-Europeans who are not Africa-identified, fear of being maltreated like "blacks" induces fear of being socially grouped with them opposite "whites." I am unsure how else advocates of multiracial identity could imagine ordinary Americans believing that "by and large, there are really only two races: black and white."[17] Anxiety among the "mixed" about racial proximity to Black Americans would not be, by Sundstrom's account, a politically responsible motivation.

The prevalence of irresponsible multiracialism will be difficult to determine in a society that discourages explicit racism. This lack of transparency is reinforced when people underestimate their own bad racial attitudes and behaviors. Research psychologists have found, for example, that a majority of white people in the United States show an "implicit bias" for "whites" relative to "blacks."[18] Although Americans who identify as racially mixed might show a similar bias, their multiracialism could be separately motivated by a positive desire for recognition of their nonblack identity. To speculate that persons who have visible African ancestry and identify as racially mixed do so from an anti-black orientation is to express a kind of skepticism that multiracial advocacy rejects.

In my use, *the term "multiracials" will designate persons who conceive of themselves as racially mixed—without the presumption that they somehow are racially mixed, except in the sense that they have mixed continental ancestry.* I proceed as if multiracials in America are rarely motivated by racism, self-loathing, or opportunism.[19] Mainstream recognition of multiracial identity is not, I argue, a threat to the social stability and political importance of Black American social identity.[20] Americans who identify as black can respectfully coexist with those who have visible African ancestry yet identify as racially mixed. We should avoid easy speculation that black-nonblack multiracials are trying to suppress their blackness or are engaging in group betrayal. There is conceptual and ethical room for multiracial identities.

At the same time, I argue that multiracial advocacy has lent credence to racialist sensibilities by evoking notions of natural racial being. An odd feature of the monoracial classification scheme in America is that most persons counted as black/Negro—from Frederick Douglass to Michelle Obama—have had substantial European as well as African ancestry.[21] **The practice of assigning persons who have visible African ancestry to a monoracial "black" grouping does not and has never tried to fully track the substantial components of their continental ancestry.** A monoracial scheme emerged because of commitment to a rigid black-white social hierarchy in a post-slavery society, not because of faulty unrecognition of purported facts, experiences, or psychologies of race.

3

Negro Blood

President John F. Kennedy announced that "race has no place in American life or law" when he sent the National Guard to desegregate the University of Alabama in 1963.[1] Fifty years later, post-racial ambition seems out of touch with the realities of intergenerational gross injustice.[2] The disconnect fuels anti-black stereotypes: if historical injustice is not a leading cause of Black American social inequality, then the greater balance of causes must be internal to the group. In effect, such conjecture shifts causal responsibility for the group's stalled progress onto forces mostly detached from legacies of inherited slavery and enforced segregation of its known descendants.[3] Yet race does have less of a place after the Civil Rights Acts of the 1960s.[4] Distinguishing between legal and social post-raciality brings this into focus.

In a legally post-racial society, law and policy no longer recognize racial identity as a relevant consideration, apart from prohibiting explicit discrimination. The US Supreme Court set the bar high in *Regents of the Univ. of California v. Bakke* (1978), which decreed that "racial and ethnic classifications of any sort are inherently suspect." Use of a "racial quota" in university admissions was ruled in violation of the Civil Rights Act of 1964, though "the goal of achieving a diverse student body" would be "sufficiently compelling to justify consideration of race . . . under some circumstances."[5] Provision for "affirmative action"—namely, to counter white anti-black phenomena surrounding inherited slavery and enforced segregation—transitioned from an obvious justice rationale intended "to promote full equality of employment opportunity" for Black Americans to a vague diversity interest applied to certain non-Anglo groups (aka "people of color").[6]

The Supreme Court set the bar higher in *Adarand Constructors, Inc v. Peña* (1995): "all racial classifications," even for "benign" purposes, were to be treated as "seldom" relevant and therefore subject to "strict scrutiny."[7] Civil rights lawyer Michelle Alexander, author of *The New Jim Crow: Mass Incarceration in the Age of Colorblindness*, interprets the impact

The Afterlife of Race. Lionel K. McPherson, Oxford University Press. © Lionel K. McPherson 2024.
DOI: 10.1093/oso/9780197626849.003.0015

in legal practice: "Demand that anyone who wants to challenge racial bias in the [criminal justice] system offer, in advance, clear proof that the racial disparities are the product of intentional racial discrimination. . . . This evidence will almost never be available."[8] Formal neutrality veiled real histories of using race classification exclusively against known descendants of American slavery and other nonwhites.[9]

A socially post-racial society, by comparison, would no longer treat racial identity as a factor in public or personal life. This is a distant prospect in the United States. Basketball legend Bill Russell recounts his skeptical wonder in the 1960s about "'I didn't notice' liberals," who professed colorblindness toward Black Americans.[10] Skin color remains socially conspicuous today; for example, a large majority of White Americans (reportedly over 75 percent) have no nonwhite friends in their "core social networks."[11] Whether this scarcity of interracial friendship is best explained by cultural difference, racial bias, or social distance does not change what de facto segregation looks and feels like. Public and personal awareness of racial identity is well documented.

Under "race," 90 percent of respondents to the 2010 US census identified as either "White" (72 percent), "Black, African Am., or Negro" (13 percent), or specific Asian national origin (5 percent); only 3 percent identified as "two or more races"; and 6 percent identified as "some other [one] race."[12] Strong patterns of monoracial self-identification are about more than mere facts of continental descent: Americans are seeing themselves and others as members of raced groups that carry social meanings. If perceptions of race had steeply declined in social importance, disparate color-conscious treatment would hardly persist. Discrimination against "blacks" among "equally qualified" job candidates, for instance, showed no change between 1989 and 2015.[13] Legal post-raciality has done little to overcome pervasive exclusion of descendants of American slavery.

It is not news that monoracial blackness in America typically has absorbed non-African ancestry. The US Census Bureau estimated in 1918 that "as high as three-fourths" of the population designated as black had European ancestors.[14] As it turns out, "mulatto" first became a census category in 1850; this lasted until a 1930 decree that persons "of mixed white and Negro blood" counted only as "Negro," before "two or more races" was added in 2000.[15] These changes were not driven by any developments in race science; boundaries of race classification shift across place, time, and politics, with at most a veneer of empirical guidance.[16] This poses a paradox about the connection between mixed continental ancestry and multiracial identity. The paradox is dissolved once we extend due respect for racial identity preference.

A version of the belief that "Negro blood" makes a person black still holds sway.[17] Persons who have obvious African ancestry will have to insist they are not black, and their non-identification will usually be viewed as a technicality. Tiger Woods, whose father was a descendant of American slavery, is widely seen as black—despite identifying as "Cablinasian" and public knowledge that his mother is Thai. Barack Obama, whose father was Kenyan, is widely seen as black—despite the occasional claim that he is not truly black because his mother was white.[18] Regardless of intrigue about black racial being, Africa-identified Americans who are not descendants of American slavery are not Black Americans in the historical sense. Few Black Americans would count as racially black if having a traceable blood relative who had no African ancestry were disqualifying: as a group, Black Americans are an Afro-European people.

Being counted in America as "Slaves," "black," or "Negro" was never about race essences or natural categories. **Equating monoracial blackness with non-northern African ancestry was a practical measure to ensure unambiguous whiteness during and after American slavery.**[19] White America's law established guidelines for denying white/European caste membership to persons known to have African ancestry: homegrown "blacks" were to comprise a group forever marked by the ascribed dishonor of slavery and subordinate personhood.[20] Those guidelines in turn socially reinforced American color caste and racial thinking.[21]

Of special note was the state of Virginia's "Racial Integrity Act" of 1924, which imposed the "one-drop rule" to assign everyone at birth to either a "white" or "colored" category. Article 5 put forth conditions for the preservation of whiteness:

> It shall thereafter be unlawful for any white person in this State to marry any save a white person. . . . [T]he term "white person" shall apply only to the person who has no trace whatsoever of any blood other than Caucasian; but persons who have one-sixteenth or less of the blood of the American Indian and have no other non-Caucasic blood shall be deemed to be white persons.[22]

Through curious erasure of race ideology, "Caucasian" has become a popular substitute for "white," as in (true-white) European.

The Racial Integrity Act's author, physician Walter Plecker, bluntly spelled out its rationale:

> There are in the State [of Virginia] from 10,000 to 20,000, possibly more, near white people, who are known to possess an intermixture of colored blood, in some cases to a slight extent it is true, but still enough to prevent them from being white.
>
> In the past it has been possible for these people to declare themselves as white, or even to have the Court so declare them. Then they have demanded admittance of their children into the white schools, and in not a few cases have intermarried with white people. . . .
>
> These persons, however, are not white in reality, nor by the new definition of this law. . . .
>
> Their children are likely to revert to the distinctly Negro type even when all apparent evidence of mixture has disappeared.[23]

That statement of race essentialism took for granted a black-white social hierarchy; in-between racial types were of lesser political importance. As head of the Virginia Bureau of Vital Statistics, Plecker oversaw his law until 1946. The state's strict defense of whiteness had legal force through 1967.

Neither the living nor the dead were immune from the spirit of race essentialism in Louisiana, where Naomi Drake spent years in control of the Bureau of Vital Statistics for New Orleans. She was finally ousted in 1965 after the public grew tired of her refusal to issue thousands of birth and death certificates for "white"-listed individuals she loosely speculated might have African ancestry.[24] Such extreme caution proved impractical in a region with a politically influential "mixed" Creole population. In addition, the demise of enforced segregation of Black Americans eased the state's commitment to tracking individuals whose blackness through American slavery was unobvious.

4

Caste Impurity

The purity doctrine of whiteness equates truly being white with having no trace-able non-European ancestors. Multiracial advocacy takes this doctrine to cast into limbo the race classification of persons who have mixed European and non-European ancestry. Race theorist David Goldberg claims, "Americans are trying to come to terms with the fact that racial purity is a thing of the past."[1] In truth, White America has long dealt with mixed continental being. The Supreme Court scoffed at routine "race mixing" of a certain gendered kind: its *Dred Scott* decision (1857) noted that white/European "men in every grade and position . . . daily and habitually acted upon" an enslaved Africa-identified people "so far inferior, that they had no rights" to protection from those men's "private pursuits," including routine sexual violence resulting in childbirth.[2] (See Part One, Section 4, "(Sub)continent-Wide Human Types?") The "mixed" children were born into slavery.

An alternative US history serves as a foil for philosophers of multira-cialism. "Oppressed and dominant communities," Linda Alcoff claims, "fail to acknowledge and accept mixed offspring, and both value a purity for racial identity."[3] This poses a false equivalence; the notion of racial purity is a norm for white racial identity, not for racial identity as such. Naomi Zack makes the same mistake when she claims that "the term 'race' always connotes pu-rity," even as she implies otherwise: "The [one-drop] regress built into the meaning of blackness [in America] and the exclusivity built into the meaning of whiteness underscore an unfairness. . . . Blacks can never undo or dilute their race."[4] Black racial being would survive admixture or impurity. In any event, multiracialism risks turning American slavery policy into an issue of race classification fairness.

Persons of mixed continental descent who have obvious African ancestry are acknowledged, as black, in America. It is especially strange to pin a ra-cial purity myth on Black Americans. Color caste ensured white/European hegemony through hypodescent, which relegated mixed-caste children to the legal status of the parent of lower caste. Visible descendants of American slavery always knew that having mixed continental ancestry was compatible

The Afterlife of Race. Lionel K. McPherson, Oxford University Press. © Lionel K. McPherson 2024.
DOI: 10.1093/oso/9780197626849.003.0016

with black/Negro identity. Sociologist E. Franklin Frazier described, for example, a "small upper class" of Black Americans whose "light skin-color was indicative not only of their white ancestry [sic], but of their descent from the Negroes who were free before the Civil War, or those who had enjoyed the advantages of having served in the houses of their masters."[5] Those "upper class" Africa-identified persons still belonged to the black caste. Racial purity was a white/European American preoccupation regarding whiteness only.

Among Black Americans, by contrast, there is an open tradition of pragmatic color consciousness: they primarily view racial identity through the lens of family and community ties, visible ancestry, and social reception. This is on display in James Weldon Johnson's Harlem Renaissance novel *The Autobiography of an Ex-Colored Man*.[6] The light-skinned, anonymous narrator was born to a black mother soon after the Civil War, eventually decides to inhabit whiteness (with retrospective regret), marries a white woman who dies during their second child's birth, and raises the children with a white racial identity. Contrary to hypodescent, Black American custom implies that persons of mixed African and European descent are functionally white when they look, grow up, and live as white. Whether unknowing "passers" might be some or other natural race ceased to be a socially salient question.[7]

The more complicated type of passing is when persons hide immediate colored ties. Bandleader Ina Ray Hutton was a real-life case: the "blonde bombshell of rhythm" died as white in 1984 and had her blackness rediscovered in 2007. Hutton's ex-colored life began under the guidance of her mother in 1925, at age 8, when she faded from Chicago's black community.[8] Generally, Black Americans were reluctant to expose active passers: the custom drew from sympathy for individuals who chose to escape explicit American anti-black oppression. Historian Allyson Hobbs writes, without irony, "Perhaps it was a larger sense of racial solidarity that compelled blacks to protect the identities of those who lived 'on the other side.'"[9] Most persons or their children who were able to pass for white/European assimilated into whiteness.

Even White Americans used to realize that purity would be a tricky standard for membership in their color caste. During the 1895 South Carolina constitutional convention, delegate George D. Tillman pleaded, "If the law is made as it now stands respectable families [will] be denied the right to intermarry among people with whom they are now associated and identified." The ramifications were sweeping: "It is a scientific fact that there is not one full-blooded Caucasian on the floor of this convention. . . . It would be a cruel injustice and the source of endless litigation, of scandal, horror, feud, and bloodshed to undertake to annul or forbid marriage for a remote, perhaps obsolete trace of Negro blood."[10] Tillman's plea, which won the day, was re-pressed within a generation. In the early twentieth century, White America legally enacted variations of the one-drop rule to protect against incursion by descendants of American slavery who might claim partial whiteness.

Reputed fear of "race mixing" had already inspired tough measures after slavery was formally abolished in 1865.[11] Legal historian Barbara Welke summarizes the gendered landscape: "Miscegenation laws . . . aimed most fundamentally to prohibit interracial unions involving white women. . . . It is telling that the case that first gave constitutional sanction to miscegena-tion law—*Pace v. Alabama* (1880)—involved a relationship between a black man and a white woman and the case long celebrated for bringing the ed-ifice of miscegenation to a close—*Loving v. Virginia* (1967)—involved a marriage between a white man and a black woman."[12] Racial terror went from a legal to an extralegal weapon. Thousands of Black American war veterans between 1877 and 1950 were targeted for lynching because, an Equal Justice Initiative report concludes, "they represented the hope and possibility of black empowerment and social equality."[13] White America's gratitude for their military service might have eased the way for intimate relationships between black males and white females—which could erode the color caste hierarchy, with white men encountering greater sexual and labor competition.[14]

Contingencies of sex between white males and black females, by compar-ison, could be adequately handled by hypodescent. Any resulting children would be legally black and, under prohibition of interracial marriage, outside the bounds of paternal obligation. Virginia's precedent-setting slavery law of 1662 addressed the dominant mixing scenario:

> WHEREAS some doubts have arisen whether children got by any Englishman upon a negro woman should be slave or free, *Be it therefore enacted and declared by this present grand assembly,* that all children borne in this country shall be held bond or free only according to the condition of the mother, *And* that if any christian shall commit fornication with a negro man or woman, he or she so offending shall pay double the fines.[15]

That law established inherited slavery via the enslaved (Africa-identified) "condition of the mother." Long after slavery was abolished, as late as the 1970s, monoracial hypodescent could disallow legal exceptions for unobvious African ancestry.

The case of Susie Guillory Phipps is often cited in support of multiracialism.[16] Phipps sued the state of Louisiana to change the designation on her birth certificate to "white" from "col."—for "colored," meaning black/Negro by the one-drop rule—which she discovered on her passport in 1977, at the age of 43.[17] Her perspective was unequivocal: "Nothing is bad about being black if you're black, but I'm white. I never was black. I was raised white. . . . My children are white. . . . My birth certificate is the only thing that says I'm black." A few years into the lawsuit, Phipps told her white husband. His response: "It didn't bother me. But it doesn't mean I'd jump up and marry a black woman or want my kids to. . . . But she's white, and if she's not black, why not right it?"[18]

By Phipps's own admission, the "col." on her birth certificate had no previous impact on her life. She was born in the 1930s, when legal blackness in Louisiana was confirmed through "common repute." This translated into a mismatch between what Phipps claimed to know about her family background and what local authorities might have known. In 1970, the state changed the law to specify that "anyone having one thirty-second or less of 'Negro blood' should not be designated as black." A Louisiana legislator offered this cryptic rationale for the change: "There were some folks of questionable origin who thought they were white—they'd been passing for white for years—but on the records they were considered black. We passed the law so they could legally pass."[19] Being faux white could suffice for unspecified legal purposes. In the official psyche, true-whiteness approached European purity.

Tillman's 1895 plea for a truer understanding of whiteness stayed lost to history. Louisiana repealed its "Negro blood" law under national pressure from the Phipps case. The update led back to "any traceable amount"—a nebulously small quantum tracking homegrown slave lineage—as "the proper way to define a Negro." Phipps's lawyer complained that "there's no way out if you look white and all your records say you're colored." The almost insurmountable "preponderance of evidence" standard for correcting one's legal race stemmed "'a surge' of requests from applicants wishing to change the description of their race on old birth records."[20] We are left to infer the black-to-white preferred direction of change.

There had been no clerical error for Phipps. The state paid to confirm her genealogical, not genetic, truth: "Authorities in Louisiana have traced Mrs. Phipps's ancestry to an 18th century French settler who had a white wife and black mistress [*sic*]. The mistress Margarita was the wife's slave. When the wife died, ownership of Margarita passed to a son. [F]ather and son became embroiled in a legal contest that left excellent records of Mrs. Phipps's lineage."[21] This information did not end her quest. By the time her case failed in a Louisiana appeals court in 1985, Phipps had spent upwards of $50,000 trying to change her legal race to "white."[22] The public story ends there.

Phipps's case cannot be enlisted for multiracialism or against the one-drop rule: the plaintiff argued that she was a white racial person by sensible application of the rule. She cared enough to protest too much, forgoing privacy, time, and money in pursuit of legal whiteness for intangible motives. A confidently white person in 1970s Louisiana might have brushed off a clerical error unnoticed for forty years. Phipps evidently was less confident than that. Her goal was certified membership in the white/European caste rather than the historical "Slaves" caste.[23]

In fairness, American race ideology-rhetoric is bewildering. The Census Bureau embraced atrocity, absurdity, and psychopathy in its 1860 population survey presented to Congress: "That a race forcibly transported to a state of slavery here, from a country without history, literature, or laws, whose people remain in barbarism, should not have been able to attain to an equality in morals with their intellectual superiors is not surprising."[24] For all its religiosity, White America was unfazed by Isaiah 5:20: "Woe unto them that call evil good, and good evil; that put darkness for light, and light for darkness; that put bitter for sweet, and sweet for bitter!"[25]

White America's "race" law was never about who persons truly are by natural race, intellectual ability, biological family, or personal identity. **The purpose of a color-conscious social hierarchy was to conserve unambiguous white/European hegemony and a black/Africa-identified bottom caste marked by American slavery.** White America had no other weighty interest in guarding entry to whiteness and legally assigning persons with traceable African ancestry to a monoracial black category. By and large, living as white has not required legal membership in the white caste or escape from legal blackness.

Inherited slavery is the heart of America's color consciousness. Caste identity circa 2020 tells us the following: "The typical black family has just 1/10th the wealth of the typical white one. In 1863, black Americans owned one half of 1 percent of the national wealth. Today it's just over 1.5 percent for roughly the same percentage of the overall population."[26] Many Americans, preferring to highlight positive change in "racial" attitudes, laud nonmaterial or token progress.[27]

To quote Martin Luther King Jr., writing a hundred years after emancipation:

> Negroes have proceeded from a premise that equality means what it says, and they have taken white Americans at their word when they talked of it as an objective. But most whites in America in 1967, including many persons of goodwill, proceed from a premise that equality is a loose expression for improvement. White America is not even psychologically organized to close the gap—essentially it seeks only to make it less painful and less obvious but in most respects to retain it. Most of the abrasions between Negroes and white liberals arise from this fact.[28]

White reactionaries are more transparent about their color-conscious values. Hence Donald Trump's 2016 presidential campaign mantras "Make America Great Again" (MAGA) and "in the good old days" championed the pre-1960s caste order.[29]

5

The Browning Thesis

Cliché notions of the United States as "a melting pot" and "a nation of immigrants" ignore certain historical facts.[1] Most of the celebrated mixing has occurred between Europe-identified persons whose known ancestors chose to come to the country. Atlantic Africans were stolen and imported as property; White law licensed their violation, extending to their descendants. It should come as no surprise that Black Americans typically have substantial European ancestry, and millions of White Americans are known to have African ancestry (roughly 10 percent in the South and 3.5 percent overall).[2] There is extensive diversity of ancestry within and across continental boundaries.

Contemporary demographic patterns have led some commentators to forecast "the browning of America" in the next few decades. The browning thesis holds that the United States will become a "majority-minority" country largely due to Hispanic and Asian immigration, shifting public focus away from the traditional black-white divide.[3] Like other forecasters, Sundstrom envisions a politically empowered nonwhite majority. Yet he distinguishes multiracialism from a vision that "racially homogenizes" nonwhites under a vague "brown" label.[4] Americans who have obvious African ancestry are not implied under that label.

The catchphrase "black and brown people" glosses over color-conscious status differences. Empirical evidence suggests that "multiracial identifications" of persons who have Hispanic or Asian backgrounds "show much less social distance from whites than from blacks, signaling the likely emergence of a black-nonblack divide that continues to separate blacks from other groups, including new nonwhite immigrants."[5] Meanwhile, the term "people of color" is rejected by Black Americans who stress the domestic priority of our justice claims as descendants of American slavery.[6] Umbrella nonwhite solidarity might rest on the belief of different peoples that their anti-discrimination interests overlap, not on empathy.[7] Black-brown coalitions would be more tactical than principled.

The Afterlife of Race. Lionel K. McPherson, Oxford University Press. © Lionel K. McPherson 2024.
DOI: 10.1093/oso/9780197626849.003.0017

High hopes for nonwhite solidarity are thus optimistic. Political scientist Claire Kim gives the example of younger Asian Americans who reject an "Asian-first" ethos by refusing to join White American opposition to "race-based" "affirmative action." Nevertheless, she cautions about assuming "the unity of nonwhite interests [when] Asian Americans are positioned differently from black people in the US racial order."[8] Upward mobility for non-black "minorities" has been judged against the bottom caste status of Black Americans.[9] Default solidarity among nonwhite Americans seems more a liberal fantasy "of color" than a potential political force.

The browning thesis presumes that current patterns of American color caste configuration will remain fixed. This assumption is unwarranted: social forces that shape racial identities are susceptible to change.[10] According to the Census Bureau, "Prior to the 1930 Census, Mexicans had been categorized as White."[11] White America could maintain a majority by accepting into its ranks as many people of Latin American descent as would be plausible. Individuals of "Hispanic origin" (that is, of New World Spanish descent) are doing their part through race self-identification: the 2010 census saw 53 percent choose "white," 36.7 percent choose "some other race" (Hispanics were 99 percent of this category overall), 6 percent choose "two or more races," and 2.5 percent choose "black."[12] The United States has a history of immigrants being viewed as less than equally white before being accepted as white. More pertinently for Hispanics—a census demographic of US government creation in the 1970s[13]—those who do not have obvious non-European ancestry could prove able to shed the group's nonwhiteness in favor of a "white ethnic" identity.[14]

Immigrant group racial transformation is the theme of historian Noel Ignatiev's *How the Irish Became White*. "The Irish who emigrated to America in the eighteenth and nineteenth centuries," he observes, "were fleeing [class-based] caste oppression and a system of landlordism that made the material conditions of the Irish peasant comparable to those of an American slave."[15] Upon arrival, they were treated little better than free Africa-identified people, segregated in urban ghettos, and limited to low-wage work. Ignatiev argues that while "white skin made the Irish eligible for membership in the white race, [they] had to earn it" in America.[16] He overstates the case somewhat. Irish immigrants rose in white subcaste status to become equally white citizens. (The Census Bureau counted all European immigrants as "white.") Italian immigrants experienced a similar process of transformation from lesser to equally white.[17] Ultimately, no Europe-identified people failed to "earn" whiteness.

Americans of (Ashkenazi) Jewish background are a less distant case. Jews were often unwelcomed in white gentile spaces. Unlike other nonblack immigrant origin groups, they had a tendency to ally themselves with Black Americans.[18] Opportunity to become equally white opened for American Jews after the Black rights movement of the 1950s and 1960s. Before 1970, the rate of Jewish intermarriage was 13 percent; by 1996, the rate had risen to 47 percent; and in intermarried households, only 33 percent of children were "raised Jewish." The upshot of these numbers: "sharp differentiation between affiliated and unaffiliated Jews, and significant differences between the in-married and intermarried . . . suggest an increasing polarization in Jewish connections."[19] This has yielded a paradox of American integration: Jewish identity morphed from a marker of quasi-racial otherness into an ethnocultural difference within the white/European American polity.[20]

Racial transformation is not a matter of natural facts. No such facts were adduced for Irish, Italians, or Jews in America; the phenomenon was social, not genetic. Social amnesia mostly erased public memory of their former status as other than equally white. Such history points to a sizable subgroup of Hispanics also being accepted as equally white, accelerated by declining fluency in Spanish and greater assimilation into mainstream Anglo-American life and culture.[21] According to the Census Bureau in 1999, "the Hispanic population is predominately [sic] White," based on self-identification "not intended to be scientific." Explanation of the "White Hispanic" construction was weak: "This group can be thought of as the Hispanic portion of the White population, or as the White portion of the Hispanic population."[22] A more honest explanation is that only Hispanics seen to fit a broadly European physical profile are socially eligible for American whiteness.[23]

In brief, future racial demographics in the United States will not match how some immigrant origin groups are seen today. Any (sub)group's transformation into White Americans would unlikely be a result of intermarriage; commonplace intermarriage would follow, not precede, transformation.[24] When people rise in caste status, they become suitable for what formerly would have been downward marriage (hypogamy) for members of a higher status group.[25] **American racial reconfigurations do not herald social post-raciality, which would require the collapse of color caste, not movement up or down the black-white bipolar hierarchy.**

6

To Be Legally Not-Black

Multiracial advocacy challenges the major races paradigm and emphasizes the psychological importance of racial identity. An immediate question is whether these aims can be brought together: mixed races could not exist unless there were unmixed major races. (On the dubious notion of pure races, see Part I, Section 9, "Renewed Race Science.") Monoracial identities were built on racialist conjecture and customary guidelines, which are unstable sources for multiracial identity. Aware of this tension, Zack protests that natural major races are thought to be real by most Americans, who "do not recognize the existence of mixed-race groups as categories distinct from the [few] major racial groups."[1] She argues that Americans who have mixed continental ancestry should be officially entitled to declare their mixed-raceness despite the nonexistence of natural major races.

Respect for racial identity preference could require state neutrality about recognition.[2] "If people who think they are monoracial have a right to racial identities, then people who are biracial or multiracial also ought to have that right," Zack contends.[3] Legal anthropologist Virginia Domínguez gets closer to the substantive issue in post–apartheid America: "To redress a legal injustice, [the Supreme] Court permits racial classification by institutions. The question is whether . . . protecting the rights of a sector of the population that has historically been subjected to systematic discrimination infringes on the rights of individuals to opt not to be racially classified and to identify themselves racially according to their own criteria."[4] Laissez-faire race self-identification could prove increasingly unreliable when attempting to track historical patterns of social inequality.

(This is a good place to reiterate that Black Americans are a people of substantial mixed continental descent and can readily identify through their distinctive social-political lineage as descendants of American slavery, without referring to a race category.)

American law has largely ceased governing racial identity. This dates to the Supreme Court's 1967 *Loving* decision that decriminalized miscegenation. Chief Justice Earl Warren wrote, "The fact that Virginia prohibits only

The Afterlife of Race. Lionel K. McPherson, Oxford University Press. © Lionel K. McPherson 2024.
DOI: 10.1093/oso/9780197626849.003.0018

interracial marriages involving white persons demonstrates that the racial classifications must stand on their own justification, as measures designed to maintain White Supremacy." He drew the following conclusion about marital license: "Under our Constitution, the freedom to marry, or not marry, a person of another race resides with the individual, and cannot be infringed by the State."[5] Specifically, the Court held that anti-miscegenation laws violated the Fourteenth Amendment's Equal Protection and Due Process Clauses.

Legal scholar Cheryl Harris analyzes how "White Supremacy," as Chief Justice Warren called it, utilized monoracial identity pre-*Loving*: "The state's official recognition of a racial identity that subordinated Blacks . . . elevated whiteness from a passive attribute to an object of law and a resource. . . . Whiteness as the embodiment of white privilege transcended mere belief or preference; it became usable property, the subject of the law's regard and protection."[6] Harris cites the legal doctrine, last reaffirmed in 1957, "that to call a white person 'Black' is to defame her" because "the allegation was likely to cause injury." Black Americans could not sue for racial defamation because White law held they had no "race reputation" to defame.[7]

Under pressure from the Black rights movement and more militant alternatives,[8] White Americans stopped arguing that whiteness should be an official basis for special rights and privileges. The prior arrangement enjoyed majority conviction or complicity well into the twentieth century.[9] For example, the Supreme Court's *Corrigan v. Buckley* decision (1926) permitted white people to enforce racially restrictive housing covenants, which across the country intensified exclusion of descendants of American slavery. The Court reversed itself in *Shelley v. Kraemer* (1948), by which time white resistance in Washington, DC, and other cities had hardened.[10] Legal anti-black housing discrimination did not end until the Civil Rights Act of 1968. The last major battle for white freedom to defy court-ordered remedies was the Boston school desegregation crisis of 1974.[11] By 1980, President Ronald Reagan's cynical race ideology-rhetoric of "colorblind" law had taken hold.[12] "Affirmative action" became an enduring pretext for post–civil rights era white resentment toward Black Americans.[13]

Generally, Black Americans have endorsed color-conscious social policy in pursuit of justice claims that race neutrality would leave stranded.[14] For visible descendants of American slavery, counting as black has been less a personal choice than a social reality that manifests in virtually every aspect of public life. The American idiom of racial identity followed monoracial convention, not the latest theories and psychologies of racial being. This basically explains why multiracial identity has struggled for formal modes of recognition: Africa-identified persons who had lighter skin were legally black along with those who had darker skin. (Discrimination "litigation could be greatly affected," the *New York Times* speculated in the run-up to the 2000 census, "if a significant number of blacks [*sic*] list themselves as being of more than one race.")[15] Recognition of racial identity is not and was never a freestanding issue in American law.

Formal recognition of multiracial identity is more credibly supported by what I call civic fairness. As an aspiring liberal democratic society, the United States would promote social toleration of different reasonable values and preferences of its members.[16] This ideal is not limited by what elite judges at various points in history have decided that the Constitution protects. Racial identity can express personal conscience. Zack unhelpfully suggests that because some persons *are* mixed race, race queries should include "a distinct category for mixed-race black and white Americans."[17] She claims that "to deny mixed-race adults or children identities as mixed is to discriminate against them on racial grounds."[18] Such trafficking in purported facts of race overreaches the goal of due respect for racial identity preference.

A balanced approach to formal recognition is already available: race queries by state and non-state institutions could routinely include "one or more" categories as an option, absent good reason not to. Good reason exceptions would depend on whether there is a compelling public interest in asking respondents to select a monoracial identity—for example, to gather information that better tracks home loan discrimination against persons who have visible African ancestry.[19] Apart from a compelling public interest, civic fairness would recommend giving respondents the option to select a racial identity of their preference or none at all.

Freedom of personal conscience also would open the door to eccentric statements of color-conscious identity. "How would you describe yourself?" queries that only imply race or ethnicity leave a loophole for respondents to check "white," "black," "Asian," etc. despite knowing they have no associated continental ancestry. This is reminiscent of English comedian Sacha Baron Cohen's "hip-hop journalist" character "Ali G," who unclearly means something other than African descent in claiming to be a black man.[20] Whether and why any individual's ancestry non-conforming identity should be accepted by ordinary members of a continental grouping does raise thornier issues, of course.

Less controversially, ancestry and color-conscious social identity can come apart when ancestry is not the intended reference.[21] Consider white abolitionist John Brown, who was sentenced to death for leading a deadly antislavery raid on Harpers Ferry, Virginia, in 1859: "Now, if it is deemed necessary that I should forfeit my life, for the furtherance of the ends of justice, and mingle my blood . . . with the blood of millions in this slave country, whose rights are disregarded by wicked, cruel, and unjust enactments,—I say, let it be done."[22] His radical commitment to the cause of Black American liberation inspired W. E. B. Du Bois to write a biography whose preface closed: "This book is at once a record of and a tribute to the man who of all Americans has perhaps come nearest to touching the real souls of black folk."[23]

Du Bois judged Brown "an honorary black man" who embodied political blackness in antebellum America.[24] Frederick Douglass had expressed a similar opinion: "His zeal in the cause of freedom was infinitely superior to mine. . . . I could speak for the slave. John Brown could fight for the slave. I could live for the slave, John Brown could *die* for the slave."[25] Persons who do not have non-northern African ancestry can be politically black. (This is a theme in Part III.)

Political values and commitments aside, multiracial identity is conceptually no less valid than monoracial identity. Race science has barely tried to explain how persons of substantial mixed African and European descent could be racially black/Negro.[26] The one-drop rule for (mono)racial blackness gained social and psychological influence that veiled its primary function as a legal contrivance, namely, to preempt Africa-identified "mixed" persons from seeking rights and privileges through claims of partial whiteness. Most descendants of American slavery could have made such claims by the time of emancipation in 1865.[27]

Racial identity is an artifact of society, not a natural extension of race categories. The sole conceptual constraint on racial identity is associated continental ancestry. **When a compelling public interest is not at stake, civic fairness supports formal recognition of racial identity that claims more than one continental source.** Appeal to a legal right of racial recognition is neither credible nor necessary.

7

Seeking Separate Social Status

Outside of formal recognition, American multiracial advocacy presses the following argument. Now that (mono)racial identities are not dictated by law, people should feel free to espouse a multiracial identity they believe fits their mixed-raceness. Lack of equal social recognition is a barrier: multiracials experience lingering suspicion that their racial identities are less worthy of respect. Another barrier is the dominant custom of identifying people by major race. At the same time, heightened public awareness of the race idea's shoddy nature has undermined monoraciality's command over common sense. Historical circumstances that socially excluded multiracial identity are long past.

A growing number of Americans openly identify as racially mixed.[1] Many persons who do not espouse a multiracial identity might prefer to if they did not fear social skepticism, with scrutiny focused on those whose African ancestry is visible. Sundstrom believes that not extending separate but equal social recognition to multiracials is akin to asking gays and lesbians to remain "in the closet"; he objects to a "demand" that multiracials "hide themselves from view in the public sphere" and practice "a public-private distinction in personal identity."[2] They might be seen and treated as black while feeling they are deeply different from regular blacks. Multiracials who are out of the closet could take pride in their nonconventional racial identity.

But Sundstrom's closet analogy is strained. Being "out" as mixed race has not carried grave threat of legal jeopardy or violent hostility, unlike homosexual identity.[3] Generally, Black Americans are not asking anyone to keep multiracial identity private: the tradition is to encourage persons whose African ancestry is visible to see themselves and identify as black; persons whose African ancestry is unobvious are mostly left to their own devices. White Americans are also not asking: the white caste loses nothing when persons whose non-European ancestry is visible identify as racially mixed. A certain scenario, however, does trigger resentment among Black Americans and White Americans alike: when persons check more than one

The Afterlife of Race. Lionel K. McPherson, Oxford University Press. © Lionel K. McPherson 2024.
DOI: 10.1093/oso/9780197626849.003.0019

race category box except when they believe there is a possible benefit, such as "affirmative action," in checking only the "black or African American" box.[4]

There is no denying that public response to multiracial self-identification is guarded in the United States, where (non-Hispanic) persons are presumed to accept whichever monoracial label appears to fit (if they have associated continental ancestry). Persons who "look black" come under greatest suspicion of being fraudulent or in denial if they do not identify under the category "black." Yet multiracialism has made inroads over the past thirty years.[5] The social legitimacy of multiracial identity is now in circulation, and the one-drop rule has morphed into vaguer guidelines for monoracial blackness. These developments highlight conceptual friction at the core of racial identity.

A tacit notion has emerged that Black Americans, despite being of Afro-European descent, are racially black—perhaps unlike multiracials whose mix includes the same components.[6] If those multiracials are thought to be "whiter" in some racial sense than average Black Americans, that could merely amount to having a larger component of European ancestry and looking "less black." Responsible multiracialism would reject the suggestion that such differences warrant a separate race category. The ethical case for recognition of multiracial identity must lie elsewhere, not in a notion of mixed-raceness that enables skin tone to serve as an implicit proxy for some continental "blood" quantum threshold.

Traditional Black American doubts about multiracialism speak to an expectation that visible descendants of American slavery will join in resisting injustice that targets their social group. For example, "the black vote" has hovered around 90 percent for the Democratic presidential candidate for over forty years—a phenomenon best explained by the Republican Party's reactionary caste politics after the end of enforced segregation of Black Americans, not by loyalty to the Democratic Party's tedious incrementalism.[7] This Black consensus expresses political solidarity under systemically unjust conditions. Historical experience has taught Black Americans that having visible African ancestry is sufficient for risk of being badly maltreated as a descendant of American slavery. Whether many Black Americans believe they truly are monoracially black/African is marginal to lived experience.

The question remains why Americans who prefer to identify as racially mixed, not black, should accede to other people's expectations. Furthermore, individuals need not identify as black to join in opposing racism. The normative answer from Black America is twofold: (1) opposition to anti-black stigma calls for identifying publicly as "black" if the conventional label fits; (2) reparations and other justice claims for Africa-identified descendants of American slavery will be harder to pursue if there are fewer declared members of the group.[8] I would agree that these legitimate concerns fall short of ethically requiring personal conscience to yield to customary guidelines for monoracial blackness.

Sundstrom is sympathetic to color-conscious political solidarity. He contends that by virtue of their white-adjacent social position, multiracials who have light skin (and visible African ancestry) "are uniquely situated to cause or perpetuate racist harm." However, he argues that their "special obligations to commit to anti-racist principles and actions" can be met without suppressing their alternative, nonblack racial identity.[9] The bottom line for multiracialism is that individuals should not, in Zack's phrase, "be compelled to belong to an association" (by race) they do not believe truly fits them.[10] Separate social recognition of multiracial identity—say, as "biracial" rather than "black"—would lower a barrier to respect for multiracial identity. But the notion of due respect is cloudy and prompts asking, "Respect for what, exactly?"

By agential respect, I mean respect for exercise of personal choice. This does not require substantive respect, that is, respect for reasons behind a personal choice. American color-conscious identity now more freely expresses individual preference. The US government decades ago stopped dictating caste membership; the "race" thing cannot be determined merely by natural facts; and color categories remain highly charged in a society where "black" has signaled lowest social status. Agential respect would have the public use a racial label that individuals prefer, whatever their reasons, in reference to their continental ancestry in full or part.

The complication is that multiracial advocacy seeks further mainstream investment in social recognition. As Zack puts it, if persons "want to identify as racially mixed, they should be allowed to do so, and respected for their choice."[11] She does not specify grounds for substantive respect for multiracial identity. Suspect motives are unsurprising and well known, which is why Sundstrom defends a "responsible" multiracialism in the first place. Observers who withhold judgment will usually be conveying impersonal doubts; substantive respect is not warranted by default when persons disclaim the label of a lower caste for which they fit customary guidelines. Since Black Americans as a social group comprise a "colored" bottom caste, most could not choose to be widely seen as not-black: they are paradigmatic "blacks" in America.[12] This Africa-identified group had included all candidates who have obvious African ancestry, which is the historical sign of American slave lineage.

Maybe it would be disrespectful if many Americans continued, in their private minds, to see as racially black those persons who have obvious African ancestry and claim they are not-black. Multiracialism seeks social recognition that involves more than personal choice and public civility: we are to see as racially mixed those persons who tell us they prefer to be identified that way in the minds of other people. I turn now to considerations that might provide a distinctive, ethically compelling rationale for why multiracial identity warrants substantive respect in a color caste society.

For the live-and-let-live approach to race self-identification I have outlined, the modest conceptual constraint is that persons must have associated continental ancestry. Tens of millions of Americans, including a vast majority of descendants of American slavery, would be eligible by that standard to identify as racially mixed. But Sundstrom is dissatisfied with widely permissive guidelines for racial being: he rejects an "individualistic libertarian conception of absolutely autonomous identity."[13] Freedom of personal conscience about race self-identification could be too free. Having biological parents who belong to different monoracial groups is supposed to be a basic warrant for multiracial identity. Working backward from that formula, Black Americans dating to slavery would have the same warrant via European-American paternity.

Recall the wholeness claim that persons who identify as racially mixed should not be pressured to adopt an identity that does not fit their complete racial self.[14] A gap between their self-conception and other people's perceptions can leave multiracials feeling alienated from themselves and from persons who accept an adjacent monoracial identity.[15] This experience of alienation could add urgency to normalizing a race category that multiracials believe better represents who they are. "If individuals cannot be identified, in the third person, as mixed race," Zack asserts, "then it is impossible for them to have mixed-race identities, in the first person.[16] Substantive respect—going deeper than the public civility of agential respect—becomes vitally important. Social affirmation of mixed-raceness would be for multiracials a background condition for feeling psychologically whole.

Racial alienation from self and others is a theme that looms large in the multiracialism literature. Alcoff writes about her childhood experience as the daughter of a (light-skinned) "mixed Spanish, Indian, and African" Panamanian father and a "white Anglo-Irish" American mother: "In Panama, my sister and I were prized for our light skin. . . . There, the mix itself did not pose any difficulties." The sisters moved with their mother to Florida: "As much as was possible, we began to pass as simply white. . . . But [our] incorporation into the white Anglo community induced feelings of self-alienation, inferiority, and a strong desire to gain recognition and acceptance."[17] The non-European components of her ancestry were unobvious, so "the mix itself" was unobvious and cannot explain her color-conscious difficulties in Florida. It appears Alcoff felt comfortable with an intermediate (multi)racial status in Panama yet uncomfortable with the light-skinned privilege that traveled with her to the United States, where she gained (mono)racial status as a member of the white/European caste.

Multiracial classification schemes can support a social hierarchy as capably as a monoracial scheme. Political scientist Michael Hanchard unpacks the situation in Brazil: "*Negro* remains a racial category many people do not want ascribed to them. . . . Focus on the numerous color categories in Brazilian racial politics can obscure [the] racial meanings [that] limit individuals' ability to simply choose their own racial category."[18] Robust bias against persons with dark skin is at odds with Alcoff's hope for a revamped "mestizo identity" that could equally include "all forms of racial mixes" and avoid "a proliferation of specific mixed identities."[19] Trying to sweep various kinds and degrees of mix into a single shared identity might intensify exclusion. Hanchard finds in both the United States and Brazil the "equation of blackness with sloth, deceit, hypersexuality and waste."[20] New-school mestizos would gain social distance from blackness, leaving intact the social norm of treating as "black" those persons of mixed African and non-African descent who are dark-skinned.

Blackness in America has all along been open to admixture—which is precisely the trouble for multiracialism. White America imposed monoracial hypodescent through the one-drop rule, particularly to govern the caste status of children born via Europe-identified males and enslaved Africa-identified females. Alcoff would instead reimagine descendants of American slavery as "mixed race persons": they might give themselves over to a "mestizo consciousness" that would be "a double vision, a conscious articulation of mixed identity, allegiances, and traditions."[21] Such hybridity is already characteristic of Black American social identity.

Du Bois famously articulated a homegrown black phenomenology in the early twentieth century: "It is a peculiar sensation, this double-consciousness, this sense of always looking at one's self through the eyes of others. . . . One ever feels his two-ness—an American, a Negro; two souls, two thoughts, two unreconciled strivings."[22] The public fact that descendants of American slavery have mixed continental ancestry has been compatible with their caste status as "blacks." Monoracial convention has failed, however, to flatten the dual nature of their distinctive black identity. Capital-"B" Black Americans are a national people of Afro-European descent, with an American historical memory of enslavement and resistance.[23]

A practical reason that Black Americans generally do not experience mixed-race alienation is that Americans who have obvious African ancestry grow up being seen as black. Compare Alcoff's "ambiguous identity" under a revamped mestizo construct: "I am not simply white nor simply Latina, and the gap that exists between my two identities . . . a gap that is cultural, racial, linguistic, and national, feels too wide and deep for me to span. . . . Peace has come for me by living that gap."[24] The ambiguity enables evading the specific question of caste identity (or subcaste identity, for example, "White Hispanic"[25]). This evasion is socially unavailable to darker-skinned "blacks" in America, where general anti-black stigma will restrict worldly access to mestizo consciousness.

Some persons might decline to state any racial or otherwise color-conscious identity. Consider the American tennis player who has obvious African ancestry and dreams of being portrayed on film by White American actress Julia Roberts: "I don't really identify myself as white or African American [or biracial]: I'm just me."[26] A stance of that type takes color-conscious identity off the table. Moreover, ahistorical individualism in a color caste society is likely to (re)introduce feelings of alienation: hopefuls will have cut themselves off from honest engagement with social reality that happens outside personal control.[27]

By contrast, Zack's vision for achieving a post-racial society does not defer the question of raced self-conception. Multiracialism would be the first step toward "a culture of microdiversity" where race categories splinter and erode: "The next step after microdiversity is racelessness," she asserts.[28] Yet Latin American custom suggests that ambition to leverage multiracial identities for social progress is probably doomed. Multiracials cannot ignore that the notion of mixed-raceness is reinforced by demarcation of lower

124 PART II. DECODING "MIXED RACE" AS AMERICAN CASTE

status monoracial groups. Racial identity in places like the United States, Brazil, and South Africa will be constrained by color caste dynamics.

Social freedom for lighter-skinned "people of color" to climb up from monoracial nonwhiteness is "the mulatto escape hatch" that worries Sundstrom. They would have to push back, as conscientious multiracials, against their access to relative skin-color privilege. **Multiracialism had advertised a remedy for racial alienation from self and others, not a cover for evasion about color-conscious social hierarchy.**

8

A Colored Breed Apart

The wholeness claim emphasizes that for some persons of mixed continental descent, social affirmation of their mixed-raceness would remove a psychological obstacle to a complete racial self. This seems to imply a tight natural relation between alleged racial being and the facts of continental descent. Monoracial guidelines such as the one-drop rule tell a different story: the components of any person's mixed continental ancestry are only a necessary and not also a sufficient condition for racial identification.

For reasons of lived genealogy, I have suggested that African ancestry equivalent at least to one great-grandparent can be substantial enough to sustain black geoancestral or racial identity through social-political lineage. Multiracialism similarly presumes that candidates will have more than merely any traceable ancestry one's racial self might prefer to identify with. But multiracialism rejects laissez-faire preferences for racial identity— which risks buying into notions of natural racial being. What else could truer criteria be thought to rest on?

American slavery law addressed the central dilemma for multiracialism, which Black Americans embody. Most realize at some level of awareness that they have European ancestry through antebellum paternity.[1] Few seem preoccupied with racial self-alienation as compared to vulnerability and distress in a society bent on devaluing their claims to equal social standing.[2] To quote Toni Morrison: racism's "hoped-for consequence was to define Black people as reactions to White presence. . . . Nobody really thought that Black people were inferior. . . . They only hoped that [we] would behave that way."[3]

There has been no exodus from Black American social identity after the 2000 census provided the option. Neither science nor popular opinion has pushed to reassign visible descendants of American slavery to a "mixed race" category. Open acknowledgment that Black Americans are a people of mixed African and European descent is met with a public shrug, whether tied or not to scientist faith that visible African ancestry is a natural and reliable sign of monoracial blackness.

The Afterlife of Race. Lionel K. McPherson, Oxford University Press. © Lionel K. McPherson 2024.
DOI: 10.1093/oso/9780197626849.003.0020

History explains why visible descendants of American slavery seem to have endorsed a monoracial identity. Their raced self-conception primarily reflects their social-political lineage defined by American caste, backed by legal designation under the US census category *Slaves*. Visible African ancestry was the color of American slavery. After emancipation, the Africa-identified caste was counted as "black" or "mulatto" under the census category *color*.[4] Nor have Black Americans treated the one-drop rule as if persons of fellow social-political lineage who pass for white/European are, in natural truth, monoracially black/African.[5] This is unsurprising since descendants of American slavery have no practical investment in Western race ideology-rhetoric.

Civil war rendered the "Slaves" label defunct (1865): an alternative conceptual framing for that caste became necessary for demographic purposes. Census forms show that "color" remained on the scene for "black" persons until 1950, when "race" alone took over for the first time. Awareness of a "black" self or social identity is for Black Americans characterized by historical memory of being Africa-identified descendants of American slavery and by lived experience of treatment as such. Although some Black Americans might feel otherwise, our social identity as a national people—despite pervasive ideology-rhetoric determined to keep "race" somehow alive—has never depended on belief in some black/African natural race phenomenon. (See Part I, Section10, "Racial Metaphysics of Distraction.") Colonial Virginia's inherited slavery law (1662) appealed only to the enslaved "condition of the mother."

The central challenge for multiracialism lies in explaining what is categorically different about multiracials who have the same substantial components of continental ancestry as most Black Americans. Zack claims, "To belong to a race in the United States involves a shared past, recognition by other large groups of racially designated people, and some kind of community or ideal of solidarity—none of which exists or ever has existed for individuals who could be designated mixed black and white race."[6] A historian points to a different possibility in Haiti: "White-mulatto solidarity was finally forged on the island. There would be just two classes of citizens, free men and slaves." That plan was finalized thirty days earlier; but on September 23, 1791, "the [National] Assembly repealed the May decree, thereby returning Haitian mulattoes to the same status as slaves."[7]

As I discussed in "Caste Impurity" (Part II, Section 4), persons in the United States who could be designated black-white biracial have belonged to a raced group. Those whose African ancestry was visible were counted by White America as "Slaves" yet began viewing themselves in the nineteenth century as an oppressed Africa-identified American people.[8] Zack's claim that "all of the mixed-race people are [identified as] black" is incorrect.[9] Most Americans of Afro-European descent who were seen to "look white" eventually were counted and counted themselves as white.[10] That explains why vanishingly few descendants of American slavery who "look white" today identify as black or Black American.[11]

In short, if the substantial components of a person's mixed continental ancestry were necessary and sufficient for racial identity, most Black Americans would be counted as mixed race. Continental color consciousness has never been about the facts of continental descent alone. White America's color caste system was in the business of servicing a slavery society that made a mockery of race science pretensions.[12]

The social reality of color caste puts advocates of multiracial identity in an awkward position to complain about "a need to be recognized for what we are racially—if anyone is anything racially."[13] That complaint implies there might be a categorical racial difference between "mixed black and white race" Americans and most Black Americans (dating to slavery). By virtue of what facts could there be racial misrecognition of "mixed race" persons? The American habit of mistaking, as it were, many persons of mixed African and European descent for racially black would clearly not be akin to confusing Sikhs and Muslims, for example, who are distinct religious peoples.[14]

Advocates of multiracial identity are quiet about the prospect of reclassifying most Black Americans as "mixed race." Sundstrom understands why fixation on the facts of continental descent would be a mistake: "Belonging to one or another race is largely, though not exclusively, tied to racialized genealogy. [D]ifferent races are not considered by widely held social mores and legal practices to mix equally. . . . Thus racial categories already include those of mixed genealogies." He cautions that "activists [who] base multiracial identity on simple genealogy do grave damage to their own cause by reinforcing [erroneous] biological conceptions of race and standards of racial purity."[15] Multiracialism has had a tendency to traffic in notions of natural race. Progress in the United States would involve more than bringing mulatto/quadroon-type distinctions under an umbrella "mixed race" category.

The problem is that Sundstrom flirts with the genealogical trap he warns against—as if multiracials form a racial group *prior to* their life experiences and alternative identity formation. "Multiracial persons," he asserts, "find themselves born into a world riven by racial fault lines, a situation they frequently carry within their bodies and identities."[16] That assertion is misleading: multiracial identity is not a default social option where "mixed race" is not a normalized category. Americans who feel compelled to declare their mixed-raceness do so despite its limited popular currency (for non-Hispanics). In a land of monoracial identities, multiracials seek social affirmation as a breed apart.

The source of the genealogical trap lies in equivocating between persons who have mixed continental ancestry and those who also would endorse a multiracial identity. From the outset I have specified "multiracials" in the latter sense. Equivocation leaves the misimpression that raced genealogy is a distinctive rationale for multiracial identity. Zack refers, for example, to "legal imposition of blackness on individuals who were of mixed race"—as if assigning them to a "black" or "Negro" grouping (that tracked American slave lineage) contravened their natural racial being and they were objectively misclassified.[17] We know that race classification has been more attentive to African ancestry than to any other components of a person's continental ancestry. "Those who bear multiracial identity," Sundstrom observes, "are relieved to various degrees of the social forces that subject their monoracial, and often darker, relations."[18] Multiracial identity would appear to be more naturally borne by lighter-skinned persons.

Responsible multiracialism is not helped by arguments circling around the notion that obvious African ancestry properly distinguishes "blacks" from "mixed race" persons who also have substantial African ancestry. The spectrum of skin tones in Black America has routinely encompassed physical profiles for persons whose African ancestry is visible if not obvious.[19] Affirming as much in her essay "Passing for White, Passing for Black," philosopher and conceptual artist Adrian Piper explores painful encounters with White Americans and Black Americans who variously perceived her—a self-described "light-skinned black" woman from Harlem—to be not truly black.[20] Piper struggles to grasp that doubts about her blackness, at least from the Black contingent, did not hinge on the European component of her ancestry. Such doubts are mainly about social-political lineage, immediate family background, default social reception, political allegiance, and cultural orientation.

Ironically or not, Piper later declared herself to be a person of African (and non-African) descent who prefers to identify as racially mixed. Her 2012 announcement of a digital self-portrait reads: "Adrian Piper has decided to retire from being black. In the future, for professional utility, you may wish to refer to her as The Artist Formerly Known as African-American."[21] Such flexibility of color-conscious identity, whether for life or art, is unavailable to (non-Hispanic) Americans whose African ancestry is obvious.

An ethical defense of multiracialism would refuse special pleading for persons whose visible African ancestry might be light enough for them to gain social affirmation as nonblack. Advocates often mention the darker, "more African" physical features of other people who have mixed continental ancestry—which in effect sets up a lighter-skinned contrast class against which to explicate multiracial identity. Responsible multiracialism will want to avoid any appearance of colorism.

9

Mixed Family Reunion

Color caste never struggled to accommodate darker-skinned descendants of American slavery who have substantial European ancestry. In general, multiracial identity is for lighter-skinned folk. Latin America made that explicit by using gradations of skin tone as a social proxy for fractional African ancestry. The United States, while socially less regimented by skin tone officially, used a mixed-continental scheme prior to 1930: Census Bureau enumerators (census takers) assigned all Africa-identified persons to a "black" or "mulatto" *color* category from 1850 to 1880. The 1890 census specified fractional Africa-identified categories up to "octoroon."[1]

The word "race" did not appear on the US census until 1900, as part of a *color or race* category.[2] Only as of 1930 did the Census Bureau adopt a single-continent scheme for closely tracking American slave lineage—assigning all Africa-identified persons to a "Negro" grouping. In any case, lighter-skinned persons have had access to relative social privilege, whether in Latin America or the US.[3] There is nothing new about American mixed African and European (or "mulatto") identity.

Multiracial and monoracial schemes calculate blackness differently yet share a logic that governs color caste: visible African ancestry triggers hypodescent, whatever the amount of European ancestry a person also has. Advocates of multiracial identity seem to accept this logic in taking for granted a (regular) "black" baseline for distinguishing (other Africa-identified) "mixed race" persons. Of course, racial identity relied on the notion of natural major races, which White America's law made compatible with the one-drop rule for (mono)racial blackness: all persons of known American "Slaves" caste lineage were leveled down to a "colored" bottom caste. The Socratic myth of the metals, as I discuss in Part I, is a precursor of racial hypodescent ideology.[4]

Continental color consciousness, however, does not require strong attachment to notions of natural race. Sociologist Edward Telles observes that in Brazil, "the term *côr* ('color') is more commonly used than 'race,'" and "the idea that each individual belongs to a racial group is less common."

The Afterlife of Race. Lionel K. McPherson, Oxford University Press. © Lionel K. McPherson 2024.
DOI: 10.1093/oso/9780197626849.003.0021

The ambiguous foundation for Brazil's complex color categories has nevertheless maintained a "primary racial boundary" between "whites" and nonwhites.[5] This *côr* boundary is a caste marker: persons who have any visible fraction of African ancestry are relegated to a social status that is further divided by proximity to "white" European physical appearance. These divisions are a proxy for genealogical distance from New World enslavement.

Devaluation of blackness, then, can attach to visible African ancestry without need for the race concept. Transatlantic antiblackness preceded race science by two centuries—through slavery that meant "permanent, violent domination of natally alienated [or 'socially dead'] and generally dishonored persons," to borrow sociologist Orlando Patterson's summary analysis.[6] Even if color categories in Brazil are not grounded in a racial logic for "Negroid" and "Caucasoid" natural kinds, the end result is a social hierarchy that guesstimates various thresholds of nonwhiteness. Brazil's example is an embarrassment to responsible multiracialism, which would disavow the notion that lighter-skinned persons are distinctively entitled to a multiracial identity. The challenge of finding an ethically compelling rationale for multiracial (fractionally Africa-identified) versus monoracial (regular Africa-identified) identity remains.

American multiracialism has called upon biological family as the best hope for answering the ethical challenge. Individuals who are steered to declare a racial identity they do not identify with might feel they are bearing false witness. Recall, along with the wholeness claim, the integrity claim that recognition and affirmation of mixed-raceness is required if persons who conceive of themselves as racially mixed are to have a racial identity they can endorse for a true self. The wholeness and integrity claims primarily tie multiracial identity to a person's biological family history. This maneuver tries to substitute raced family in place of natural race per se. A distinction between genealogical facts and the facts of continental descent searches for a meaningful difference.

Toward that end, Zack privileges racial identity that closely tracks "biological family relations."[7] She attributes a troubled type of self-conception to race impurity: "If designated black Americans are not racially pure—and most are not—then [their] attempts to identify the self on a foundation of family history, where the individual identifies with black forebears, will be seriously frustrated by the presence of oppressive white forebears." For certain persons, "the American problem of mixed race creates crises of personal identity."[8] Africa-identified descendants of American slavery might psychologically suffer from raced family fissure, whether they realize it or not.

Zack suggests that multiracial identity could be a positive treatment option for Black Americans. The therapy would involve psychological reconciliation with one's White American forefathers who perpetrated sexual violence against one's enslaved foremothers, whose ensuing children were born into slavery and routinely disowned by the white side of the biological family.[9] Through declaring their complete and true multiracial selves, Black Americans could claim their partial European ancestry and overcome traumatic "interracial" family histories of abuse and estrangement.[10] Setting aside doubts about the value of such therapy, I question assumptions behind the race impurity problem.

If I, for example, learned that my great-great grandfather was a Confederate general or that a nineteenth-century Natchitoches Tribe chief was a distant cousin, these findings would make no appreciable difference to me as a Black American of substantial mixed continental descent. My biological family history does not represent what I care about in my social-political lineage vis-à-vis persons I might particularly care to identify with. I am an Africa-identified descendant of American slavery; I am slightly curious about my African (and Native American) origins and incurious about my European origins. The notion that Black Americans typically have been racially misclassified is puzzling to me—though I do not think of anyone as determinately being "racially" anything. Maybe my stance toward unknown or suppressed genealogical facts is idiosyncratic.[11]

The basic moral of my personal story still holds: family as a social unit is not always and importantly defined by biological relations.[12] American multiracialism, by contrast, presupposes the importance of one's raced genealogy to raced self-conception, then uses this in defense of an overarching claim that "mixed race" persons have "a need to be recognized for what [they] are racially."[13] The facts of one's continental ancestry ultimately bear the load of that circular reasoning: the genealogical facts are raced according to one's continental origins. Vague appeal to "family history" adds nothing extra. Without presupposing the importance of brute "biological family relations," there would not be the appearance of a mystery about how one could endorse a racial identity that fails to allude to the purported race of some traceable lineal ancestors. **Acknowledging forebears from different continents is not naturally connected with endorsing a ("mixed") racial identity that incorporates theirs.**

The dilemma for multiracialism's family argument comes down to this: while (biological) family ties often play an important role in raced self-conception, connecting multiracial identity to all one's forebears is a roundabout way to reintroduce natural major races. Psychological wholeness and integrity would be attained by encompassing all a person racially is supposed to be, transmitted through the brute biology of all substantial components of continental ancestry. Sundstrom tries to avoid "racialized genealogy" that would reduce race to biological family; he disclaims the notion that multiracials "need to recognize all the members of [their] interracial families."[14] Instead, he appeals to the deep affective nature of certain family ties.

Sundstrom is especially concerned about criticism that "white mothers" have spearheaded multiracialism on behalf of their nonwhite children. He cites "thick relationships of family belonging that are nurtured rather than rejected in the post-civil rights era." But his explanations do little to counter the criticism, if only because they seem irrelevant to a defense of multiracialism. "The demand to recognize [our] mothers is a moral response," Sundstrom asserts, "itself grounded in our obligations to care for those who are connected to us by bonds of love."[15] For persons who have substantial mixed continental ancestry and a biological mother who belongs to a monoracial group, endorsing a multiracial identity that alludes to her purported race would be a filial obligation.

Why a duty of care to certain family members would require endorsing a racial identity that incorporates theirs, Sundstrom does not explain. Are we to believe, for example, that "transracially" adopted persons are obliged to endorse a racial identity that alludes to the race of their adoptive mothers? Sundstrom suggests no argument for why loving care toward one's mother, biological or adoptive, should be expressed through one's racial identity. We also get no indication of why mothers, and not also fathers, are singled out as crucial to raced familial bonds.

Multiracialism fosters the impression that at least for persons whose biological parents are different purported races, not endorsing a multiracial identity is tantamount to hiding or ignoring part of who they truly are. This raises the bar for multiracial identity from a legitimate personal choice to a duty to self and others. "The demand for the recognition of multiracial identity," Sundstrom asserts, "ought to be grounded in children's recognition of particular responsibilities (psychological, social, economic, and political) to their mothers and grandmothers," given "obligation to the memory of our mothers."[16] Heartfelt sentiment cannot mask the lack of a distinctive, ethically compelling rationale for multiracial versus monoracial Africa-identified identity in the United States.

10

Slavery Subcaste Drama

As descendants of American slavery, Black Americans typically have substantial mixed African and European ancestry. Colonial Virginia's inherited slavery law of 1662 preempted "any Englishman" from legally recognizing his children born to enslaved Africa-identified women. The underlying English legal doctrine *partus sequitur ventrem* ("that which is born follows the womb")—originally meant to apply to "all tame and domestic animals"— was retrofitted to determine the "Free" versus "Slaves" status of American children.[1] That doctrine eventually morphed into a "race" rule for marking "black" status via the African "blood" of enslaved Americans.

There was no racial mystery in inherited slavery. Black Americans, without paradox, can broadly acknowledge lineal ancestors of European descent; the ancestors who mattered for hypodescent were of American slave lineage. This historical reality is compatible with granting that persons who have mixed continental ancestry should be entitled to identify as "racially" mixed on that basis regardless of the ethics of color caste identity.

All told, my case study of "mixed race" as American caste is about a slavery society that later invested heavily in race ideology-rhetoric, namely, to distract from continuous subjugation and exclusion of the homegrown "black" population. The foundational color-conscious division in America was between Europe-identified "Free White" persons and Africa-identified "Slaves"; in between were "All Other Free Persons" (mainly of native lineage). From 1790 to 1950, Census Bureau enumerators monitored, house to house, this bipolar caste formation. The "Slaves" caste was relabeled by "color" after the Civil War.

Except for the period 1930 to 1990, the United States has always officially recognized persons of mixed Afro-European descent: "black" or "mulatto" was their description from 1850 to 1880, which became "black, mulatto, quadroon, octoroon" in 1890 and retreated to "black (Negro), mulatto" in 1910. The Census Bureau today strangely claims that "the term 'Negro' was used to refer to full-blooded individuals and the term 'of Negro descent' was used to refer to 'Mulattos.'"[2] By that measure, a vast majority of descendants

The Afterlife of Race. Lionel K. McPherson, Oxford University Press. © Lionel K. McPherson 2024.
DOI: 10.1093/oso/9780197626849.003.0022

of American slavery would have counted as "mulatto" rather than "black." One-drop racial blackness did not become a scientistic census category until 1930.

Early census forms make explicit that the US government was never trying to track the demographics of some "race" thing. Nor has the government cared about counting the country's inhabitants according to racial inner selves or feelings of racial belonging. The American Founders established a tyrannical caste distinction grounded in the Anglo-American institution of inherited slavery.[3]

Africa-identified persons comprised a conspicuous out-group—born into a condition of human property, from generation to generation, through White America's violence and law. After emancipation came Reconstruction and its defeat.[4] The rest is active history: from enforced segregation of slavery's visible descendants to their mass incarceration, accompanied by the permanent wealthlessness of Black America.

PART III
NON-EXCLUSIONARY BLACK (AMERICAN) SOLIDARITY

The reality of equality will require extensive adjustments in the way of life of some of the white majority. Many of our former supporters will fall by the wayside as the movement presses against financial privilege. Others will withdraw as long-established cultural privileges are threatened. During this period we will have to depend on that creative minority of true believers.

—Martin Luther King Jr., "Racism and the White Backlash" (1967)*

1

Social-Political Lineage Matters

Needless confusion and intrigue surround public understanding of Black American social identity. A great majority of Africa-identified persons in the United States are of homegrown background. In the historical sense, "Black American" refers to descendants of American slavery; they typically have substantial mixed African and European ancestry, and they were almost the only "blacks" in America prior to the 1980s.[1] These are facts of social-political lineage and geographic descent.

Hence, whatever natural or social thing may be meant by "race," Black Americans comprise an Africa-identified national people. They were labeled "Slaves" by the US Census Bureau from 1790 to 1860, with "Free Colored" persons added in 1820.[2] After emancipation, they were relabeled "black" (or "mulatto") by "color." Circa 2020, as 11 percent of the US population, their visible descendants owned less than 2 percent of the national wealth. This wealth gap is enduring evidence of a society that was divided between Africa-identified enslaved persons and Europe-identified free persons—a division that defines American color caste.[3] The modern idea of race as continental groups has been a device of moral and intellectual distraction from gross injustice.

Persistent race inquiry sustains what I call the race ideology-rhetoric complex. The core drama in the United States circles around whether some "race" thing alone might factor into the permanent social disadvantage of descendants of American slavery. Parts I and II of this book provide a multidisciplinary philosophical assessment of the race concept, particularly in relation to "American slavery or American caste," as abolitionist James McCune Smith put it in his introduction to Frederick Douglass's second autobiography.[4] Douglass "self-identified as a citizen of the USA," a historian explains, "and rejected all arguments that African-Americans had any racial, national or spiritual connection with African peoples"—"not only because he believed ['schemes for colonization and emigration'] distracted from the struggle against slavery in the USA, but also because he was convinced that Anglo-American civilization provided far greater opportunities for

The Afterlife of Race. Lionel K. McPherson, Oxford University Press. © Lionel K. McPherson 2024.
DOI: 10.1093/oso/9780197626849.003.0023

individual and collective betterment than relocation to Africa."[5] In that light, Douglass opposed pan-Africanism.[6] He was a proponent of black specificity in America for descendants of American slavery.

An American public fiction has emerged that "blacks," whether of homegrown or immigrant background, are basically a same people because visible African ancestry indicates they belong to "the black race."[7] In truth, they more modestly have in common a general susceptibility to anti-black prejudice. Consider the fabled question of whether someone is "black enough."[8] For descendants of American slavery, blackness queries mainly concern (doubts about) solidarity with persons of fellow social-political lineage. By contrast, blackness queries about persons of immigrant background concern their stateside social and political identification with Black Americans.[9] Notions of black racial being are unilluminating in either case.

The "race" thing in America was never about some shared elusive factor. US census forms show that "black" did not become an official "color" designation until 1850, followed by "color or race" in 1900—labeling maneuvers that disappeared slavery while tacitly tracking homegrown slave lineage. Biologistic race science and ideology in the twentieth century merely shored up (super)naturalist faith in the existence of major races.[10]

Some philosophers have tried to update faith in natural race. Quayshawn Spencer's "OMB race theory," for instance, treats the US Office of Management and Budget's "White" and "Black or African American" demographic categories as proxy for "biologically real" races—not, as official color categories were and largely are meant, as euphemistic shorthand for the Europe-identified "Free White" caste and the Africa-identified "Slaves" caste.[11] Foundational American caste today translates into stark "racial" disparities in the world's wealthiest country, which has by far the world's highest incarceration rate.[12] Under "race," the Census Bureau hides the historical Black American specificity of such disparities by flattening all Africa-identified peoples into generic blackness despite national origin. Unexplained, also under "race," the Census Bureau recognizes only specific and uncolored national origin identities for Asia-identified peoples, except those of "the Middle East." (On census "color or race" perplexities, see Part I, Section 11, "Enter 'Geoancestry.'")

I have argued for shunning race intrigue and, in the United States, openly acknowledging a deep social-political distinction between Black Americans and Africa-identified persons of immigrant background.[13] The distinction helps clarify two overlapping but different kinds of justice themes. There is the lower case-"b" kind that addresses general bias against Africa-identified persons. There is also the capital-"B" kind that belongs to the American "Slaves"/"black" caste tradition of righteous struggle for social equality. The Jim Crow Museum of Racist Memorabilia, for example, displays a wide range of everyday propaganda that targeted descendants of American slavery, though some depictions would apply to Africa-identified persons overall.[14] (The black immigrant population was less than 1 percent when White America overturned Jim Crow segregation laws in the 1960s.)[15]

White America imposed inherited slavery and enforced segregation on the homegrown Africa-identified population—ruling over a "colored" bottom caste forever associated with deprivation, disadvantage, and vulnerability. The post–civil rights era is marked by high residential and school segregation that rivals Jim Crow times, with Black America's wealthlessness a common denominator.[16] In 2016, a study finds, "black child households had just one cent for every dollar held by non-Hispanic white child households," and the disparity has a highly negative impact on "intergenerational mobility."[17] Another study finds hierarchical color consciousness operating across the socioeconomic spectrum: "Black men consistently earn less than white men, regardless of whether they're raised poor or rich. . . . No such income gap exists between black and white women raised in similar households."[18] The intersection of gender and color caste can appear inverted when Black America is viewed through the lens of patriarchal norms.[19] Such inversion is a distinctive feature of the social culture of descendants of American slavery.[20]

Part III of this book has two central aims. The first is to demonstrate that the core of Black American social identity has been consciously political (not "racial"). Visible descendants of American slavery—increasingly aware of the inefficacy of civil rights law and the two-party political system regarding their own people's attainment of social equality—have begun to insist that reparations are indispensable to any progressive American politics.[21] Think a domestic Marshall Plan dedicated to repairing the "Slaves"/"black" caste.[22] Whether races can be said to exist in some intelligible sense or other has little to do with explaining the fate of this Africa-identified national people. In general, neither racial metaphysics nor race science will advance our understanding of a society's color-conscious social hierarchy.

The second central aim of Part III is to demonstrate that the Black American justice cause (aka "the Cause") will welcome support from persons of conscience regardless of their color-conscious identities. **Everyone in the United States has morally compelling reasons to support, and should support, Black equality politics that strives to end American color caste.** This inclusive verdict deviates from the Black radical tradition: "Black nationalism, as manifested in the nineteenth-century United States," historian Wilson Moses writes, "was a racial nationalism, premised on the assumption that membership in a race could function as the basis of a national identity."[23] I arrive at a post-nationalist politics because that is where my reflective judgment leads me, particularly in response to philosopher Tommie Shelby's important argument for Black political, rather than racial or cultural, solidarity.[24]

After the George Floyd protests that swept the United States in 2020, "interracial" support for Black political solidarity no longer seems wildly naive as an actionable proposition.[25] Those protests—sparked by the on-camera, broad daylight, slow motion police killing of another defenseless Black American—were joined for months by many younger White Americans.[26] There is growing unease that state-governed violence excessively directed against "blacks" belongs to a vicious routine of refusal to redress Black America's caste plight.[27] The future for corrective justice seems less dim for descendants of American slavery.

2

Plessy on Being Colored

Through *Plessy v. Ferguson* (1896), the US Supreme Court created the shameless legal fiction of "equal, but separate, accommodations for the white and colored races." Plaintiff Homer Plessy was "of mixed descent, in the proportion of seven eighths Caucasian and one eighth African blood; [and his] colored blood was not discernible." He claimed to be "entitled to every recognition, right, privilege and immunity secured to the citizens of the United States of the white race."[1]

Clearly, American slave lineage was the fundamental issue: Plessy had mostly European ancestry and did not "look black" to the Court. Justice John Marshall Harlan rebuked his colleagues for acting in flagrant bad faith: "The thing to accomplish was, under the guise of giving equal accommodation for whites and blacks, to compel the latter to keep to themselves while traveling in railroad passenger coaches."[2] White America had moved from asserting a right to enslave Africa-identified persons to a right to absolutely exclude them.

Harlan noted that by the Louisiana statute in dispute, members of "the Chinese race" were barred from citizenship but "can ride in the same passenger coach with white citizens." A "colored citizen," the jurist wryly explained, "does not object, nor, perhaps, would he object to separate coaches for his race if his rights under the law were recognized."[3] Instead, the Supreme Court endorsed keeping Black Americans to inferior public facilities, when provided any.[4] Harlan, though, embodied the potential of White American conscience: a Unionist born into a Kentucky slaveholding family, he had opposed President Abraham Lincoln's Emancipation Proclamation as "an unconstitutional crusade to rid the nation of slavery."[5] The early Harlan failed to grasp the absurdity of constitutionalism that enabled a geographic people's prerogative to enslave persons of a kind who have known ancestry from a different geographic region.

White America's "racial" logic led the *Plessy* Court to reserve the term "colored" for persons who had any visible or reputational fraction of African ancestry, presumed less than full. The contemporary liberal variant "people

The Afterlife of Race. Lionel K. McPherson, Oxford University Press. © Lionel K. McPherson 2024.
DOI: 10.1093/oso/9780197626849.003.0024

of color" (for certain non-Anglo groups) ignores the foundational caste division—"Free White" versus "Slaves"—constructed through more than two hundred years of enslaving Africa-identified persons in America. **Replacing "colored" with the variant "of color" falsely implies roughly comparable American caste status and lived experience across nonwhite national peoples—which obscures the distinctive history and circumstances of Black Americans.**[6] No amount of patronizing language (e.g., "Oppression Olympics," "BIPOC") will detract from the moral urgency and priority, in the United States, of the "racial" justice claims of Africa-identified descendants of American slavery.[7]

Unfortunately, the idealism of Harlan's lone dissent has become infamous: "In the eye of the law, there is in this country no superior, dominant, ruling class of citizens. There is no caste here. Our Constitution is color-blind."[8] *Plessy*, Harlan predicted, would be "quite as pernicious as" *Dred Scott* (1857), where the Court reaffirmed that "the descendants of Africans who were imported into this country and sold as slaves were not included nor intended to be included under the word 'citizens' in the Constitution."[9] His personal verdict was blunt: "The thin disguise of 'equal' accommodations . . . will not mislead anyone, nor atone for the wrong this day done."[10] Persuading White America of reasonable merits of the case was unnecessary for purposes of Jim Crow segregation law.[11]

At the height of the Black rights movement seventy years later, *Plessy* inspired Alabama governor George Wallace's inaugural address:

> Today I have stood, where once Jefferson Davis stood, and took an oath to my people. It is very appropriate that from this Cradle of the Confederacy, this very Heart of the Great Anglo-Saxon Southland, that today we sound the drum for freedom as have our generations of forebears before us done. . . . I say . . . segregation now . . . segregation tomorrow . . . segregation forever.[12]

Such visions of white freedom only require social tribalism, not continental races.[13] Reportedly, some notion of "Anglo Saxons as a 'chosen people'" goes back to early medieval times—"continuing and reinventing a tradition in Christian historiography, which represented individual Christian nations after patterns established by Old Testament authors."[14] The American Founders, however, were not in the business of reformulated Hebrew mythology about different European peoples in the New World.

The US Constitution originally protected a pan-European conception of "Free White" persons while ensuring the subjugation and exploitation of an Africa-identified "Slaves" caste.[15] Explicitly tying "race prejudice" to "the institution of slavery," Harlan put faith in constitutional amendments that "obliterated the race line from our systems of governments" (even as White consensus would leave the material ramifications of slavery mostly intact).[16] His "color-blind" aspiration in the historical context of White law was later twisted by the Supreme Court's sham scrutiny about some "race" thing (in efforts to protect legacies of social inequality produced through inherited slavery and enforced segregation). The result has been unending legal obstruction of the possibility of social equality for descendants of American slavery as a national people.[17]

Plessy added social contempt to material injury done to Black America, as Harlan spelled out: "We boast of the freedom enjoyed by our people above all other peoples. But it is difficult to reconcile that boast with a state of the law which, practically, puts the brand of servitude and degradation upon a large class of our fellow citizens."[18] He was preoccupied with the aftermath of inherited slavery, as were his Supreme Court colleagues, for contrary motivations. Haunted by *Dred Scott* and his own slaveholding past,[19] Harlan treated the Court's mendacity as an affront to any respectable institution of law. He did not claim a theory of law or legal interpretation to justify reading the Constitution and its amendments in a far more favorable moral light than intended by their framers or tolerated by the White body politic.

After the legal demise of Jim Crow segregation by the late 1960s, policing color-conscious identity became almost defunct. Customary guidelines for American blackness have loosened enough that persons of mixed continental descent whose African ancestry is visible are free to adopt a nonblack identity. Being Black American has become an elective social identity (which is not to suggest it typically feels like a choice, particularly for the darker skinned). The arrival of elective Black American social identity strengthens, not weakens, Black political consciousness: proud volunteers will be more committed to the Cause than Africa-identified hostages.

3

Du Bois's Normative Negro

"The truth is that there are no races: there is nothing in the world that can do all we ask 'race' to do for us," philosopher Anthony Appiah concludes in his essay "The Uncompleted Argument: Du Bois and the Illusion of Race."[1] Appiah's "truth" is provocative, but it overreaches. There and in subsequent work, he casts suspicion that Black American social identity depends on commitment to the notion that every member of each racial group has some essential quality distinctive to their group. The error of this suspicion is instructive and, ironically, an error Appiah shares with the early Du Bois. The race concept ensnares even critical thinking about color-conscious identity. (I am resuming the discussion of Du Bois in Part I, Section 10, "Racial Metaphysics of Distraction.")

Du Bois knew better than to believe there were important innate differences between (sub)continental groupings of people. In "The Conservation of Races" (1897), he rejected racialism that placed Negroes at the bottom of a natural hierarchy of human beings regarding mind/brain. Du Bois recognized that "the grosser physical differences of color, hair, and bone go but a short way toward explaining" the contributions that different races, including Negroes, have made to world culture.[2] Of course, physical differences could hardly explain broad cultural developments. He did believe, though, that different racial peoples are suited to making different cultural contributions.

The early Du Bois fell halfway into race essentialism. He appealed to "subtle forces [that] have generally followed the natural cleavage of common blood" and "have divided human beings into races."[3] Such races would have to depend on some unifying property to yield the respective membership of each group. Presumably, the property would be genetically heritable via having sufficient non-northern African ancestry. Appiah argues that Du Bois imagined there is a black racial people—across epochs, continents, languages, and subcultures—whose natural endowment, like that of other races, renders them distinctive contributors to world culture as revealed through their common history.

The Afterlife of Race. Lionel K. McPherson, Oxford University Press. © Lionel K. McPherson 2024.
DOI: 10.1093/oso/9780197626849.003.0025

That notion should have seemed implausible to Du Bois. As he observed, "Race differences have followed mainly physical race lines, yet no mere physical distinctions would really define or explain the deeper differences—the cohesiveness and continuity of these groups." The question, then, is what kind of heritable property would deliver racial "cohesiveness and continuity." He implied that the different races have different natural psychological dispositions: "The deeper differences are spiritual, psychical, differences—undoubtedly based on the physical, but infinitely transcending them."[4] The source of such differences remained mysterious. Many descendants of American slavery, including Du Bois, were conspicuously of mixed African and European descent.

My present purposes do not concern trying to better grasp how the early Du Bois understood the idea of race. (Philosophers have paid inordinate attention to the topic—an assessment the later Du Bois would share.) The pertinent issue now is why he would go to metaphysical lengths to propose a variation of race essentialism. Two main motivations stand out: to resist Western race ideology that pseudo-rationalized the extreme oppression of blacks/Negroes on grounds of their alleged natural inferiority, and to articulate a foundation for Black solidarity. I have argued that his flirtation with racial metaphysics was marginal to his color-conscious values and politics.

In *Dusk of Dawn* (1940), Du Bois gave up his earlier attraction to race essentialism. He acknowledged that "neither my father nor my father's father ever saw Africa or knew its meaning or cared overmuch for it," while his "mother's folk were closer and yet their direct connection, in culture and race, became tenuous." What Black Americans had in common, "the real essence of [their] kinship," was less "the badge of color" than the "social heritage of slavery; the discrimination and insult; and this heritage binds together not simply the children of Africa, but extends through yellow Asia and into the South Seas."[5] This type of "essence" would be found in a color-conscious experience of Western domination, not in nature. Appiah questions such an expansive approach to social identity: "How can something [Du Bois] shares with the whole nonwhite world bind him to only a part of it?"[6]

Although Du Bois's line of thought in the *Dusk of Dawn* passage is compressed, we can discern what he meant. As a "Negro," he was expressing solidarity with other non-European peoples subjected to Western aggression and said by Western race ideology to be naturally inferior to the "white" European. (Recall his receptiveness to B. R. Ambedkar's vision of Afro-Dalit solidarity.)[7] More specifically, Du Bois was expressing solidarity with persons whose African ancestry was turned into a color-conscious device of distraction from their subjugation. Most specifically, he was expressing solidarity among Africa-identified descendants of American slavery, whose bottom caste status under White America's law marked their national people.

Du Bois's non-Western cosmopolitanism and pan-Africanism were compatible with prioritizing, in the United States, a color-conscious politics geared toward social equality for Black Americans. In his words, "White America has crucified, enslaved, and oppressed the Negro group and holds them still, especially in the South, in a legalized position of inferior caste."[8] Unsurprisingly, there will be complexities and tensions in how non-Western, pan-African, and Black American modes of solidarity interact.[9] But to downplay the moral and political significance of Black American specificity would misconstrue a Du Boisian vision of color-conscious equality in the United States.[10]

I believe Appiah misses the point when claiming that "Du Bois' experience of 'discrimination and insult' in his American childhood and as an adult citizen of the industrialized world was different in character from that experienced by, say, Kwame Nkrumah in colonized West Africa." Self-evident differences highlight rather than undermine how Du Bois thought about color-conscious solidarity. The common theme is subjugation of Africa-identified and other non-European peoples by certain national European (e.g., Britain, France) and settler European (e.g., the United States, Canada) nations—which represents a shared type of "insult" and not merely, as Appiah thinks, "the *badge* of insult."[11]

Color-conscious ties might well be weak among Africa-identified peoples globally. Africa-identified descendants of American slavery, however, share historical memory and lived experience of color caste that eventually employed race ideology to pseudo-rationalize atrocity for purposes of exploitation. Again: "More than eight-in-ten black adults" (the percentage of descendants of American slavery) "say the legacy of slavery affects the position of black people in America today."[12] Capital-"B" Black Americans are an involuntarily formed national people of mixed continental descent. The race concept's flat blackness gets in the way of grasping this, if far less so for "blacks" of immigrant background.[13] Being Black American consists in a distinctive social-political lineage that predates, by almost three centuries, black immigration to America.

Du Bois suggested that in addition to the historical sense of being a home-grown Africa-identified American people, there is a normative sense. Visible descendants of American slavery continue to identify as black in relation to their experience of caste, regardless of any "race" thing of theirs (other than having non-northern African ancestry). The Census Bureau, starting in 1870, renamed the "Slaves" caste by superimposing the "color" designation "black" (once all Americans were "free" after the Civil War)—which became a "color or race" designation from 1900 to 1930, went back to "color" alone in 1940, became "race" alone in 1950, returned to "color or race" in 1970, etc. But what if many Black Americans in the twenty-first century were to put faith in the professed ideal of a "post-racial" society?[14] That could threaten Black solidarity, leading to cultural loss and political miscalculation were they to count on "color-blind" empathy to deliver social equality for their people.

Anticipating the dream of a post-racial society, Du Bois went on the offensive in "The Conservation of Races":

> If I strive as a Negro, am I not perpetuating the very cleft that threatens and separates Black and White America? . . . Does my black blood place upon me more obligation to assert my nationality than German, or Irish or Italian blood would?
>
> It is such incessant self-questioning . . . that is making the present period a time of vacillation and contradiction for the American Negro; combined race action is stifled, race responsibility is shirked.
>
> . . . We are Americans, not only by birth and by citizenship, but by our political ideals, our language, our religion. Farther than that, our Americanism does not go. At that point, we are Negroes, members of a vast historic race. . . . [I]t is our duty to conserve our physical powers, our intellectual endowments, our spiritual ideals; as a race we must strive by race organization, by race solidarity, by race unity.[15]

The lived experience of Africa-identified descendants of American slavery would ground being racially black in America. In turn, cultural and political value attached to that blackness would ground a duty to identify as Black Americans and join in solidarity as a national people.

Du Bois wanted to end "incessant self-questioning" about who Black Americans are by preempting any serious doubt about the descriptive basis of their social identity (namely, blackness rooted in American slavery and segregation) and the normative basis of their solidarity (namely, American Negro "racial" duty). Some bare natural fact of race could not establish that members of any racial group have a membership duty of political solidarity. Hence he introduced the normative claim that Black Americans have necessary contributions to make if social equality is ever to approximate realization for their national people, which also would benefit "the Negro people" globally.[16]

4

Black Specificity in America

Underlying Du Bois's conception of a Black American people is his concern that too many members of the group will not join in solidarity, especially the most talented and privileged.[1] Prospective nonparticipants might reject identifying as Black American because they prefer a "multiracial" identity; or, while accepting the continental color label "black," they might seek assimilation to cultural whiteness; or, though affirming a Black American cultural orientation, they might be indifferent to political solidarity. Du Bois worried that rampant individualism would threaten the group's collective agency.

White America has had a fondness for "good Negroes" apparently willing to "define themselves as subservient to antebellum traditions or subject to the pacifying structure" of American slavery and segregation.[2] There is also a more principled approach to Black nonresistance, which educator and public intellectual Booker T. Washington articulated: "To those of my race who depend on bettering their condition in a foreign land or who underestimate the importance of cultivating friendly relations with the Southern white man . . . I would say: 'Cast down your bucket where you are.'"[3] The accommodationist road to group uplift would be slow and forbearing. By contrast, the tradition of solidarity among descendants of American slavery is about actively resisting injustice.[4]

While insider threats to Black (American) solidarity have been real, Du Bois overestimated temptations to disidentify as Black or repudiate the group's politics.[5] Few members have deemed the incentives worth the trouble: he underestimated social-political bonds that through intergenerational lived experience forge Black social identity.[6] As Shelby observes, Black Americans "frequently call upon, even pressure, one another to become a more unified collective agent for social change."[7]

Nevertheless, norms of Black solidarity have left room for individual preference. There is no custom, for example, of openly criticizing ordinary folk who pass for "white" or decline to identify as "black."[8] Wary of such laxity, the early Du Bois sought a nonelective basis for Black social identity, which was race on an essentialist understanding. Racial blackness that grounded

The Afterlife of Race. Lionel K. McPherson, Oxford University Press. © Lionel K. McPherson 2024.
DOI: 10.1093/oso/9780197626849.003.0026

racial duty would bind visible descendants of American slavery to Black social identity, Black solidarity, and "the Negro people." For these normative
purposes, the early Du Bois reduced blackness to some (super)natural "race"
thing. The racial being of Black Americans, a people of Afro-European descent, would have to "transcend" the total facts of their continental descent.

The Census Bureau had explicitly tracked slave lineage before making the
"Slaves" caste legally coincide with "black" demographic being. Nearly all
Africa-identified persons in America, treated by White law until the 1960s
as subordinate to Europe-identified persons, were descendants of American
slavery. Despite Du Bois's emphasis on that "social heritage" in shaping
Black social identity, he harbored doubt about the group's commitment to a
politics of resistance.[9] The evidence shows, more than a hundred years after
"Conservation," that being a homegrown Africa-identified American still
gives rise to a sense of common fate through historical memory and lived
experience.[10] The nexus of residential segregation, household wealth, pre-
college education, and mass incarceration is a blatant manifestation.[11] There
is no need for investment in any notion of race to explain this. It seems that
Appiah amplifies Du Bois's error, in the service of a competing conclusion.

Many Black Americans, Appiah argues, accept a version of the one-
drop rule for racial blackness. He notes that many persons in America
who would be black by this rule do not "look black," are unaware they have
African ancestry, and believe they are racially white, that is, of (nearly) full
European descent. Yet these persons are treated as "rare in relation to the
African American population," which Appiah deems important: "The result is a norm of solidarity such that African Americans very often have a
reason for identity-based generosity to people they believe [are] white. If
they acted on the norm based on the one-drop rule, their identity-based generosity would be regularly directed toward people they regard as whites."[12]
Such incoherence would undermine policies such as "race conscious" college admissions—by substituting flat blackness in place of being an Africa-
identified descendant of American slavery.[13] Appiah derives his worry
through ahistorical thought experiment.

Racial blackness, not mere African ancestry, was a conjured secondary
quality of enslaved Americans. To pseudo-rationalize a social hierarchy of
brute domination, White America superimposed an Africa-identified "race"
thing on the "Slaves" caste. Inherited slavery in America preceded both
the modern idea of race as continental groups (est. 1684) and the advent
of scientistic race ideology (est. 1777).[14] *A capitalist institution of inherited*

slavery never depended on belief in race. As of 1662, forty-three years after the first African captives landed in Virginia, the free or slave "condition of the mother," and nothing else, legally determined the freedom or enslavement of her children.[15] Not until the 1850 census did the US government start pretending to track the racial blackness of Africa-identified Americans.

Along the lines of Toni Morrison's characterization of "the very serious function of racism," I have argued that drama about some racial X factor provides endless "distraction" from subjugation, exploitation, and nonrepair.[16] Americans who have ancestry that traces to Atlantic Africa, while typically also having European ancestry, were made "racially" the same as non-northern Africans, who were themselves made "racially" black across the vast subcontinent. The race ideology-rhetoric complex shifts attention to flawed beliefs and attitudes about continental races . . . as a principal cause of European attachment to New World slavery. Here lies the game of pseudo-rationalization.

I have posed a rhetorical question throughout this book: What could any modern thinker have thought about human beings that would be morally relevant to enslaving them and their children, in perpetuity? The notion that some people, indicated by continental ancestry, are unworthy of freedom is patently absurd. There never were, because there never could be, honest efforts to justify the intergenerational atrocity of inherited slavery. Preoccupation with moral justification is a red herring, as if there might possibly have been something about "Slaves"/"blacks" themselves to explain their categorical dehumanization under "Free White" law.[17] I encourage readers to consult *The American Slave Code in Theory and Practice* (1853) and *American Slavery as It Is: Testimony of a Thousand Witnesses* (1839) rather than simply condemn in abstraction a nation's willful evil that defies moral comprehension.[18]

White America's post–civil rights approach to blackness erases the slavery and relies solely on declared African ancestry. A result, for example, is this: "According to Professor [Henry Louis] Gates, more than two thirds of all Harvard's black students were either the children or grandchildren of West Indians or Africans. Very few, he said, of Harvard's black students were the descendants of American slaves."[19] Meager redress policies intended for the American "Slaves"/"black" caste—achieved through their people's political activism and personal sacrifice—disproportionately benefit individuals whose families were permitted to migrate to the United States, a vast majority after 1980.[20]

Race fixation by historically White institutions, particularly the Census Bureau and the Supreme Court, is a primary source of public confusion.

Demystification is simple enough: the relevant "blacks" regarding reparations for descendants of American slavery are Africa-identified descendants of American slavery. Evasive talk of "minorities," "diversity," "underrepresentation," and the like only clouds the point.[21] Explicit racism, as sociologist William Julius Wilson suggested forty years ago, has been of "declining significance" in explaining the caste disadvantage and the collective identity of Black Americans after the 1960s.[22] Since then, moreover, the continental color label "black" has become a fifth less than 99 percent reliable in referring to them as compared to other Africa-identified persons in America.

Appiah needs the faulty assumption that Black American identity relies on belief in (super)natural race to drive his argument about deep incoherence in the identity. More careful observers of American caste understand that "one drop" of "colored blood" was figurative for visible African ancestry as a telltale sign of homegrown slave lineage.[23] **The normalization of black "racial" identity in America is a prevalent effect (sponsored by the US government), not a contributing cause, of the bottom caste status of descendants of American slavery.** Du Bois, after early preoccupation with trying to metaphysically specify racial blackness, was satisfied loosely referring to Africa-identified Americans as "we who are dark"[24]—by virtue of having traceable African ancestry, in domestic context of the former "Slaves" caste, as the target of antiblack ideology that is foundational to a "Free White" civic religion.

In sum, Black (American) social identity is not merely a function of living in America, having visible African ancestry, and being susceptible to some similar kinds of color-conscious maltreatment.[25] Capital "B"-Black Americans are descendants of American slavery—whose social-political lineage continues to matter a lot more than belonging to a continental race grouping. Persons of Africa-identified immigrant background have been proud to agree, affirming their national origin lineages while embracing an "immigrant America" willing to accept them.[26] The "immigrant America" construct includes neither Native Americans nor descendants of American slavery.

5

Geoancestry and Elective Identity

A nonelective aspect of the blackness of Black Americans consists in non-northern African ancestry that marks their homegrown lineage in the former "Slaves" caste. We can understand this without the race concept's intervention. Indeed, the notion that Africa-identified descendants of American slavery comprise a subgroup of the black/Negro race is misleading. Nonelective blackness—as in visible African ancestry—is not equivalent to and does not presuppose any "race" thing.

Among persons who believe in some reality of race, a basic problem is specifying criteria that adequately match ordinary racial ascriptions. If some biological account of race criteria is given, there often will be mismatches between how persons are racially classified and how they are seen as looking. Such mismatches in America were the byproduct of a legal regime that governed pervasive reproductive mixing between Europe-identified males and enslaved Africa-identified females. The visible result of slavery hypodescent was a wide physical profile for homegrown "blacks."

If biological criteria for monoracial black/Negro raciality were to capture having any substantial African ancestry (say, equivalent at least to one great-grandparent), many Americans who have had a white social identity would count as racially black.[1] If biological criteria for monoracial blackness were to capture having mainly African ancestry, many Americans who have had a black social identity would count as racially white. Even if biological criteria could be revised to correspond more tightly to black social identities in America, this would not represent objective facts of race.[2] Rather than tracking some elusive black/Negro "race" thing, the chosen guidelines would be chasing ordinary racial ascriptions tied to geographic descent.

Another hurdle for biological criteria for race is that ordinary racial ascriptions are not globally uniform. In the United States, for example, visible African ancestry widely translates as monoracially black; in Brazil, different shades of African ancestry can translate as either monoracially black ("*preto*") or mixed race (e.g., "*pardo*" for brown).[3] Accounts of race as a social construction run into a similar issue: guidelines for social race vary across

The Afterlife of Race. Lionel K. McPherson, Oxford University Press. © Lionel K. McPherson 2024.
DOI: 10.1093/oso/9780197626849.003.0027

societies and can be unstable within a society, while vaguely attached to notions of natural race. In short, boundaries of racial membership must rely on ambiguous and contested meanings of "race." The only global constant for membership in a particular racial grouping is ancestry from the associated base continent—which is not a sufficient condition, as curiously "Caucasian" northern Africa illustrates.

I argued in Part I for moving on from the idea of race in favor of the concept of *geoancestry*. My case study has been the American "Slaves"/"black" caste: how Black Americans, a people of Afro-European descent, could belong to a uniform black race is a permanent riddle. The empirical-conceptual haze is not a special feature of having traceable ancestors from different continents. "Know it when we see it" guidelines for ascribing racial identity retain a hold over the American imagination to keep insinuating themselves in complicated stories about the reality of race. "Race" adherents of all stripes assume or set forth various accounts of the word's meaning, which are supposed to steer toward some unobvious continental basis for racial groupings.

No matter some meaning(s) of "race," Africa-identified descendants of American slavery constitute a national people, neither native nor immigrant. Questions of racial being are irrelevant to viewing them as members of a conventional Africa-identified grouping. Continental color consciousness can be rendered unmysterious. **Geoancestry is a color-conscious mode of thought giving expression to the continental place(s) a person's traceable ancestors originated from, in full or part.** Inquiry into the true nature of a "race" thing was always superfluous. The modern idea of race became an ideological-rhetorical device of distraction from European transatlantic slavery and colonialism.

Discarding race intrigue, the geoancestry concept stands for continental groupings that reflect Western practices of attaching importance to visible continental ancestry. A global Africa-identified grouping tracks visible African ancestry without presumption that its members have only or primarily African ancestry. Under this conception of geoancestral blackness, Black Americans constitute a subgroup of Africa-identified peoples and as such would have a "black" social identity.

Partisans on either side of the natural race versus social race divide agree that abandoning search for the meaning(s) of "race" risks important conceptual, informational, political, or existential loss. Rejecting this opinion, I showed in Part I that continental color consciousness does not actually depend on the race concept. A simple neologism like "geoancestry," coined for the uncomplicated purpose of alluding to color-conscious geographic descent, is more transparent than "race" and helpfully unfamiliar, with greater mental distance from racialist baggage.

When individuals self-identify as members of a geoancestral group, they are acceding to customary criteria for the corresponding color-conscious identity. This does not entail robust cultural identification with their continental grouping (e.g., Africa-identified). But geoancestral subgroup membership in connection with a national identity (e.g., Black American, Haitian, or Nigerian) will often involve a cultural dimension expressed through cuisine, music, worldview, etc. Globally or nationally, the geoancestry concept lends straightforward support to a non-racial sense of Africa-identified social identity. Nothing of substance would be lost were the idea of race to fade away along with its ideology-rhetoric of whatever persuasion.

I have proposed that Black American social identity has become elective: descendants of American slavery who fit an Africa-identified profile and (as typical) are of Afro-European descent are no longer bound by rigid social policy and practice to accept the continental color label "black." Since 2000, the Census Bureau has acknowledged public latitude for individuals to self-identify through their geographic "origins" of preference—via "one or more" or "some other race." Previously, "color" had been good enough for government purposes; only since 1990 has "race" settled in as a freestanding demographic thing.[4] Respondents themselves, not Census Bureau enumerators, have filled out census forms since the 1960s.

Despite the US government's contemporary sponsorship of self-identification by race, the geoancestry concept strongly recommends reframing that query: boxes for (sub)continental and national origins would suffice. (Census forms do this for different national peoples of Asia.) The Census Bureau already allows respondents to "mark one or more boxes." The early returns are anticlimactic. Mass exodus from one-box blackness shows no signs of materializing. Of the total US population, persons who self-identify as "black or African American in combination" with "white" rose from 0.3 percent in 2000 to 0.6 percent in 2010. Persons who self-identify as "white in combination" with any other "race" rose from 1.9 percent in 2000 to 2.4 percent in 2010.[5] So there also has been no mass exodus from one-box whiteness.

The tendency to equate color-conscious social identity with race classification sustains intrigue about the natural or social reality of race. Yet identifying as a descendant of American slavery neither depends on any metaphysics of race nor presupposes "racial" identity. Being Africa-identified in America was never reducible to facts of continental descent. The very purpose of the one-drop rule for blackness was to resolve American caste membership for persons of mixed European and (enslaved) African descent. The rule's eventual scientific pretensions hardly masked its primary role as a color-conscious heuristic for legally and socially marking Americans of "Slaves"/ "black" caste lineage.

Racial being for Black Americans was never truly about some "race" thing. Only their visible or reputed African ancestry was raced. The enlightened rejoinder is not that Black Americans are, strictly speaking, racially mixed (or "some other race"). A "race"-preserving gambit runs the risk of doubling down on essentialism—as if color-conscious identities are supposed to represent true racial being by the total facts of substantial continental descent. There is anyway not much content to racial identity, apart from associated continental ancestry in social context.

Thus, I argued in Part II, having traceable African and non-African ancestry does not provide any distinctive rationale for identifying as multiracial or disidentifying as Black American. I also argued that persons should be free to pursue their own conceptions of the good, which includes the freedom to adopt color-conscious identities of their preference given their (sub)continental origins. By this principle, no one has a social duty to self-identify through color as Black American or black—which is as true for persons who fit the paradigmatic "black" profile, namely, they have traceable non-northern African ancestry only. Continental "race" labels are ultimately a Western thing.

Contrary to the fears of Black Americans who care about Black (American) solidarity, there has been no rush to exit from Black American social identity. The push for greater mainstream recognition of multiracial identities highlights that being Black is no longer an ascribed color-conscious identity that Americans of homegrown background who have visible African ancestry are virtually obliged to accept. The option to disidentify as Black (American) or black only strengthens the appearance of commitment to electively identifying as such, in a geoancestral sense.

6

The Very Idea of Black Solidarity

This section is a prelude for addressing whether Black Americans ought to join in solidarity based on a common social identity. The view that they should belongs to a long tradition of American political thought. "Classical black nationalism"—which as Wilson Moses describes it "originated in the 1700s, reached its first peak in the 1850s, underwent a decline toward the end of the Civil War, and peaked again in the 1920s"—championed "the creation of an autonomous black nation-state, with definite geographical boundaries."[1] Modern variants drop the vision of a separate nation-state for Black Americans.

What prompts black nationalism is the profound historical injustice Black Americans have suffered as an Africa-identified national people. My interpretation of these circumstances in terms of American caste—from inherited slavery, to enforced segregation, to mass incarceration—puts the focus squarely on subjugation, exploitation, and nonrepair rather than on race ideology-rhetoric that distracts from systemic maltreatment. There is still a place for flatter black solidarity in response to anti-black bias and discrimination. Yet black nationalism recognizes the specificity and special moral standing of the Black American justice cause, which fundamentally is about the unreconstructed aftermaths of American slavery and segregation.[2]

That Black Americans are seen as somehow racially black is epiphenomenal to the blackness of their social identity as members of the American "Slaves"/"black" caste. My engagement with the question of whether Black Americans have a special obligation to join in group solidarity reaches only to persons who are Africa-identified descendants of American slavery, not to all Americans for whom the label "black" could be applied as a "color or race" designation. Appeal to any alleged fact of membership in the same race is unnecessary and unwelcome for my purposes.

I will not be interested in further establishing justification for Black political solidarity. Shelby, who elaborates the philosophical foundations in *We Who Are Dark*, argues that Black solidarity requires neither a "black racial

The Afterlife of Race. Lionel K. McPherson, Oxford University Press. © Lionel K. McPherson 2024.
DOI: 10.1093/oso/9780197626849.003.0028

identity [with] its origins in the ideological fiction of 'race'" nor a thick "collective cultural consciousness."[3] Against Appiah, Shelby and I together have argued that "dubious criteria for assigning racial membership" do not threaten the legitimacy of Black Americans' aim "to maintain, and perhaps strengthen, their political solidarity."[4] If Black Americans as a people no longer comprised a bottom caste group (effectively signaling the end of American caste), the question of their color-conscious solidarity would be far less important. Shelby and I have stated that we "support the efforts of [Black] Americans to seek the full liberties and equality that justice requires, and we regard [B]lack political solidarity as a practical and legitimate way to advance this cause."[5] There is nothing racially mysterious or politically obscure about such efforts.

Thus I construe political blackness in a nonracial sense. I set aside the notion, popular among reactionaries and liberals alike, of striving for public indifference to color-conscious identities—an ahistorical ideal often misread into Martin Luther King Jr.'s dream of a time when people "will not be judged by the color of their skin, but by the content of their character."[6] Nor am I entertaining speculation about whether Black solidarity might be counterproductive to bringing about a nonracist country. The aim of Black solidarity is not to fight implicit racial bias, promote racial harmony, or foster progressive coalitions. The form of Black solidarity that I am referring to is *political*. **The value of Black political solidarity is instrumental with respect to social equality for Black America.** It would be nice if strengthening Black America would also, as I think is plausible, strengthen America across lines of color, despite early instability caused by a substantial redistribution of resources and opportunities to build up Black America for the first time in history.

Black political solidarity is not tantamount to "the politics of difference." To sum up the latter: race, gender, sexual orientation, etc. can yield a sense of deep "difference" that represents "a highly valued preference that many persons and groups would [like to] have accommodated and recognized as the basis of their participation in civic, political, and economic life."[7] Philosopher Lucius Outlaw sides with this politics when he writes, "As was Du Bois, I remain convinced that both struggles against racism and invidious ethnocentrism, as well as struggles on the part of persons of various races and ethnicities to preserve, enhance, and share their [distinctive] 'messages' with all humans, require the conservation of races"—with emphasis on "black people" who comprise "the African race."[8] By contrast, I take the blackness of Black equality politics to be separable from goals such as preserving ethnic community.[9] My point is not to criticize nonpolitical forms of Black solidarity but to clarify that I construe the political form narrowly, in the context of the United States, and not as part of any broader Black or black cultural agenda.

In black nationalist thought, there tends to be ambiguity over the question of whether Black Americans have an obligation to join in group solidarity. Consider this statement by Outlaw: "Whether or not an [Africa-identified] individual can enjoy a relatively unrestricted and flourishing life is tied to the well-being of the group; the well-being of the group requires concerted action ... within the context of a shared [black] identity."[10] Even Shelby, who rejects the notion that a black cultural identity is required for Black political solidarity, takes the orienting question to be, "What political principles can blacks reasonably expect all other blacks, because they are black, to commit to as a basis for group action?"[11] He claims: "Of course we all, whether black or not, have an obligation to resist racial injustice. . . . But blacks would argu-ably have an obligation to pursue their antiracism *through* black solidarity if in its absence racial justice could not be achieved."[12] Principally, the standard euphemism "racial" justice refers to descendants of American slavery, who are the "blacks" Shelby mainly has in mind, largely regarding legacies of slavery and segregation in the United States.

The ambiguity I am locating concerns what is meant by the claim that Black Americans *ought to* join in solidarity—which might translate either as a requirement or a recommendation. Our question is not about whether Black solidarity is worth recommending to Black Americans who are pre-pared to commit to the Cause. Rather, the narrower question is whether Black Americans have a special obligation to join in political solidarity to ad-vance justice for the group—an obligation to "all other blacks, because they are black," as Shelby puts it. What in the nature of blackness could an obliga-tion of such kind rest on?

My test case is Black Americans who might be solidarity skeptics. I am not asking whether all persons who would be counted in a "black" race cat-egory in the United States would thereby have an obligation to join in Black solidarity. I am asking whether Black Americans (as members of a home-grown Africa-identified people under a caste system that governed slavery) might have some such obligation. The pertinent skeptics are not ideologi-cally opposed to Black solidarity, but they reject the view that commitment to group solidarity is a requirement of individuals. So the test case deals with the possibility that there could be Black Americans who would opt out of joining in Black solidarity—where opting out would not be a breach of obligation *as* Americans who are Africa-identified. What ethical mistake, if any, might these solidarity skeptics be making?

Getting a straight answer about whether Black Americans are supposed to have an obligation to join in Black solidarity proves difficult. Shelby avoids clarifying whether he means "should" in the strong sense as an ethical requirement (and thus non-optional) or in the moderate sense as an ethical recommendation (and thus optional). He prefers to "leave open the question of whether a commitment to black political solidarity is strictly obligatory, for answering it would require resolving the difficult empirical question of whether such solidarity is absolutely necessary to achieve racial justice." Instead, Shelby limits the issue to "what should and should not be required of those who *choose* to fight antiblack racism through black political solidarity."[13] This changes the subject from whether Black Americans have an obligation to join in Black solidarity . . . to which kinds of action should be taken by those already committed to Black solidarity.

Shelby implies that whether Black Americans have something like an obligation to join in political solidarity turns on the empirical matter of whether their commitment is necessary for advancing the Cause. I will argue that this way of framing the issue—by making obligation derivative from practical necessity—begs the underlying question of why in the first place Black Americans might have an obligation to join in Black solidarity. Generally, the fact that justice could use your service for a cause does not entail your obligation to serve that cause. You might have had nothing to do with the bad situation, or you might already be burdened by it.

For comparison, assume that world hunger could be solved if everyone with disposable income contributed their fair share to help solve it. We know that some able persons will not do their part, which leaves others to do more than what would have been their fair share. Since potential contributors are not defined by factors other than ability to contribute, conscientious individuals will be compelled to pick up the slack: they would be doubly burdened by their commitment to help. Whether all able persons have an obligation to support the world hunger cause is unobvious to start with, let alone beyond what an equitable scheme would require. The claim that practical necessity alone can entail obligation to support a worthy cause takes ingenuity to try to defend.

Proponents of Black political solidarity are characterized by willingness to commit to the Black American justice cause. This is not the case for skeptics: philosophical effort is called upon to explain why they should reverse their reluctance to join in group solidarity. Shelby's observation that a believer "could rightly be criticized for failing to live up to obligations she has voluntarily accepted as a member of that solidarity group" sidesteps the issue of whether skeptics would be wrong for declining to join the group to begin with. Shelby "leaves open the possibility that it is permissible for blacks to work for racial justice through some other means, whether group-based or not."[14] What about those who would rather not concern themselves with "racial" justice, period? They might be dedicated to family well-being, self-actualization, and professional life, for instance, even if they would welcome any justice gains for fellow descendants of American slavery.

To be clear, my query in this scenario applies only to Black Americans. Shelby takes his account of political solidarity to apply to all "blacks" in the United States, that is, persons who satisfy local "thin social criteria for being classified as black."[15] Yet there is no disagreement between us if he means principally Black American black political solidarity. I can envisage no powerful objection to a secondary, more expansively black politics in the United States.

7

Bound by Colored Politics?

In the light of American history, I take for granted that Black political solidarity can be instrumental toward social equality. The Black rights movement of the 1950s and 1960s, which sparked the end of "legal" segregation in
White America, is the most famous case.[1] Yet an instrumental rationale will
not garner much interest today from Black Americans who are inclined to
be solidarity skeptics despite (let's imagine) their general concern to do the
right thing. They would need convincing not simply that Black solidarity is
good for the group and most of its members. Skeptics would also need to be
convinced that as individuals who are Black American, they have a special
obligation to the group.

Of the nationalist notion that Black Americans have distinctive non-
optional reasons to engage in Black collective action, I am philosophically
doubtful. Black nationalist discourses lead me to distinguish four types of
reason that are offered to explain why Black Americans should recognize a
special obligation to join in Black political solidarity. These reasons concern
(1) responsibility to a tradition of great value; (2) the need for Black solidarity; (3) enlightened self-interest as a Black American; and (4) fairness as a
beneficiary of Black collective action.

First, there is an active tradition of Black political solidarity. The "Black
vote" is a striking example. In each presidential election since 1964, nearly
90 percent of Black Americans voted for the Democratic candidate; from
1936 through 1960, nearly 70 percent did.[2] A succinct explanation for ex-
ceptionally strong Democratic Party identification by Black voters is that
Democratic candidates typically have been less hostile than Republicans to
Black progress. Shelby offers a deeper explanation: "The common experience
of antiblack racism has for centuries provided a firm and well-recognized
basis for mutual identification and special concern among blacks. This
shared experience partially accounts for the bonds that exist today, for blacks
understand one another's burdens and empathize with each other on this
basis. [T]he legacy of collective struggle to remove this burden is a cherished

The Afterlife of Race. Lionel K. McPherson, Oxford University Press. © Lionel K. McPherson 2024.
DOI: 10.1093/oso/9780197626849.003.0029

inheritance for many black Americans."[3] Solidarity skeptics, however, prefer not to accept this group inheritance.

To ethically simplify, assume that skeptics would have supported Black solidarity prior to the demise of enforced segregation in the 1960s. Solidarity proponents cannot reasonably demand that skeptics get with the collective political program (e.g., Democratic Party fidelity) because most Black Americans have done so.[4] Nor is it plausible that a group's valuable tradition could be endlessly owed the commitment of individual members. Skeptics will want to know why they, by virtue of their historical American blackness, should be expected to participate. They need not doubt there are reasons for Black Americans to value a tradition of solidarity to doubt that those reasons should compel any respectable member of the group to act accordingly.

Second, there certainly remains a need for Black political solidarity. I argued in the previous section that, in general, practical necessity does not entail obligation to serve a particular justice cause. My argument now goes further. Even if solidarity proponents can claim necessity, this would not show that Black Americans have an obligation as "blacks" to join in group solidarity. Americans overall—Black, White, Asian, or other—have a civic duty to support the Cause, given the moral urgency and priority of Black America's corrective justice claims. No special reason attaches to Black Americans by practical necessity alone that obligates their commitment to Black solidarity.

King Jr. realized that ending "legal" segregation would not dismantle color caste: Blacks being free and equal at last would "have to depend on that creative minority of true believers" among White Americans.[5] More than fifty years later, White America's hostility to "racial" justice for Black America has grown.[6] Shelby suggests that without the pressure of organized Black politics, Whites will continue to prevent reform policies that Black social equality would require.[7] He thinks that since Black Americans are more willing to take up their justice cause, they should do so, lest Black progress stall or backslide. Black political leadership after the Black rights movement is not a source of progressive hope.[8]

Solidarity skeptics cannot easily be dismissed as selfish individualists unwilling to do their Black duty. They have a good argument on their side: to project onto members of a subordinated group non-optional ethical reasons of membership to join in group solidarity constitutes a double burden. Although committing to group advancement at personal expense might be a commendable choice for individuals to make, demanding their commitment to Black solidarity—"because they are black"—is morally perverse. White America, in the broad scheme of civic life, has been responsible for or complicit in maintaining Black America's entrenched social inequality. So the heaviest burden of corrective justice would fairly fall on White Americans; comprising the economically and politically dominant group, they have the wherewithal to renounce their intergenerational attachment to de facto benefits of Black social disadvantage.[9] Denial that this color-conscious hierarchy exists—with roots in American slavery and segregation—is contradicted by history and data I reference throughout *The Afterlife of Race*. Caste is relational and relative, and thus cannot (nor is meant to) be cured through universal "rising tide" policies that might improve working wages and social conditions.

US President William H. Taft, the later Chief Justice of the United States, could not have been clearer about the White establishment's commitment to caste hierarchy under Jim Crow segregation: "The fear that in some way or other a social equality between the races shall be enforced by law or brought about by political measures really has no foundation except in the imagination of those who fear such a result."[10] His vision of post-slavery American caste has proven true for more than a hundred years. The first female to serve on the Supreme Court, Sandra Day O'Connor, "called Taft a 'great Chief Justice . . . who deserves almost as much credit as [proslavery Chief Justice John] Marshall for the Court's modern-day role but who does not often receive the recognition.'"[11]

Many White Americans have convinced themselves that Black Americans are due little after Black rights legislation of the 1960s. As these naysayers tell it: the country has done enough, through "racial" anti-discrimination law and token preferential treatment for certain "minorities," to address the collective social disadvantage of descendants of American slavery; today's Whites are only marginally implicated in past exploitation and exclusion of those "blacks," so the country is not morally obliged to support concrete policies of repair for Black America; or descendants of American slavery should seek equality of opportunity, not "racially" comparable outcomes that would be antithetical to the recent quest for post-racial meritocracy.[12] Such is the context in which I am examining the need for Black political solidarity. There has been a dearth of non-Black persons, including those of post–Jim Crow generations, who prioritize a social equality agenda for "blacks" as descendants of American slavery.[13]

Shelby argues that Black Americans' obligation to join in political solidarity "would follow from the principle that if one wills the end, one also wills the necessary means," given that the means are morally permissible—the end being Black social equality.[14] But some Whites and other non-Blacks also affirm the Cause, and the principle would apply no less to them. **While Black Americans recognize that the practical burdens of Black equality politics will fall more heavily on them, this does not translate into a duty for Blacks as such to commit themselves to Black political solidarity.** Rational willing is color neutral.

Third, there is no denying that Black political solidarity can serve enlightened self-interest. The argument for special obligation on this front roughly goes as follows. Black Americans remain vulnerable to caste disadvantage and maltreatment; few will not be negatively affected, at least concerning extended family.[15] As individuals, then, they can expect to have more reason than not to join in Black solidarity. Rational failure to act accordingly would count as an ethical failing, given that people should try to improve the quality of their own lives, especially when this would also promote justice. In that spirit, political scientist Michael Dawson has theorized a "black utility heuristic": by his empirical analysis, "it is efficient for individual African Americans to use their perceptions of the interests of African Americans as a group as a proxy for their own interests."[16] Since the goal is Black social equality, Black political solidarity does not reduce to identity group politics.

For Shelby, Black political solidarity is to be distinguished from "broader forms of anti-racist solidarity and coalition building." His account provides a rationale for the exclusionary blackness of the Black political solidarity collective, so to speak, as well as a rationale for joining: "Although a joint commitment to fighting racial injustice in all its forms can help create interracial solidarity, it is often the shared experience of *specific* forms of racial injustice that creates the strongest motivation to act and the most enduring bonds among victims of racism." The harms Shelby cites range from "racist ideology, with its images of blacks as lazy, stupid, incompetent, hypersexual, and disposed to gratuitous acts of violence," to "severe social problems—joblessness, alarmingly high rates of incarceration, concentrated poverty, failing schools, a violent drug trade—that plague some black communities."[17] As it turns out, even solidly middle-class Black communities are subject to systemic economic marginalization.[18]

The problem for the prudential argument for Black political solidarity is that serving the Cause is unlikely to be in every conscientious Black person's best interests. Prudential calculation requires not merely that a certain course of action would yield some benefit for the person. There is value of some kind, I assume, for Black Americans engaged in political solidarity; at a minimum, they affirm moral and political entitlement to social equality that their caste continues to be denied. I am particularly sympathetic to philosopher Bernard Boxill's argument that black self-respect calls for protest and self-defense against anti-black ideology and injustice.[19] Yet self-respect can come at the cost of personal benefit.

Prudential calculation of self-interest requires, simply put, net gains for the agent in the long run. As philosopher Joseph Butler observed in the eighteenth century: "When we sit down in a cool hour, we can neither justify to ourselves this or any other pursuit, till we are convinced that it will be for our happiness, or at least not contrary to it."[20] Butler was making not only a psychological claim about motivation but also a normative claim about the credible demands of moral conscience.

We can imagine solidarity skeptics who are, along with close Black family and friends, prosperously stationed in life. Their prudential calculation about whether to participate in Black politics could forecast a net loss for them. For example, during the 1990 US Senate campaign in North Carolina between segregationist Republican incumbent Jesse Helms and Democrat challenger Harvey Gantt, basketball legend Michael Jordan declined to endorse Gantt since "Republicans buy sneakers, too." (In 2020, Jordan's billion-dollar brand pledged $100 million to fight anti-black racism.)[21] Additionally, some Black Americans who are not well-off might also reach a negative prudential verdict about their own risk-to-reward circumstances.[22] If obligation to join in Black political solidarity depends on calculations of enlightened self-interest, some skeptics would not have such an obligation.

Fourth, there is a general line of thought that fairness requires expected beneficiaries of collective action to contribute to the effort. Proponents of Black political solidarity proceed on the assumption that the unjust social circumstances of Black Americans as a group will substantially improve only if members commit to political solidarity. The ethical charge now is that solidarity skeptics do not do their share in the collective struggle: they are likely to benefit from a Black political agenda (re law enforcement, redlining, tax policy, health care, etc.[23]) while making no effort to support the Cause. The basic "free rider" objection, then, is that nonparticipants would unfairly gain from the fruits of other persons' labor or risk.[24] Skeptics stand accused of failing to meet an obligation not only as "blacks" but also as immediate beneficiaries of Black politics by virtue of group membership.

Although a free rider argument for Black political solidarity is tempting, I do not believe it will work. Revisit my prior objection to Black duty by practical necessity. Skeptics can argue that since Black Americans should never have been subjected to caste abuse and wealthlessness that are testament to their foundational American status, those who serve the Cause go beyond what duty requires when they could be trying to live their own best lives. The fact that collective social and electoral politics have represented an important group norm does not turn abiding by the norm into a matter of special obligation that individuals have to the group. To claim that those who would disengage from post-1960s Black politics are free riding on Black progress seems, again, a morally perverse imposition on members of a bottom caste group. I do not deny, however, that moral suasion might be a legitimate tool to inspire Black allegiance to the Cause—and non-Black allegiance as well.[25]

The bottom line is that I reject the notion that Black Americans, "because they are black," have non-optional reasons to join in Black political solidarity. My stance is at odds with the black nationalist tradition at large. But I am aligned with the nationalist tradition in affirming that Black Americans have ethically good reasons to support Black politics in unapologetic pursuit of Black social equality. Volunteers, not hostages, are needed for the Cause. Black political solidarity skeptics, already burdened as members of the American "Slaves"/"black" caste, are individualistically entitled to exercise the freedom to go their own way (without doing positive harm to the group).

8

Anti-Caste Black Solidarity for All

I have distinguished the legitimacy of Black (American) political solidarity from Black special obligation to join in that solidarity. While
I endorse the former, I am skeptical about the latter. To conclude Part III,
I round out the argument by resolving the fraught issue of whether being
black—that is, an Africa-identified person who has non-northern African
ancestry, typically visible—ought to be a necessary condition for being in
Black solidarity. The black nationalist tradition answers in the affirmative.

The exclusionary "racial" feature of black nationalist thought has been a
bête noire to nonblack persons, real or apocryphal, interested in supporting
the Black American justice cause. This feature of black nationalism now
seems to me mistaken. My revisionist conclusion comes from a place of
close sympathy with Tommie Shelby's account of "pragmatic black nationalism" as political, not racial or ethnocultural, solidarity. Instead of cagey
defense of the exclusionary feature, Shelby should be led by his own argument to reject it. Being in Black political solidarity does not require being
an Africa-identified person. This revision suits a twenty-first-century Black
politics that is decentralized in leadership and less tied to taking pride in
social identity principally defined by (nonvoluntary) blackness.[1] Yet the
animating vision of Black social equality remains the same.[2] **My view can be
described as Black post-nationalist: Black political solidarity should welcome support from everyone moved, regardless of their social identity,
by the moral urgency and priority of corrective justice for descendants of
American slavery, which would finally dismantle color caste in the United
States.[3]**

The "racial" wealth gap has carried over from inherited slavery to enforced
segregation to de facto segregation.[4] Social inequality is an inherited condition for Black America under continuous legal and political terms set by
White America.[5] Caste applies at a social group level, which today allows
wide variation in the socioeconomic status of individuals within and across
the color-conscious divide. To quote David Dinkins, the first Black mayor of
New York City (1990): "A white man with a million dollars is a millionaire.

The Afterlife of Race. Lionel K. McPherson, Oxford University Press. © Lionel K. McPherson 2024.
DOI: 10.1093/oso/9780197626849.003.0030

A black man with a million dollars is a nigger with a million dollars."[6] So goes American caste, now more than 150 years after emancipation.

Shelby's view that Black political solidarity is for and exclusively by Africa-identified persons is one I used to share. In a co-written article, we explain how "two modes of blackness" define that solidarity: "The racial mode specifies a necessary qualification (viz., being black according to the prevailing social criteria) for receiving the benefits—the group loyalty, trust, and special concern—that political solidarity generally entails; and the political mode determines what basic commitments a black person typically must accept [to count as] a member in good standing."[7] The lingering question is why should Black political solidarity require that participants have African ancestry and a black social identity if, as Shelby and I agree, having a politics of Black social equality is the critical factor. After all, "group loyalty, trust, and special concern" are not themselves political ends but common building blocks of Black resistance to White America's unceasing varieties of caste dominance.

For my part, I had conceived of geoancestral blackness as a conceptual constraint for being in Black political solidarity—the black in "Black." Conflating *solidarity by* and *solidarity for* was the source of my confusion. Prompted by puzzled queries from students, I realized the error of my defense of the exclusionary feature: a squarely political account of Black political solidarity, which dispenses with race intrigue, lacks a principled rationale for excluding non-Black participants as "members" of the solidarity collective.[8] (The familiar contrast with "allies" would suggest that black racial identity is a criterion for membership.)[9] From a revisionist perspective, Black political solidarity is marked not by a person's color-conscious identity but by distinct political commitment to Black American social equality.[10] Concrete Black equality politics, unlike black identity politics, rejects a social identity criterion for Black (American) solidarity fellowship.

Shelby might be convinced by my post-nationalist revision, which better fits his account of the political nature of Black solidarity he argues is philosophically sound. (Accounts of black nationalism that build in a cultural component are his main target.)[11] But his argument reads otherwise because of an exclusionary take on what it means to be a member of the political solidarity collective, which he considers a pragmatic constraint. He writes: "Although blacks should surely work with antiracist nonblacks against racism and other forms of social injustice, there is no principled reason why blacks must give up their solidaristic commitment to each other to do so."[12] I believe Shelby would agree that their solidarity in the United States principally concerns historical injustice and current antiblackness directed against Africa-identified descendants of American slavery. At any rate, neither of the two pragmatic considerations he appeals to for restricting the solidarity collective to "blacks" stand up to further review.

First, Shelby argues that "antiblack racial injustice" is "unique as a form of racial subjection" in the New World.[13] While I agree in spirit, I would drop the "racial" framing and clarify that slavery is the root of transatlantic anti-black injustice. Antiblackness in White America marked out a home-grown "Slaves" caste as a "colored" people to exploit, abuse, and exclude, with accompanying ideology-rhetoric.[14] As Shelby would agree, social-political lineage ultimately matters more than beliefs and attitudes about any general "race" thing supposedly relevant to sane thoughts that slavery and segregation (via color caste) might be morally tolerable. Black political solidarity is not a normatively "racial" form of solidarity.

But Shelby pivots to non sequitur in defense of the exclusionary feature: shared experiences of American antiblackness produce "the strongest motivation to act," he observes, where the "additional motivational impetus is needed to overcome the moral complacency and conservative resistance that inhibit political reform in the racial arena."[15] While I agree with the spirit of that as well, the motivational point is descriptive and predictive—which does not bolster a social identity criterion for Black political solidarity membership. Non-Black persons who would support Black equality politics are already motivated to join with descendants of American slavery. Exclusionary Black solidarity fellowship would only discourage non-Black true believers, already in short supply, from publicly promoting the Cause.

Second, Shelby defends the exclusionary feature by appeal to doubts about non-Black commitment to the goal of Black social equality. "The black experience with racism in America makes it difficult," he observes, "for many blacks to fully trust nonblacks when it comes to fighting against racism."[16] A hundred years of postbellum evidence that liberal Whites were more attached to White America's national unity and privileged access to resources than to transformative "racial" progress led Black Power advocates, for example, to reject White involvement in the Black freedom struggle.[17] Moreover, Shelby points out, "other ethnoracial minority groups have solidaristic commitments of their own, which have sometimes been used to exploit the economic and political disadvantages of African Americans as a group"—with "whites in power sometimes favor[ing] these other groups over blacks."[18] Non-specific politics "of color" can be a diversion from, if not an obstacle to, policy focused on Black (American) progress.

There is no proud history of Asian, Latino, or African immigrant groups contributing to Black America's struggle for equality. In effect, "black and brown" solidarity has mainly consisted of non-Black groups pursuing their interests under the auspices of "racial" justice, "diversity, equity, and inclusion," and "antiracism"—by amorphous association with corrective justice claims grounded in American slavery and segregation.[19] For their part, Black Americans have an unparalleled record of actively taking up justice causes outside their national people, to no expected benefit of their own.[20] These Africa-identified descendants of American slavery are, at the same time, a global symbol of perpetual state-sponsored abuse and contempt.[21]

John Lennon and Yoko Ono did not mince words in a 1972 song about the nature of America's Africa-identified bottom caste: "Woman is the nigger of the world / Think about it, do something about it // . . . Woman is the slave to the slaves / Yes she is, if you believe me, you better scream about it."[22] The objectified "nigger," still serving contrastive purpose when no longer unfree as heritable property, has to remain in a state of disrespect and disrepair. Lennon-Ono hoped that their song would create a stir and raise feminist consciousness. "Woman," even as the right hand of European supremacy, could become a figurative "slave" who, as literal master, had enjoyed legal right to enslave "the nigger."[23] Neither the White nor the Asian American woman, of course, was ever akin to the American slave—whose children would be born into slavery, until whenever the "Free White" institution might end.[24]

I agree with Shelby that abundant non-Black support for the Black American justice cause can neither be expected nor relied upon.[25] Those modest expectations, however, do not warrant exclusionary Black solidarity fellowship. The persons to be excluded would be the very ones trying to help make social equality for descendants of American slavery a political imperative for the first time in the country's history. If the day does arrive when that imperative garners support from a substantial minority of non-Black true believers, it will mark the beginning of the end of American caste. Who might not prioritize this foundational, forever overdue, anti-caste goal, and why, is worth asking, regardless of color.

Yet I am doubtful about the importance Shelby assigns to Black in-group trust. Substance and strategies of Black politics, particularly in an age of social media, will be publicly and sometimes heatedly contested within the group.[26] There is no organizational avenue for emergence of a Black national leadership to build community consensus.[27] A politics focused on Black social equality will nevertheless ward off, by virtue of its black specificity, solidarity pretenders who have predominant other agendas. Trust in solidarity fellowship has always been a challenge, in-group or out-group.[28] Active support for concrete policies to advance justice for Black America—without contingency on some other, grander social transformation—is an adequate sign of solidarity fellowship under normal circumstances of political engagement in twenty-first-century America.

The specificity of Black post-nationalism might feel unsettling to social justice advocates. In the post–civil rights era of "diversity" politics, American caste can give the impression there is something counterproductive or even offensive about Black Americans asserting distinct justice claims to social equality.[29] Meanwhile, various other causes seem publicly entitled to their dominant-issue voters.[30] It is up to Black Americans to resist being shamed for insisting on the moral urgency and priority of their people's vital interests. This will involve rejecting the notion that Black Americans, on call, should be the moral conscience of the country—and owe first allegiance to noble principles in service of other people's social justice projects.[31] "Interracial" solidarity as an "all lives" affair is an impossible substitute for Black political solidarity.

In that light, non-exclusionary Black (American) political solidarity would not leave the terms of Black politics hostage to values or preferences of prospective non-Black fellows. Black post-nationalism endorses Black self-determination of proximate political goals and strategies. Putting much stock in Black in-group trust and loyalty, though, virtually announces vulnerability to infiltration, as the tragic history of radical Black activism illustrates.[32] Chuck D of hip hop group Public Enemy summed up the reality of in-group divided loyalties: "Every brother ain't a brother 'cause a color / Just as well could be undercover."[33] The lesson applies to Black persons of any gender.

Black political solidarity cannot be channeled into political party fealty that rests on default "lesser evil" logic. Any major party platform worth progressive support should be made to answer to Black America's corrective justice agenda, not vice versa.[34] The alternative for the past fifty years has been "racial" symbolism within establishment parameters, bounded by Taft's ambition to elevate the Supreme Court as "an 'expounder of national principle.'"[35] Opposition to social equality for descendants of American slavery has remained such a principle—formerly explicit, now implicit.[36]

9

Parting Political Thoughts

In his *Autobiography*, Malcolm X recounts a White college student who asked what she could do to help the Cause: "Nothing," he replied.[1] Years later, he expressed "regret" for his dismissive stance. "I wish that now I [could] tell her . . . what I tell white people now when they present themselves as being sincere." It was: "Where the really sincere white people have got to do their 'proving' of themselves is not among the black *victims*, but out on the battle lines of where America's racism really is—and that's in their own home communities." They would not be needed in his "own particular Black Nationalist organization, the organization of Afro-American Unity," with its combined political and ethnocultural missions.[2]

Tentatively, Malcolm X had revised his Black nationalism in favor of non-exclusionary political solidarity. I have taken the next step, which is to positively endorse Black political solidarity fellowship for all persons, regardless of color, who prioritize the political struggle to steer the United States to materially transcend American slavery and end its color caste hierarchy.

The challenge for any truly progressive American politics is, as it has always been, how to support transformative progress toward Black social equality in the wake of majoritarian non-Black resistance, complicity, and apathy. This will require rejecting incrementalist or universal policies tacitly premised on leaving Black America in a better bottom caste place.[3] **There can be no way around directly addressing Black America's intergenerational social disadvantage and disrepair rooted in inherited slavery and enforced segregation, whose consequences have been made to appear inevitable and even normal through centuries of White America's anti-black rhetoric and social practice.** The American caste hierarchy will not dissipate through sheer passage of color-conscious time, improved attitudes about raciality, "intersectional" gender politics, "people of color" mantras, "class-first" dogma, pan-African zeal, or flat black consciousness.

Non-exclusionary Black (American) political solidarity means an unapologetic, post-partisan, and de-raced commitment to Black politics, which

The Afterlife of Race. Lionel K. McPherson, Oxford University Press. © Lionel K. McPherson 2024.
DOI: 10.1093/oso/9780197626849.003.0031

calls everyone who cares about the profound corrective justice circumstances of descendants of American slavery. Without such commitment, American color caste will endure—and Black American social equality will remain a distant dream.

Forgiveness on Layaway

Reflections from South Africa

*Updated from a keynote address delivered at the University of the
Witwatersrand, Johannesburg, June 2011.*

There is a consensus in the United States that enslaving Africa-identified persons was wrong for more than two hundred years and that the Jim Crow system of apartheid for the next one hundred years was also "a moral problem." National myth holds that false beliefs and bad feelings about "blacks" were mostly to blame for vicious behaviors by "whites." We are to think that British settlers in America thus saw fit, under their law, to import captives for any use without limit, whose children would be born into slavery, from generation to generation—a "forever" mark of the American caste system, which Abraham Lincoln later defended because of "the superior position assigned to the white race."[1]

In the tradition of Frederick Douglass and Toni Morrison, I have explained that there never was an honest question (at least in the modern era) about moral justification for inherited slavery.[2] Western slavery societies simply did not care about the lives of persons of (enslaved) African descent, who were treated as mere utilities for manual, sexual, and reproductive labor. After the European transatlantic slave trade was well under way, race (super)naturalism provided fraudulent ideological support for sustaining the most extreme form of human exploitation, in the generally successful hope of turning exceptional profits.

The public pretense in the United States is that "racial" justice has been deeply "complicated." This allows that morally respectable slaveholders—including presidents, members of Congress, and Supreme Court judges—could exist and be worthy of veneration.[3] There is supposed to be no absurdity in such possibility—unlike, say, the notion of basically noble Holocaust-era Nazis and their complicitors.[4] As the statement of guarded apology goes: "Mistakes were made." However, Americans learn little about these "mistakes" since American slavery and segregation are hardly taught in schools. Endless "controversy" is stirred by proponents of whitewashing

The Afterlife of Race. Lionel K. McPherson, Oxford University Press. © Lionel K. McPherson 2024.
DOI: 10.1093/oso/9780197626849.003.0032

America's history, reproducing willful ignorance of the depth, scope, and contemporary relevance of subjugation of the "Slaves"/"black" caste, also known as Black Americans.[5]

White America is not opposed in principle to acknowledging historical injustice through official apologies and reparations. For example, Japanese American victims of relocation and internment during World War II received a formal apology from the US government under President Ronald Reagan in 1988, with subsequent reparations paid in the amount of $1.6 billion.[6] In 2008, the United States House of Representatives did pass a "symbolic resolution" that serves as the first apology of any kind by the federal government for the country's history of violence, exploitation, and exclusion targeting Black Americans. The introduction to this apology defies sober comprehension: "The fact is slavery and Jim Crow are stains upon what is the greatest Nation on the face of the Earth and the greatest government ever conceived by man."[7] For White America, inherited slavery and enforced segregation, followed by their nonrepair and mass incarceration, do not disqualify an exceptionalist nation from moral and political greatness.

Tangible redress, including reparations, for Black America has been ruled out in White establishment discourse on the following alleged grounds: unfairness to the ever-growing majority of individuals who were not around to contribute to color-conscious "sins of the past" (setting aside questions of material inheritance); sheer expense (as compared to annual foreign aid of around $40 billion circa 2020); political infeasibility (due to White refusal); or constitutional barriers (made self-fulfilling by the Supreme Court).[8] The US approach to reckoning with historical injustice against descendants of American slavery illustrates what I will call *half-hearted reconciliation*, at best.

Other countries have made better faith efforts to acknowledge historical injustice. South Africa convened the Truth and Reconciliation Commission (TRC) soon after the end of apartheid. The primary mandate of the TRC, led by Archbishop Desmond Tutu, was to gather testimony about human rights crimes committed under the system of white-minority rule and official racial segregation. The TRC is widely seen, in South Africa and internationally, as a practical if imperfect model of how to promote "national unity and reconciliation."[9] Perpetrators could seek amnesty for their political crimes and would be granted amnesty, if they made full disclosure about their role in the crimes; no demonstration of remorse was necessary. Applications for amnesty involving "gross" human rights violations—such as abductions, killing, and torture—were dealt with in public hearings.[10]

I will bracket discussion of what resources and opportunities descendants of American slavery are justly due in efforts to repair intergenerational social disadvantage that burdens them individually or collectively.[11] My present agenda is to explore, through comparison of the United States and South Africa, a burgeoning philosophical literature and applied practice of political apology. Although I am not hostile to the spirit of forgiveness, I reject emphasis on intangible modes of "restorative justice"—particularly under conditions of vast social inequality rooted in historical injustice.

* * *

Apology ordinarily functions in the interpersonal domain, between specific individuals. Philosopher Nick Smith broadly describes the moral dynamics: "An apology can recognize that we have been harmed, helping us to understand what happened and why. The person apologizing accepts blame for our injury and she explains why her actions were wrong. This validates the victim's beliefs, and she can begin or resume a relationship based on these shared values."[12] Smith does not make a distinction here between interpersonal and political apology. He presupposes a core domain of apology that is interpersonal—grounded in a certain conception of reactive moral attitudes such as praise, blame, gratitude, and resentment.

The moral logic of reactive attitudes has attached the value of apology to a forgiveness-or-resentment/retribution dichotomy. As Smith goes on to propose, "When we think of apologies in these respects, we can appreciate why personal and political relationships may hinge on them. . . . If we think of all of the festering injuries that cause so much pain in our intimate lives as well as our global conflicts, apologies often seem like the best means of cleaning and stitching those wounds."[13] I beg to differ. Outside a justice framework that starkly counsels either forgiveness or retribution, the role of apology is more modest.[14]

An apology is a moral performance in words, which may or may not be heartfelt. Descriptive inquiry examines how apologies function in actual social contexts. Smith is impressed by evidence, for example, that in an adversarial legal system, "a few words of contrition, regardless of their sincerity by any measure, can dramatically decrease the likelihood of costly litigation."[15] By contrast, normative inquiry examines how apologies are morally supposed to function. Advocates of reactive attitudes logic seem especially eager for apology to yield forgiveness. The normative question is whether and why an apology ought to be accepted, and what acceptance ought to involve.

Apologies by themselves are often an insufficient means of addressing wrongs. Many Americans believe as much in a country that supports the highest incarceration rate in the world (while "violent and property crimes have plunged since the 1990s").[16] I contend that trying to disconnect apology from repair is morally irresponsible. Political apology is mainly important as a precondition for corrective justice. Compare the 2008 "symbolic" US House resolution apologizing to Black Americans for inherited slavery and enforced segregation—which "prohibits the use of the resolution to justify the issuance of reparation payments to the descendants of slaves," with "assurances that the nonbinding resolution would have little real-world effect."[17] Explicit refusal to redress historical injustice renders the symbolism ridiculous and insulting.

Generally, I am concerned about the gap between acknowledging and rectifying wrongs, which can be so wide as to call into question the sincerity of an apology. This gap is encouraged by implying that retribution naturally follows non-forgiveness—which suggests, I think misleadingly, that nonviolent progress is virtually impossible without forgiveness. I will focus on the relation between apology and forgiveness in the political domain, where my reservations about interpersonal forgiveness are amplified.

Philosopher Charles Griswold has set forth a comprehensive account of apology and forgiveness. A candidate for forgiveness would have to meet six conditions: (1) through apology, acknowledge her responsibility for wrongdoing; (2) repudiate herself for doing wrong; (3) express regret to the victims; (4) commit to becoming a person who does not wrong others; (5) show her understanding, from the victim's perspective, of the damage done by her wrongdoing; and (6) explain how she came to do wrong, and how she is becoming worthy of moral esteem.[18] On the other side of the equation are six conditions that a person who forgives would have to meet: these mainly concern renouncing claims to revenge, letting go of resentment, and communicating to the wrongdoer that her apology is accepted and that she is forgiven.[19] Griswold does not exactly argue that when a wrongdoer has met the conditions for forgiveness, the person wronged must forgive. Instead, he characterizes forgiveness as a "virtue"; and on occasions when forgiveness "should" be granted, doing so reflects what the wrongdoer is "due"; yet the wrongdoer has no "right" to forgiveness.[20]

I am unsure why forgiveness per se would be a virtue, why a person who is wronged would typically have a moral stake in forgiving the wrongdoer, and why a wrongdoer would be due forgiveness. My reservations hold for any account of apology and forgiveness tied to traditional, retributivist-oriented conceptions of moral desert. Griswold claims that his approach to forgiveness is secular, and that he is only trying "to understand the notion and its conceptual structure."[21] This is unconvincing—especially since he follows Joseph Butler, the 18th-century Christian bishop and philosopher, in construing the resentment that forgiveness would overcome as "a species of moral hatred . . . aroused by the perception of what we take to be an un-warranted injury . . . a reactive as well as retributive passion that instinctively seeks to exact a due measure of punishment."[22] Fear that angry moral judgment will motivate retribution looms large in the philosophical literature on forgiveness.[23]

The normative challenge I am raising is not answered by interpreting forgiveness merely as "reconciliation" (a term that presupposes there were "friendly relations" in the first place). Reconciliation requires social uptake, which will depend on reactive moral attitudes variously had by people in different cultures.[24] The type of forgiveness under review is supposed to be a distinctive phenomenon of worldwide application—and thereby not reducible to practical benefits of striving for social harmony or attaining inner peace. Contemporary theorists cast forgiveness along the lines of "a corrective attitude that replaces an initial attitude of resentment that we no longer find worthy."[25] Yet mere indifference seems adequate to that corrective task. Speaking autobiographically, I cannot recall entertaining the notion of "forgiving" someone who has wronged me; nor have I sought "forgiveness" from persons I believe I have treated poorly. I do endorse apology as a marker of significant moral or personal transgression against others. While I appreciate having my apology accepted, I am not hoping for moral redemption of my character or worldly relief from other people's metaphysical visions of just deserts.

Apology, I argue, appropriately precedes and supersedes forgiveness. The purpose of interpersonal apology is to acknowledge and take responsibility for one's wrongful behavior, to signify that one cares about the persons wronged, and to try to make this understood to them. The message conveyed by apology is that they were and are worthy of respect, despite one's mistreatment of them. When wrongs have caused material harm, an apology that does not express commitment to material repair would be ill-suited to promoting friendly relations going forward. Refusal to undertake material repair casts doubt on the apologizer's sincerity and risks generating further resentment among the persons wronged. My overriding question is this: What does forgiveness critically add concerning the normative value of apology?

Griswold's answer is highly abstract: "Forgiveness points toward a . . . notion of transcendence within time and the circle of sympathy. It assumes a background picture of human life as temporal and mortal, embodied, emotive, and interdependent or social, recognizes the pervasiveness of suffering. . . . Forgiveness is a model virtue for the project of reconciliation with imperfection."[26] This sounds like a version of the thesis that since we are all sinners, we share an underlying moral interest in forgiveness. To err is human, I agree, but we are not all similarly in need of moral redemption in this corporeal world.

How to think about transcending the human condition belongs more to discussion of spirituality. I wonder first and foremost about political contexts in which, for instance, mothers took their children to join extralegal White mobs to watch Black males being lynched (with images captured in souvenir postcards), or state agents took liberty to orchestrate the assassination of activists for the Black American justice cause (not all of whom were "blacks").[27] Let's say apologies eventually do come. "At a minimum," Griswold claims, "the commitment to forgiveness entails that neither [the wrongdoer nor the wronged] will interfere with the other again as they go forward." Although non-interference does not require forgiveness, he maintains that forgiveness in the "paradigm case" requires reconciliation: "Interpersonal forgiveness is a necessary condition of reconciliation in the stronger sense of affirmation and friendship; but not of mere acceptance in the minimal sense of the term."[28] Forgiveness, after all, would not be necessary for the persons wronged to move on with their lives without harboring designs of revenge or succumbing to resentment.

My question remains what morally recommends forgiveness as a distinctive secular phenomenon. Where real fellowship did not exist, forgiveness that might yield strong reconciliation does not seem particularly called for. Where real fellowship has existed—with estranged family, friends, or community members, for example—forgiveness again seems gratuitous: persons who are positively invested in a relationship already have a non-transcendental source of motivation to seek strong reconciliation. A heartfelt apology in either case seems sufficient to begin establishing or restoring positive social relations. With due respect to Archbishop Tutu, a more just and peaceful future is possible without forgiveness. Apologies can be sincerely made and accepted without hinging on inner spiritual transformation.

I will not further pursue my normative doubts about forgiveness in secular life. It is enough for me to argue that the "paradigm case," interpersonal forgiveness, is inadequate vis-à-vis historical injustice. Tutu's argument for the amnesty provision negotiated by post-apartheid South Africa's Truth and Reconciliation Commission provides a textbook example of the theory and practice of politically prioritizing forgiveness:

> One might go on to say that perhaps justice fails to be done only if the concept we entertain of justice is retributive justice, whose chief goal is to be punitive. . . .
>
> We contend that there is another kind of justice, restorative justice, which was characteristic of traditional African jurisprudence. . . . In the spirit of *ubuntu*, the central concern is the healing of breaches, the redressing of imbalances, the restoration of broken relationships, a seeking to rehabilitate both the victim and the perpetrator, who should be given the opportunity to be reintegrated into the community he has injured by his offense. . . .
>
> Once amnesty is granted . . . then the criminal and civil liability of the erstwhile perpetrator, and of the state in the case of its servants, are expunged. . . . This means that the victim loses the right to sue for civil damages in compensation from the perpetrator, and if it is a former state official, then the state will have been indemnified against liability for damages.
>
> That is indeed a very high price to ask the victims to pay. That happens to have been the price those who negotiated our relatively peaceful transition from repression to democracy believed the nation had to ask of victims.[29]

The amnesty "ask" did not rely on popular approval among Black South Africans but was answered for them under President Nelson Mandela's authority.[30] Tutu's endorsement of blanket amnesty highlights how a forgiveness-or-resentment/retribution dichotomy can be deeply problematic in political contexts.

Retributivism, with its fixation on the metaphysics of desert and imposing suffering on wrongdoers, is Tutu's straw man. This paves the way for his kitchen-sink defense of the Truth and Reconciliation Commission's mandate. He presents "restorative justice" as the prime alternative to retributive justice, which seeks to punish simply based on what wrongdoers are supposed to morally deserve. Locating that debate in African tradition, he implies that *ubuntu* supports victims losing any legal right to pursue criminal or civil claims (as if reserving the right would be motivated by retribution). He asserts that victims will have to accept the blanket amnesty "price" South Africa's new leadership negotiated, ostensibly on the victims' behalf, with the outgoing apartheid government. In short, Tutu runs together justice and realpolitik rationales for the TRC's decision to forswear both criminal prosecution and tangible redress.

The actual effectiveness of the TRC's process depends on who is asked what. Philosopher Margaret Walker cites research on a "central question" of ambiguous importance: "Did the acceptance of the truth (as represented by the TRC's report) lead to reconciliation among South Africans?"[31] Answers, understood in terms of "racially reconciled attitudes," showed that "truth-acceptance" had a "remarkably strong" impact on the attitudes of White South Africans toward Black South Africans, but not vice versa.[32] Nevertheless, Walker believes the TRC "embodied a set of core ideals that had been developed in the previous twenty years of restorative theory and practice in ordinary criminal justice and some civil contexts." Her optimistic verdict does not readily fit two of the ideals she lists: "Restorative justice aims above all to repair the harm caused by wrong, crime, and violence"; "Restorative justice makes central the experiences and needs (material, emotional, and moral) of victims."[33] Indeed, the TRC redefined "reparation" to include emotional and moral "healing" and "restoration," to the exclusion of material repair.[34] This probably helps explain the research finding that post-apartheid "acceptance of the truth [had] virtually no impact on black attitudes toward whites."[35]

In effect, restorative justice has treated social inequality wrought by historical injustice as a problem to be indefinitely deferred. Travel journalist Pippa de Bruyn gives a sobering synopsis of post-TRC South Africa: "The 22,000 victims of gross human rights violations had to wait until April 2003 to hear that each would receive a onetime payment of R30,000 [~$4,000 USD], a decision that was greeted with dismay by the victims. In contrast, big businesses (and most whites) were relieved to hear that the government had rejected the TRC's proposed tax surcharge on corporations. . . . [President Thabo] Mbeki emphasized that the TRC was not expected to bring about reconciliation but was 'an important contributor to the larger process of building a new South Africa.'"[36] The road from historical injustice to material repair is typically not paved with transformative policy to promote social equality. In South Africa, de Bruyn dryly concludes, "It appears that forgiveness has taken place, albeit in individual hearts."[37] There is a thin line between forgiveness and nonviolent resignation.

A 2021 report by the World Inequality Lab finds "no evidence that wealth inequality has decreased since the end of apartheid."[38] For Griswold, such lack of material progress toward social equality is not an embarrassment to forgiveness. His account insulates forgiveness, and moral sentiments it would transform, from politics of political apology. The forgiveness theorist's transcendental moralization of justice shuns mixing nonmaterial and material repair. Victims would learn that "forgiveness or the withholding thereof ought to respond to its own reasons and conditions" and should not be "held hostage to money," as Griswold puts it.[39] Reasonable concern for reparations would be left to the mercy of skeptics who hold political apology hostage to noncommitment to enact corrective justice.

We are expected to see virtue in the notion that while reparations would be a vulgar requirement for doing right, it is not vulgar to insinuate that victims would be crudely exploiting their grievances in pursuit of social equality. Griswold articulates familiar assumptions and sensibilities surrounding attention on (Afro-)forgiveness in the political domain. The orientation of Western forgiveness thought evades a simple truth: morality must have a commitment to justice, which is the very thing that generally enables an apology to be meaningful, especially in the case of historical injustice.[40]

* * *

I close by sketching a positive account of meaningful political apology. My argument has been that when apology is taken to be a primary component of reconciliation in response to historical injustice, the political domain becomes distorted by an interpersonal approach to morality. This line of criticism might seem to suggest that apology has no serious place within a corrective justice framework, which I am not implying. Apology, I argue, can have an important role in the political domain—albeit in an attenuated sense not freighted with many of the issues that arise in the ordinary interpersonal domain.

On my view, the standard case of political apology is promissory: apology is a precondition for taking good faith, substantive steps toward rectifying historical injustice that continues to be a deep source of social inequality. This type of view directs attention away from words, or symbolism, and morally interior contrition. Instead, reckoning with gross injustice is directed to actual deeds in relation to the individuals and broader class of people who were wronged, whose circumstances cannot be justly addressed as "universal" symptoms of social disadvantage such as poverty or student debt.[41] Real reconciliation aims at moral *and* material repair. Apologies that are merely a moral performance in words—unaccompanied by commitment to material repair—cannot and should not be expected to do the substantive work of reconciliation, regardless of how sincere the apologizers believe they are.

The major role of apology is thus provisional and secondary. When victims of historical injustice accept an apology, this marks their willingness to move forward with the wrongdoers or historical wrongdoing class, without active resentment. There is no credulous leap of faith since the wrongdoing class would have committed to materially rectifying wrongs for which they accept a significant degree of ownership—whether they are directly responsible agents, indirect beneficiaries, or collective custodians of nonrepair.

Reconciliation can be considered a normative byproduct. If members of the historical wrongdoing class do what they can to reverse the unjust burdens of vast social inequality, then reconciliation is much more likely to occur than not, where its occurrence would be appropriate as a matter of morality and justice. But reconciliation would not be a normative justification for corrective measures since those measures would be justified by direct appeal to what morality and justice recommend. Simply put, reconciliation can reasonably be expected to follow—and does not precede—significant progress in promoting corrective justice and social equality.

The typical obstacles to going down the path of corrective justice are practical. These obstacles highlight the dangers of an interpersonal, transcendental emphasis on apology and forgiveness. The historical wrongdoing class, having made its apologies, might well hope that apology will suffice to "restore" moral relations—and that no further commitment to promoting corrective justice is called for, especially when that would come at significant personal or group expense. There is no denying there are zero-sum dynamics in the material realm of limited resources and opportunities. But apologies for historical injustice that decline commitment to reversing gross injustice will and should ring hollow among the class of wronged people.

"By their deeds, you shall know them" is biblically famous for a reason. Half-hearted reconciliation—characterized by the elevation of words over tangibles, along with a convenient focus on transcendental progress—is a predictable outcome when beneficiaries or custodians of historical injustice remain unwilling to cede their collective position of great social advantage. Concession to unjust political reality, however, cannot be justice. The United States and South Africa illustrate that weak efforts at "racial" reconciliation will leave intact vast "black" social inequality, making half a mockery of color-conscious progress.

Notes

Part I. Section 1

* Toni Morrison, "A Humanist View" (speech, Portland State University, May 30, 1975), Portland State Library Special Collections, 31, 48–36, 45 [06:50–43:01], accessed February 14, 2021, https://soundcloud.com/portland-state-library/portland-state-black-studies-1.

1. See, e.g., Thomas Piketty, "Slave and Colonial Societies," chs. 6–7, in *Capital and Ideology*, trans. Arthur Goldhammer (Cambridge, MA: Harvard University Press, 2020), 203–303; Clint Smith, *How the Word Is Passed: A Reckoning with the History of Slavery Across America* (New York: Little, Brown, 2021); and Ellora Derenoncourt et al. [Chi Hyun Kim, Moritz Kuhn, and Moritz Schularick], "Wealth of Two Nations: The U.S. Racial Wealth Gap, 1860–2020," National Bureau of Economic Research, Working Paper 30101, June 2022, 1–40, http://dx.doi.org/10.3386/w30101.

2. See, e.g., Ian Haney López, *Dog Whistle Politics: How Coded Racial Appeals Have Reinvented Racism and Wrecked the Middle Class* (New York: Oxford University Press, 2014), 3–5; and Eduardo Bonilla-Silva, "The Linguistics of Color Blind Racism: How to Talk Nasty about Blacks without Sounding 'Racist,'" *Critical Sociology* 28, no. 1–2 (2002): 42–43 [41–64], https://doi.org/10.1177/08969205020280010501. Compare, e.g., Jason Stanley, *How Propaganda Works* (Princeton: Princeton University Press, 2015), 41; David Pilgrim, *Understanding Jim Crow: Using Racist Memorabilia to Teach Tolerance and Promote Social Justice*, fore. Henry Louis Gates Jr. (Oakland: PM Press, 2015), 34–35; and Jim Crow Museum of Racist Memorabilia (online), "Racist Cartoons," https://www.ferris.edu/HTMLS/news/jimcrow/cartoons/homepage.htm.

3. Compare, e.g., Sarah Wallace Goodman, "'Good Citizens' in Democratic Hard Times," *Annals of the American Academy of Political and Social Science* 699, no. 1 (2022): 68–78, https://doi.org/10.1177/00027162211069729; Ta-Nehisi Coates, "'Better Is Good': Obama on Reparations, Civil Rights, and the Art of the Possible," *Atlantic*, December 21, 2016, https://www.theatlantic.com/politics/archive/2016/12/ta-nehisi-coates-obama-transcript-ii/511133/; and "The Question of Slavery," *New York Times*, January 16, 1863, https://www.nytimes.com/1863/01/16/archives/the-question-of-slavery.html. **For "Negro rights" survey findings that "over time, trends on normative questions have risen more sharply than have those on social contact,"** see Mildred A. Schwartz, *Trends in White Attitudes toward Negroes* (Chicago: National Opinion Research Center, 1967), 17.

4. Compare, e.g., Theodore Dwight Weld, *American Slavery As It Is: Testimony of a Thousand Witnesses* (New York: American Anti-Slavery Society, 1839).

5. Frederick Douglass, *Oration, Delivered in Corinthian Hall, Rochester, by Frederick Douglass, July 5th, 1852* (Rochester, NY: Lee, Mann, 1852), 20, 17–19. See also Frederick Douglass, "The Last Flogging," in *My Bondage and My Freedom*, intro. James McCune Smith (New York: Miller, Orton and Mulligan, 1855), 233–49.

6. See, e.g., William Goodell, *The American Slave Code in Theory and Practice: Its Distinctive Features Shown by Its Statutes, Judicial Decisions, and Illustrative Facts* (New York: American and Foreign Anti-Slavery Society, 1853), 15–17. A selection of section titles: "Slaves Cannot Marry"; "Education Prohibited"; "No Access to the Judiciary, and No Honest Provision for Testing the Claims of the Enslaved to Freedom." Control over the enslaved's rational agency is obviously implicit in these limitations and hindrances.

7. See, e.g., Huaping Lu-Adler, "Kant and Slavery—Or Why He Never Became a Racial Egalitarian," *Critical Philosophy of Race* 10, no. 2 (2022): 263–94, https://www.muse.jhu.edu/article/866171; Heike Raphael-Hernandez and Pia Wiegmink, "German Entanglements in Transatlantic Slavery: An Introduction," *Atlantic Studies* 14, no. 4 (2017): 419–35, https://doi.org/10.1080/14788810.2017.1366009; and Julia Roth, "Sugar and Slaves: The Augsburg Welser as Conquerors of America and Colonial Foundational Myths," *Atlantic Studies* 14, no. 4 (2017): 436–56, https://doi.org/10.1080/14788810.2017.1365279. Compare, e.g., Pauline Kleingeld, "Kant's Second Thoughts on Race," *Philosophical Quarterly* 57, no. 229 (2007): 583–84 [573–92], https://doi.org/10.1111/j.1467-9213.2007.498.x. On "ambiguities" in "Kant's reflections on the role of race and commercial relations," see Lea Ypi, "Commerce and Colonialism in Kant's Philosophy of History," in *Kant and Colonialism: Historical and Critical Perspectives*, ed. Katrin Flikschuh and Lea Ypi (Oxford: Oxford University Press, 2014), 99 [99–126]. On Kant's ostensible "absolutism about lying," see Seana Valentine Shiffrin, "Lies and the Murderer Next Door," in Seana Valentine Shiffrin, *Speech Matters: On Lying, Morality, and the Law* (Princeton: Princeton University Press, 2014), 6 [5–46].

8. On the "mutual indispensability" of "the history of science to the philosophy of science"—which, despite the dogma of interpretive charity, invites comparison with political, social, and economic history re moral and political philosophy—see George E. Smith, "Revisiting Accepted Science: The Indispensability of the History of Science," *Monist* 93, no. 4 (2010): 545 [545–79], https://doi.org/10.5840/monist201093432. The modified version of Smith's thesis became apparent when we co-taught a seminar on the "moral history" of Western slavery. Historian David Brion Davis's work was our primary guide. See, e.g., David Brion Davis, *The Problem of Slavery in Western Culture* (Ithaca: Cornell University Press, 1966).

9. Franz Boas, *The Mind of Primitive Man*, rev. ed. (1911; New York: Macmillan, 1938), 29.

10. See, e.g., Reginald Horsman, *Race and Manifest Destiny: The Origins of American Racial Anglo-Saxonism* (Cambridge, MA: Harvard University Press, 1981), 1–3; Charles Hirschman, "America's Melting Pot Reconsidered," *Annual Review of Sociology* 9 (1983): 406–7 [397–423], https://doi.org/10.1146/annurev.so.09.080183.002145; and Quentin Skinner, "John Milton and the Politics of Slavery," *Prose Studies* 23, no. 1 (2000): 1 [1–22], https://doi.org/10.1080/01440350008586692.

11. Compare, e.g., Tala Hadavi, "Support for a Program to Pay Reparations to Descendants of Slaves Is Gaining Momentum, But Could Come with a $12 Trillion Price Tag," CNBC.com, August 12, 2020, https://www.cnbc.com/2020/08/12/slavery-reparations-cost-us-government-10-to-12-trillion.html; Valerie Russ, "Controversial Movement Is Challenging Traditional Black Organizations with 'Project Takeover,'" *Philadelphia Inquirer*, November 15, 2019, https://www.inquirer.com/news/civil-rights-comcast-ados-naacp-takeover-chapters-new-jersey-harriet-20191115.html; and P. R. Lockhart, "Calls for Reparations Are as Old as Emancipation. Will Global Powers Finally Listen?," NBCNews.com, December 26, 2021, https://www.nbcnews.com/news/nbcblk/calls-reparations-are-old-emancipation-will-global-powers-finally-list-rcna9800.

12. William Howard Taft, "The South and the National Government," intro. Andrew Carnegie (speech, North Carolina Society, New York, December 7, 1908), 11–12 [9–16]. Taft explained that "Negroes should be given an opportunity equally with whites . . . to meet the requirements of eligibility which the State Legislatures in their wisdom shall lay down in order to secure the safe exercise of the electoral franchise": the federal government would not stop allowing mass disenfranchisement of Africa-identified descendants of American slavery.

13. On Supreme Court Justice Oliver Wendell Holmes's *Giles v. Harris* (1903) opinion that reinforced *Plessy v. Ferguson* (1896) via "the sweeping doctrinal principle that equity cannot enforce 'political rights,'" see Richard H. Pildes, "Democracy, Anti-Democracy, and the Canon," *Constitutional Commentary* 17 (2000): 305–7 [295–319], http://dx.doi.org/10.2139/ssrn.224731. See also Robert L. Allen, "The Bakke Case and Affirmative Action," *Black Scholar* 9, no. 1 (1977): 9–16, https://www.jstor.org/stable/41066954; William Yeomans, "Parents Tried to Desegregate Their Schools. The Roberts Court Said No," *Nation*, September 24, 2015, https://www.thenation.com/article/archive/parents-tried-to-desegregate-their-schools-the-roberts-court-said-no/; and Richard H. King, "Rights and Slavery, Race and Racism: Leo Strauss, the Straussians, and the American Dilemma," *Modern Intellectual History* 5, no. 1 (2008): 55–82, http://dx.doi.org/10.1017/s1479244307001539.

14. Compare, e.g., R. A. Huttenback, "The British Empire as a 'White Man's Country'—Racial Attitudes and Immigration Legislation in the Colonies of White Settlement," *Journal of British Studies* 13, no. 1 (1973): 108–37, https://doi.org/10.1086/385652; Julian Go, "'Racism' and Colonialism: Meanings of Difference and Ruling Practices in America's Pacific Empire," *Qualitative Sociology* 27, no. 1 (2004): 35–58, https://doi.org/10.1023/B:QUAS.0000015543.66075.b4; and Mélanie Lamotte, "Before Race Mattered: What Archives Tell Us about Early Encounters in the French Colonies," University of Cambridge Research, November 16, 2016, https://www.cam.ac.uk/research/features/before-race-mattered-what-archives-tell-us-about-early-encounters-in-the-french-colonies.

15. For a sociological-historical account of fundamental relations between Western slavery, freedom, and personhood, see Orlando Patterson, *Freedom in the Making of Western Culture* (New York: Basic Books, 1991). Compare, e.g., James R. Barrett and David Roediger, "Inbetween Peoples: Race, Nationality and the 'New Immigrant' Working Class," *Journal of American Ethnic History* 16, no. 3 (1997): 3–4 [3–44], http://www.jstor.org/stable/27502194.

16. See, e.g., Douglas S. Massey and Nancy A. Denton, *American Apartheid: Segregation and the Making of the Underclass* (Cambridge, MA: Harvard University Press, 1993); Richard Rothstein, *The Color of Law: A Forgotten History of How Our Government Segregated America* (New York: Liveright, 2017); Sheryll Cashin, *White Space, Black Hood: Opportunity Hoarding and Segregation in the Age of Inequality* (Boston: Beacon Press, 2021); and Robert P. Jones, "Self-Segregation: Why It's So Hard for Whites to Understand Ferguson," *Atlantic*, August 21, 2014, https://www.theatlantic.com/natio nal/archive/2014/08/self-segregation-why-its-hard-for-whites-to-understand-fergu son/378928/. Compare, e.g., Kerner Commission, *Report of the National Advisory Commission on Civil Disorders* (Washington, DC: Government Printing Office, 1968); and Alice George, "The 1968 Kerner Commission Got It Right, But Nobody Listened," *Smithsonian*, March 1, 2018, https://www.smithsonianmag.com/smithson ian-institution/1968-kerner-commission-got-it-right-nobody-listened-180968318/.

17. Taft, "The South and the National Government," 10. Compare "Respek," Borat goes wine tasting in Mississippi, *Da Ali G Show* (TV series), HBO, 03:40–04:16, July 18, 2004.

18. On the "persistent and desperate struggles of the American Negro against slavery"—which "took eight forms, none of which have yet received anything like the treatment they deserve: (1) The purchase of freedom; (2) strikes; (3) sabotage; (4) suicide and self-mutilation; (5) flight—to communities of runaways, to the French, Indians, Canadians, Dutch, Spanish, Mexicans, British armies; (6) enlist- ment in federal forces . . . ; (7) anti-slavery agitation—talking, writing—(Douglass, Tubman, Walker, Still, Steward and a host of others); (8) slave revolts"—see Herbert Aptheker, "American Negro Slave Revolts," *Science and Society* 1, no. 4 (1937): 512 [512–38], http://www.jstor.org/stable/40399115.

19. See Antonin Scalia, "Originalism: The Lesser Evil," *University of Cincinnati Law Review* 57, no. 3 (1989): 849 [849–65]. On Taft's reactionary goal "to build up the [Supreme Court's] authority as an 'expounder of national principle,'" see also Erick Trickey, "Chief Justice, Not President, Was William Howard Taft's Dream Job," *Smithsonian*, December 5, 2016, https://www.smithsonianmag.com/history/chief- justice-not-president-was-william-howard-tafts-dream-job-180961279/. Compare Jamal Greene, "The Age of Scalia," *Harvard Law Review* 130, no. 1 (2016): 180–82 [144–84], http://www.jstor.org/stable/44072404; and Corey Robin, *The Reactionary Mind: Conservatism from Edmund Burke to Donald Trump*, 2nd ed. (New York: Oxford University Press, 2018), 28–29.

20. *Plessy v. Ferguson*, 163 U.S. 537, 544 (1896).

21. See Jerrold M. Packer, *American Nightmare: The History of Jim Crow* (New York: St. Martin's Press, 2002); and C. Vann Woodward, *The Strange Career of Jim Crow* (New York: Oxford University Press, 1955).

22. Morrison, "Humanist View," 36:23–36:45. See also Toni Morrison, "Black Matters," in *Playing in the Dark: Whiteness and the Literary Imagination* (Cambridge, MA: Harvard University Press, 1992), 6–12 [1–28].

23. See, e.g., Adam Shapiro, "The Dangerous Resurgence in Race Science," *American Scientist*, January 29, 2020, https://www.americanscientist.org/blog/macroscope/the- dangerous-resurgence-in-race-science. Compare, e.g., Stephen Ceci and Wendy M.

Williams, "Should Scientists Study Race and IQ? YES: The Scientific Truth Must Be Pursued," *Nature* 457 (2009): 788–89, https://www.nature.com/articles/457788a.

24. See, e.g., Equal Justice Initiative, "A History of Racial Injustice: Essays on People and Events in American History," Equal Justice Initiative, https://eji.org/news/tag/history-of-racial-injustice/. See also Daina Ramey Berry, *The Price for Their Pound of Flesh: The Value of the Enslaved, from Womb to Grave, in the Building of a Nation* (Boston: Beacon Press, 2017); Ronald W. Walters, "The Impact of Slavery on 20th- and 21st-Century Black Progress," *Journal of African American History* 97, no. 1–2 (2012): 110–30, https://doi.org/10.5323/jafriamerhist.97.1-2.0110; Raj Chetty et al. [Nathaniel Hendren, Maggie R. Jones, and Sonya R. Porter], "Race and Economic Opportunity in the United States: An Intergenerational Perspective," *Quarterly Journal of Economics* 135, no. 2 (2020): 711–83, https://doi.org/10.1093/qje/qjz042; and Anthony W. Marx, *Making Race and Nation: A Comparison of South Africa, the United States, and Brazil* (Cambridge: Cambridge University Press, 1998).

25. American Anti-Slavery Society, *Caste* (New York: R. G. Williams, ~1839], 1–2, 7. Compare, e.g., S. Chandrasekhar, "Caste, Class, and Color in India," *Scientific Monthly* 62, no. 2 (1946): 152 [151–57], https://www.jstor.org/stable/18862; and Sydney H. Schanberg, "Untouchability Persists Despite 40 Years of Opposition," *New York Times*, November 18, 1970, https://www.nytimes.com/1970/11/18/archives/untouchability-persists-despite-40-years-of-opposition.html.

26. Goodell, *American Slave Code*, "Slaves Can Possess Nothing," iv, 89–104. On "the least known of the front-rank abolitionists"—who "devoted his waking hours to combating slavery and promoting human rights for nearly forty years"—see Steve Gowler, "Radical Orthodoxy: William Goodell and the Abolition of American Slavery," *New England Quarterly* 91, no. 4 (2018): 592 [592–624], https://doi.org/10.1162/tneq_a_00705. See also National Abolition Hall of Fame and Museum, "Inductees," https://www.nationalabolitionhalloffameandmuseum.org/inductees.html.

27. Natasha S. Alford, "Sen. Kamala Harris Gets Backlash Over Question About Reparations for African-Americans," *Grio*, February 25, 2019, https://thegrio.com/2019/02/25/sen-kamala-harris-gets-backlash-over-question-about-reparations-for-african-americans/.

28. H. L. Mencken, "Mencken's Reply to La Monte's Third Letter," in Robert Rives La Monte and H. L. Mencken, *Men versus the Man: A Correspondence between Robert Rives La Monte, Socialist and H. L. Mencken, Individualist* (New York: Henry Holt, 1910), 115–16 [107–22]. Compare, e.g., Gyanendra Pandey, *A History of Prejudice: Race, Caste, and Difference in India and the United States* (New York: Cambridge University Press, 2013).

29. See Mencken, "Reply to La Monte's Third Letter," 116–17.

30. Quoted in Isaac Goldberg, *The Man Mencken: A Biographical and Critical Survey* (New York: Simon and Schuster, 1925), 293.

31. See, e.g., Martha J. Cutter, "'As White as Most White Women': Racial Passing in Advertisements for Runaway Slaves and the Origins of a Multivalent Term," *American Studies* 54, no. 4 (2016): 75–76, [73–97], http://www.jstor.org/stable/44982355.

32. See US Census Bureau, "History: Through the Decades: Questionnaires," accessed November 14, 2021, https://www.census.gov/history/www/through_the_decades/questionnaires/.

33. See William Waller Hening, ed., Act XII, "Negro Womens Children to Serve According to the Condition of the Mother," *Laws of Virginia*, December 1662, in *The Statutes at Large; Being a Collection of All the Laws of Virginia, From the First Session of the Legislature, in the Year 1619*, vol. 2 (Richmond: Samuel Pleasants, 1809–23), 170. Compare François Bernier, "A New Division of the Earth" (1684), in *The Idea of Race*, ed. Robert Bernasconi and Tommy L. Lott (Indianapolis: Hackett, 2000), 1 [1–4].

34. See, e.g., Thomas D. Morris, *Southern Slavery and the Law, 1619–1860* (Chapel Hill: University of North Carolina Press, 1996), 44–45.

35. Plato, *The Republic*, trans. G. M. A. Grube, 2nd ed. (Indianapolis: Hackett, 1992), Bk. 3, 414c–415d (91–92). See also Alan Garfinkel, *Forms of Explanation: Rethinking the Questions in Social Theory* (New Haven: Yale University Press, 1981), 106–7.

36. See James McCune Smith, "Introduction," in Douglass, *My Bondage and My Freedom*, xxiii [xvii–xxxi]. **Smith was reflecting on the "joys" of enthusiastic reception by "British and Irish audiences" and "social circles" who "had never drank of the bitter waters of American caste" (xxii–xxiii). Douglass was literally "a fugitive" abroad; his supporters in England negotiated and paid to "render him entirely & Legally free" from the White American family who had held him.** See "History Resources: Buying Frederick Douglass's Freedom, 1846," Gilder Lehrman Institute of American History, 2014, https://www.gilderlehrman.org/history-resources/spotlight-primary-source/buying-frederick-douglass's-freedom-1846.

37. Gunnar Myrdal, *An American Dilemma: The Negro Problem and Modern Democracy* (New York: Harper and Brothers, 1944), 54. See also Allison Davis, Burleigh B. Gardner, and Mary R. Gardner, *Deep South: A Social Anthropological Study of Caste and Class* (Chicago: University of Chicago Press, 1941).

38. See Isabel Wilkerson, *Caste: The Origins of Our Discontents* (New York: Random House, 2020), e.g., 105–7, 250–56; and W. E. B. Du Bois, "Caste in America: That Is the Root of the Trouble," *Des Moines Register and Leader*, October 19, 1904, 5 (W. E. B. Du Bois Papers, MS 312, Special Collections and University Archives, University of Massachusetts Amherst Libraries).

39. B. R. Ambedkar, *Annihilation of Caste: The Annotated Critical Edition*, ed. S. Anand, intro. Arundhati Roy, "The Doctor and the Saint" (London: Verso, 2014), 236. See also Charles W. Mills, *The Racial Contract* (Ithaca: Cornell University Press, 1997), 11.

40. See, e.g., Anindya Sekhar Purakayastha, "W. E. B. Du Bois, B. R. Ambedkar and the History of Afro-Dalit Solidarity," *Sanglap* 6, no. 1 (2019): 22–23 [20–36], http://sanglap-journal.in/index.php/sanglap/article/view/116; and Wilkerson, *Caste*, 25–27. See also Yogita Goyal, "On Transnational Analogy: Thinking Race and Caste with W. E. B. Du Bois and Rabindranath Tagore," *Atlantic Studies* 16, no. 1 (2019): 54–71, https://doi.org/10.1080/14788810.2018.1477653.

41. George Peiris Malalasekera and Kulatissa Nanda Jayatilleke, preface to *Buddhism and the Race Question* (Paris: UNESCO, 1958). **On Martin Luther King Jr.'s dawning acceptance that, as introduced during his 1959 tour in India, he was "a**

fellow untouchable from the United States of America," compare Isabel Wilkerson, "America's 'Untouchables': The Silent Power of the Caste System," *Guardian*, July 28, 2020, https://www.theguardian.com/world/2020/jul/28/untouchables-caste-system-us-race-martin-luther-king-india.

42. See Akeel Bilgrami, *Secularism, Identity, and Enchantment* (Cambridge, MA: Harvard University Press, 2014), 110.

43. See Cornel West, *Race Matters* (Boston: Beacon Press, 1993). See also Farah Stockman, "'We're Self-Interested': The Growing Identity Debate in Black America," *New York Times*, November 8, 2019, https://www.nytimes.com/2019/11/08/us/slavery-black-immigrants-ados.html.

44. See, e.g., Joshua Glasgow, Sally Haslanger, Chike Jeffers, and Quayshawn Spencer, *What Is Race?: Four Philosophical Views* (New York: Oxford University Press, 2019); and Paul C. Taylor, "What Races Are: The Metaphysics of Critical Race Theory," in *Race: A Philosophical Introduction*, 2nd ed. (Cambridge: Polity Press, 2013), 68–119.

45. The scientistic "modern" idea of race (as continental groups) has continuities with but can be distinguished from a general pre-Enlightenment idea of ancestry groups (e.g., tribes, nations) characterized by essential differences. See, e.g., Taunya Lovell Banks, "Dangerous Woman: Elizabeth Key's Freedom Suit—Subjecthood and Racialized Identity in Seventeenth Century Colonial Virginia," *Akron Law Review* 41, no. 3 (2008): 799–800, 831–33 [799–837], http://dx.doi.org/10.2139/ssrn.672 121. See also Geraldine Heng, *The Invention of Race in the European Middle Ages* (Cambridge: Cambridge University Press, 2018), 3–4; and Benjamin Isaac, *The Invention of Racism in Classical Antiquity* (Princeton: Princeton University Press, 2004), 37–38.

46. See, e.g., Katarzyna Bryc et al. [Eric Y. Durand, J. Michael Macpherson, David Reich, and Joanna L. Mountain], "The Genetic Ancestry of African Americans, Latinos, and European Americans Across the United States," *American Journal of Human Genetics* 96, no. 1 (2015): 42 [37–53], https://doi.org/10.1016/j.ajhg.2014.11.010.

47. Juliana Menasce Horowitz, Anna Brown, and Kiana Cox, "Race in America 2019," Pew Research Center, April 9, 2019, https://www.pewsocialtrends.org/2019/04/09/race-in-america-2019/.

48. For illustration of how disaggregation of "blacks" (that is, Africa-identified persons) by national origin can be socially and politically important, see Tod G. Hamilton, "Understanding Social and Economic Disparities Between Black Immigrants and Black Americans: Toward a More Comprehensive Framework," in *Immigration and the Remaking of Black America*, foreword Douglas S. Massey (New York: Russell Sage Foundation, 2019), 45–172.

49. See, e.g., Michael C. Campbell and Sarah A. Tishkoff, "African Genetic Diversity: Implications for Human Demographic History, Modern Human Origins, and Complex Disease Mapping," *Annual Review of Genomics and Human Genetics* 9 (2008): 403–33, https://doi.org/10.1146/annurev.genom.9.081307.164258.

50. Steven J. Micheletti et al. [Kasia Bryc, Samantha G. Ancona Esselmann, William A. Freyman, Meghan E. Moreno, G. David Poznik, and Anjali J. Shastri], "Genetic Consequences of the Transatlantic Slave Trade in the Americas," *American Journal*

of Human Genetics 107 (2020): 265, 270 [265–77], https://doi.org/10.1016/
j.ajhg.2020.06.012.

51. See Johann Friedrich Blumenbach, *On the Natural Variety of Mankind* (1795), in *Idea of Race*, 27–28 [27–37]. Compare, e.g., "Human Races," in *Encyclopedia of Genetics, Genomics, Proteomics and Informatics* (Dordrecht: Springer, 2008), https://doi.org/10.1007/978-1-4020-6754-9_7931.

52. See, e.g., Naomi Zack, "Introduction to Part III: Metaphysics and Philosophy of Science," in *The Oxford Handbook of Philosophy and Race*, ed. Naomi Zack (New York: Oxford University Press, 2017), 135–38.

53. See, e.g., Nina G. Jablonski, "Skin Colors and Their Variable Meanings," in *Living Color: The Biological and Social Meaning of Skin Color* (Berkeley: University of California Press, 2012), 157–68.

Section 2

1. Martin W. Lewis and Kären E. Wigen, *The Myth of Continents: A Critique of Metageography* (Berkeley: University of California Press, 1997), 21.

2. National Geographic Education Resource Library, s.v. "Continent," https://www.nationalgeographic.org/encyclopedia/Continent/.

3. Compare, e.g., Ernst Mayr, *Systematics and the Origin of Species from the Viewpoint of a Zoologist* (New York: Columbia University Press, 1942), 103–5.

4. Walker F. Connor, "Myths of Hemispheric, Continental, Regional, and State Unity," *Political Science Quarterly* 84, no. 4 (1969): 560 [555–82], https://doi.org/10.2307/2147125.

5. Peter Heylyn, *Cosmographie in Foure Bookes. Contayning the Chorographie and Historie of the Whole World* (London: Henry Seile, 1652), 22.

6. Herodotus, *The Histories*, trans. George Rawlinson, intro. Rosalind Thomas (New York: Knopf, 1997), 4.45 (321).

7. Compare, e.g., Anahad O'Connor, "Fake Meat vs. Real Meat," *New York Times*, December 3, 2019, https://www.nytimes.com/2019/12/03/well/eat/fake-meat-vs-real-meat.html.

8. See, e.g., Ta-Nehisi Coates, "My President Was Black," *Atlantic*, January/February 2017, https://www.theatlantic.com/magazine/archive/2017/01/my-president-was-black/508793/.

9. *Jacobellis v. Ohio*, 378 U.S. 184, 197 (1964).

10. M. F. Ashley Montagu, "The Origin of the Concept of 'Race,'" in *Man's Most Dangerous Myth: The Fallacy of Race*, 2nd ed. (New York: Columbia University Press, 1945), 1–2 [1–26]. See also M. F. Ashley Montagu, "The Concept of Race in the Human Species in the Light of Genetics," *Journal of Heredity* 32, no. 8 (1941): 243–48, https://doi.org/10.1093/oxfordjournals.jhered.a105051.

11. Nicholas T. Rinehart, "Black Beethoven and the Racial Politics of Music History," *Transition* 112 (2013): 117–18 [117–30], http://dx.doi.org/10.17613/M62R8C.

12. Rinehart, "Black Beethoven," 121–22.

13. See, e.g., Verna M. Keith and Cedric Herring, "Skin Tone and Stratification in the Black Community," *American Journal of Sociology* 97, no. 3 (1991): 766–73 [760–78], https://doi.org/10.1086/229819.

14. David Wallace, May 27, 2016 (8:47 AM), comment on "On Campus Visits: A Job Candidate's Critique," Daily Nous (blog), May 26, 2016, http://dailynous.com/2016/05/26/on-campus-visits-a-job-candidates-critique/#comments. Compare, e.g., Keith Wailoo, "Genetic Marker of Segregation: Sickle Cell Anemia, Thalassemia, and Racial Ideology in American Medical Writing 1920–1950," *History and Philosophy of the Life Sciences* 18, no. 3 (1996): 305–20, http://www.jstor.org/stable/23331960.

15. See, e.g., Martin Ericsson, "What Happened to 'Race' in Race Biology? The Swedish State Institute for Race Biology, 1936–1960," *Scandinavian Journal of History* 46, no. 1 (2021): 125–48, https://doi.org/10.1080/03468755.2020.1778520.

16. On "how East Asians became yellow" in the Western world after the seventeenth century, when they had been described as white by Europeans since the fourteenth century, see Michael Keevak, *Becoming Yellow: A Short History of Racial Thinking* (Princeton: Princeton University Press, 2011), 1.

17. Modern intelligence testing wasn't invented until the early 1900s, well after the formation of scientistic conjecture about natural intelligence differences between Europeans and Africans. See, e.g., Lewis M. Terman, *The Measurement of Intelligence* (Boston: Houghton Mifflin, 1916), 91–92.

18. For effort to measure presupposed "distinctive points of difference between the Caucasian and the Negro brain," see Robert Bennett Bean, "Some Racial Peculiarities of the Negro Brain," *American Journal of Anatomy* 5, no. 4 (1906): 353 [353–432], https://doi.org/10.1002/aja.1000050402.

19. See, e.g., William Z. Ripley, "The Three European Races," in *The Races of Europe: A Sociological Study* (New York: D. Appleton, 1899), 103–30. On IQ testing that scored the "Nordic" European race highest, with many "Alpine" and "Mediterranean" types scored below "the average negro," see Matthew Frye Jacobson, *Barbarian Virtues: The United States Encounters Foreign Peoples at Home and Abroad 1876–1917* (New York: Hill and Wang, 2000), 169–70.

20. Italian disbelief about natural racial mentality might reflect educational and cultural policy. On "the least known and most mystified part of Italian national history," whereby "most Italians are not aware of what really happened between 1885 and 1943 in the four African regions that Italy conquered through force and maintained through terror" under cover of a "civilizing mission," see Angelo Del Boca, "The Obligations of Italy Toward Libya," in *Italian Colonialism*, ed. Ruth Ben-Ghiat and Mia Fuller (New York: Palgrave Macmillan, 2005), 195 [195–202].

21. (Super)naturalist race ideology has not been obsessed with indigenous peoples; barely theorized defenses of conquest (and then assimilation) were adequate. To that extent, Native Americans and Aboriginal Australians bear an adjacent relation to the scientistic three "major races" scheme. See, e.g., Carla D. Pratt, "Tribal Kulturkampf: The Role of Race Ideology in Constructing Native American Identity," *Seton Hall Law Review* 35, no. 4 (2005): 1259 [1241–60], https://scholarship.shu.edu/shlr/vol35/iss4/4; and Kay Anderson, "Rethinking 'Race' from Australia," in *Race and the Crisis of Humanism* (London: Routledge, 2007), 109–45.

Section 3

1. Compare, e.g., Michael O. Hardimon, *Rethinking Race: The Case for Deflationary Realism* (Cambridge, MA: Harvard University Press, 2017), 180–81, n. 13.
2. See, e.g., John Maynard, "The Legacy of Jack Johnson on Aboriginal Australia," *Research in the Sociology of Sport* 7 (2013): 148, 153–55 [147–59], https://doi.org/10.1108/S1476-2854(2013)0000007012; and Daryl Adair, "Australia's Role in the Making of Jack Johnson, Labelled the Greatest by Muhammad Ali," Australian Broadcasting Corporation (ABC) News, May 27, 2018, https://www.abc.net.au/news/2018-05-28/boxer-jack-johnson-redemption-after-trump-pardon/9806368. **I thank Aboriginal scholars who attended lectures I gave at Macquarie University in 2017 (generously hosted by Albert Atkin and Adam Hochman) for first telling me about "global black" Australian Aboriginal identity.**
3. See, e.g., Myra Willard, *History of the White Australia Policy to 1920* (Melbourne: Melbourne University Press, 1923); and Gwenda Tavan, *The Long, Slow Death of White Australia* (Melbourne: Scribe, 2005).
4. **An "Acknowledgment of Country" dedication reads, for example, "We acknowledge Aboriginal and Torres Strait Islander peoples as the First Australians and Traditional Custodians of the lands where we live, learn and work."** See, e.g., Joey Watson, "How the Acknowledgment of Country Became a Core National Custom—and Why It Matters," Australian Broadcasting Corporation (ABC) Radio National, March 17, 2020, https://www.abc.net.au/news/2020-03-18/history-indigenous-acknowledgment-of-country-national-custom/12029886. **On "*Terra nullius*—meaning 'land belonging to no-one'—. . . the legal concept [*sic*] used by the British government to justify the settlement of Australia,"** see "Challenging Terra Nullius," National Library of Australia, accessed February 4, 2022, https://www.nla.gov.au/digital-classroom/senior-secondary/cook-and-pacific/cook-legend-and-legacy/challenging-terra.
5. Frederick Law Olmsted, *A Journey in the Seaboard Slave States; With Remarks on Their Economy* (New York: Dix and Edwards, 1856), 179. **For the hypothetical that Virginia could "buy . . . all the slaves now in the State, send them to Africa, provide each family of them five hundred dollars to start with when they reached there, and leave still a surplus which, divided among the present white population of the State, would give between two and three thousand dollars to each family,"** see Olmsted, *Seaboard Slave States*, 171. Compare Paul A. David et al. [Herbert G. Gutman, Richard Sutch, Peter Temin, and Gavin Wright], *Reckoning with Slavery: A Critical Study in the Quantitative History of American Negro Slavery* (New York: Oxford University Press, 1976); and Derrick Bell, "The Space Traders," in *Faces at the Bottom of the Well: The Permanence of Racism* (New York: Basic Books, 1992), 158–94.
6. Hillary Rodham Clinton, *It Takes a Village: And Other Lessons Children Teach Us* (New York: Simon and Schuster, 1996), 61–62. Compare Douglas A. Blackmon, *Slavery by Another Name: The Re-Enslavement of Black People in America from the Civil War to World War II* (New York: Doubleday, 2008), 53–57; and Alex Lichtenstein, *Twice the Work of Free Labor: The Political Economy of Convict Labor in the New South* (London: Verso, 1996), xvii–xviii.

7. See, e.g., Harriet A. Washington, *Medical Apartheid: The Dark History of Medical Experimentation on Black Americans from Colonial Times to the Present* (New York: Doubleday, 2006); Billy Hawkins, *The New Plantation: Black Athletes, College Sports, and Predominantly White NCAA Institutions* (New York: Palgrave Macmillan, 2010); Caitlin Rosenthal, *Accounting for Slavery: Masters and Management* (Cambridge, MA: Harvard University Press, 2018), 179–80; and Campbell Robertson, "A City Where Policing, Discrimination and Raising Revenue Went Hand in Hand," *New York Times*, March 4, 2015, https://www.nytimes.com/2015/03/05/us/us-deta ils-a-persistent-pattern-of-police-discrimination-in-a-small-missouri-city.html.

8. See Mary Mederios Kent, "Immigration and America's Black Population," *Population Bulletin* 62, no. 4 (2007): 4–6 [3–16], https://www.prb.org/wp-content/uploads/ 2007/12/62.4immigration.pdf. Compare David Scott FitzGerald and David Cook-Martín, "The United States: Paragon of Liberal Democracy and Racism," in *Culling the Masses: The Democratic Origins of Racist Immigration Policy in the Americas* (Cambridge, MA: Harvard University Press, 2014), 82–140.

9. Ira de A. Reid, "Negro Immigration to the United States," *Social Forces* 16, no. 3 (1938): 411 [411–17], https://doi.org/10.2307/2570817.

10. **On politically low-key Black American attitudes toward immigration,** see Niambi Michele Carter, "Conflicted Nativism: An Empirical View," in *American While Black: African Americans, Immigration, and the Limits of Citizenship* (New York: Oxford University Press, 2019), 136–61. Compare, e.g., Michelle Alexander, *The New Jim Crow: Mass Incarceration in the Age of Colorblindness* (New York: New Press, 2010), 20–26.

11. See "Reparations and the Elusive Definition of Black Identity" (transcript), Code Switch, NPR, April 15, 2020, https://www.npr.org/transcripts/834027120. See also, e.g., Wesley Lowery, "Which Black Americans Should Get Reparations?" *Washington Post*, September 18, 2019, https://www.washingtonpost.com/national/which-americ ans-should-get-reparations/2019/09/18/271cf744-cab1-11e9-a4f3-c081a126de70_st ory.html. **Moral/political calls for reparations for Black Americans are principally justified by appeal to hundreds of years of extreme oppression and social disadvantage produced through American slavery and segregation—not by appeal to harms of racism, "racial" discrimination, or color prejudice as generally encountered by Africa-identified persons in White America after the 1960s.**

12. See Samara Lynn, "Controversial Group ADOS Divides Black Americans in Fight for Economic Equality," ABCNews.com, January 19, 2020, https://abcnews.go.com/US/ controversial-group-ados-divides-black-americans-fight-economic/story?id=66832 680. Compare, e.g., Jesse Washington (Associated Press), "Some Blacks Insist: 'I'm Not African-American,'" NBCNews.com, February 5, 2012, http://www.nbcnews. com/id/46264191/ns/us_news-life/t/some-blacks-insist-im-not-african-american/.

13. Compare, e.g., K. Anthony Appiah, "Race, Culture, Identity: Misunderstood Connections," in K. Anthony Appiah and Amy Gutmann, *Color Conscious: The Political Morality of Race* (Princeton: Princeton University Press, 1996), 98–99 [30–105].

14. **On White Americans making ranked distinctions between (black) Africa-identified national peoples, with Black Americans lowest,** see Lauretta Charlton,

"Study Examines Why Black Americans Remain Scarce in Executive Suites," *New York Times*, December 9, 2019, https://www.nytimes.com/2019/12/09/us/black-in-corpor ate-america-report.html. See also Eleanor Marie Lawrence Brown, "An Alternative View of Immigrant Exceptionalism, Particularly as It Relates to Blacks: A Response to Chua and Rubenfeld," *California Law Review* 103, no. 4 (2015): 993, 1013–16 [989–1017], http://www.jstor.org/stable/24758493. Compare, e.g., Onoso Imoagene, *Beyond Expectations: Second-Generation Nigerians in the United States and Britain* (Oakland: University of California Press, 2017).

15. Keishel Williams, "For Some Children of Immigrants, 'African American' Doesn't Fit Their Unique, Black Experiences in the US," Insider, July 16, 2021, https://www.insi der.com/children-of-immigrants-detail-struggles-with-blackness-in-america-2021- 7. Compare, e.g., Mary C. Waters, "Encountering American Race Relations," in *Black Identities: West Indian Immigrant Dreams and American Realities* (New York: Russell Sage Foundation, 2000), 140–91; Kevin J. A. Thomas, "Racial Identity and the Political Ideologies of Afro-Caribbean Immigrants," *Review of Black Political Economy* 45, no. 1 (2018): 22–39, https://doi.org/10.1177/0034644618770762; and Nimo M. Adbdi, "Immigrant Hierarchies and White Supremacy: Race and Representation among East African Immigrant Muslims," *International Journal of Qualitative Studies in Education* 33, no. 2 (2020): 274–84, https://doi.org/10.1080/09518398.2019.1681542.

16. On "Du Bois's analysis of Americans' refusal to come to terms with slavery's persistent political implications and his demonstration of the central role played by African Americans in the making of the United States," which "indicate the urgency of seriously considering the idea of reparations as one element of a third attempt at democratic reconstruction," see Lawrie Balfour, "Unreconstructed Democracy: W. E. B. Du Bois and the Case for Reparations," *American Political Science Review* 97, no. 1 (2003): 40 [33–44], https://doi.org/10.1017/S0003055403000509. See also Ron Eyerman, *Cultural Trauma: Slavery and the Formation of African American Identity* (Cambridge: Cambridge University Press, 2001); and Lionel K. McPherson and Tommie Shelby, "Blackness and Blood: Interpreting African American Identity," *Philosophy and Public Affairs* 32, no. 2 (2004): 171–92, https://doi.org/10.1111/ j.1088-4963.2004.00010.x.

17. On "the struggle" in the twentieth century "to continue the healthy and politically necessary existence of a relatively autonomous Afro-American identity," see Ernest Allen Jr., "Afro-American Identity: Reflections on the Pre-Civil War Era," *Contributions in Black Studies* 7, no. 1 (1985): 93 [45–93], https://scholarworks. umass.edu/cibs/vol7/iss1/4. On the "absence of black disaggregation [that] elicits a larger set of questions pertaining to solidified racial group formation, bloc voting, and generalized self-identification," see Christina M. Greer, *Black Ethnics: Race, Immigration, and the Pursuit of the American Dream* (New York: Oxford University Press, 2013), 3.

18. Philip Gourevitch, "The Optimist," *New Yorker*, February 23, 2003, https://www. newyorker.com/magazine/2003/03/03/the-optimist-3. Earlier that year, Annan's classmates at Macalester College picketed a local movie theater chain for segregating its affiliate theaters in the South. See Herta Pitman, "Macalester Students

Picket Jim Crow Policies, 1961," Macalester College Archives, November 16, 2017, https://dwlibrary.macalester.edu/archives/wp/?p=828.

19. The N-word was a local slur meant for a specific people: "Historically, *nigger* defined, limited, and mocked African Americans. It was a term of exclusion, a verbal justification for discrimination. Whether used as a noun, verb, or adjective, it reinforced the stereotype of the lazy, stupid, dirty, worthless parasite." David Pilgrim, "Nigger and Caricature," Jim Crow Museum (online), September 2001 (rev. 2012), https://www.ferris.edu/HTMLS/news/jimcrow/caricature/homepage.htm. Compare "Clayton Bigsby," Black white supremacist, *Chappelle's Show* (TV series), Comedy Central, January 22, 2003.

20. See, e.g., Carter G. Woodson, "The Beginnings of the Miscegenation of the Whites and Blacks," *Journal of Negro History* 3, no. 4 (1918): 335–53, https://doi.org/10.2307/2713814; Leslie B. Rout Jr., *The African Experience in Spanish America: 1502 to the Present Day* (Cambridge: Cambridge University Press, 1976); and Peggy Pascoe, *What Comes Naturally: Miscegenation Law and the Making of Race in America* (New York: Oxford University Press, 2009). See also David Eltis and David Richardson, *Atlas of the Transatlantic Slave Trade*, foreword David Brion Davis, afterword David W. Blight (New Haven: Yale University Press, 2010).

21. See, e.g., Press Trust of India, "Indian Population Originated in 3 Migration Waves from Africa, Iran and Asia," *Business Standard*, May 11, 2017, https://www.business-standard.com/article/current-affairs/indian-population-originated-in-3-migration-waves-from-africa-iran-asia-117051100378_1.html; Antonio F. Pardiñas et al. [José Luis Martínez, Agustín Roca, Eva García-Vazquez, and Belén López], "Over the Sands and Far Away: Interpreting an Iberian Mitochondrial Lineage with Ancient Western African Origins," *American Journal of Human Biology* 26, no. 6 (2014): 777–83, https://doi.org/10.1002/ajhb.22601; and Iosif Lazaridis, Dani Nadel, and David Reich, "Genomic Insights into the Origin of Farming in the Ancient Near East," *Nature* 536 (2016): 419–24, https://doi.org/10.1038/nature19310.

22. See *Congressional Record*, 89th Cong., 2nd sess., 1966, 112, pt. 20, 27523 [27523–25]. Compare Mark A. Stoler, "What Did He Really Say? The 'Aiken Formula' for Vietnam Revisited," *Vermont History* 46, no. 2 (1978): 100–108, https://vermonthistory.org/journal/misc/AikenVietnam.pdf.

23. See, e.g., Anthony Appiah, "The Uncompleted Argument: Du Bois and the Illusion of Race," in *"Race," Writing, and Difference*, ed. Henry Louis Gates Jr. (Chicago: University of Chicago Press, 1986), 21–37; Appiah, "Race, Culture, Identity"; and Naomi Zack, *Philosophy of Science and Race* (New York: Routledge, 2002).

24. See, e.g., Quayshawn Spencer, "How to Be a Biological Racial Realist," in *What Is Race?: Four Philosophical Views*, ed. Joshua Glasgow, Sally Haslanger, Chike Jeffers, and Quayshawn Spencer (New York: Oxford University Press, 2019), 73–110; and Sophia Efstathiou, "How Ordinary Race Concepts Get to Be Usable in Biomedical Science: An Account of Founded Race Concepts," *Philosophy of Science* 79, no. 5 (2012): 701–13, https://doi.org/10.1086/667901. Compare Everett Mendelsohn, "The Biological Sciences in the Nineteenth Century: Some Problems and Sources," *History of Science* 3, no. 1 (1964): 39–59, https://doi.org/10.1177/007327536400300103.

25. "Human Races," in *Encyclopedia of Genetics*, https://doi.org/10.1007/978-1-4020-6754-9_7931. Compare, e.g., Equal Justice Initiative, "Racial Eugenics," under *History of Racial Injustice*, October 1, 2013, https://eji.org/news/history-racial-injustice-racial-eugenics/.

26. See, e.g., Fawn M. Brodie, "Thomas Jefferson's Unknown Grandchildren: A Study in Historical Silences," *American Heritage* 27, no. 6 (1976): 23–33, 94–99, https://www.americanheritage.com/thomas-jeffersons-unknown-grandchildren; and Farah Stockman, "Monticello Is Done Avoiding Jefferson's Relationship with Sally Hemings," *New York Times*, June 16, 2018, https://www.nytimes.com/2018/06/16/us/sally-hemings-exhibit-monticello.html. Compare, e.g., "Seduction, Rape, Concubinage," in *We Are Your Sisters: Black Women in the Nineteenth Century*, ed. Dorothy Sterling (New York: W. W. Norton, 1984), 18–31.

27. See, e.g., Frances Frank Marcus, "Louisiana Repeals Black Blood Law," *New York Times*, July 6, 1983, https://www.nytimes.com/1983/07/06/us/louisiana-repeals-black-blood-law.html.

28. See, e.g., "On Views of Race and Inequality, Blacks and Whites Are Worlds Apart," Pew Research Center, June 27, 2016, http://www.pewsocialtrends.org/2016/06/27/on-views-of-race-and-inequality-blacks-and-whites-are-worlds-apart/.

Section 4

1. Robert Bernasconi and Tommy L. Lott, introduction to "A New Division of the Earth" (1684), by François Bernier, in *The Idea of Race*, ed. Robert Bernasconi and Tommy L. Lott (Indianapolis: Hackett, 2000), 1 [1–4].

2. **Racial aesthetic judgments, if not Bernier's own, have often been vicious. On their connection to visions of race superiority in "eighteenth-century European debates regarding blackness,"** see Robin Mitchell, *Vénus Noire: Black Women and Colonial Fantasies in Nineteenth-Century France* (Athens: University of Georgia Press, 2020), 37. See also George L. Mosse, *Toward the Final Solution: A History of European Racism* (New York: Howard Fertig, 1978), 11–12.

3. Bernier, "New Division of the Earth," 1–2.

4. Immanuel Kant, "Of the Different Human Races" (1777), in *Idea of Race*, 9, 11–12 [8–22]. Compare, e.g., Emmanuel Chukwudi Eze, ed., *Race and the Enlightenment: A Reader* (Cambridge, MA: Blackwell, 1997), 65; and Maurice Jackson, *Let This Voice Be Heard: Anthony Benezet, Father of Atlantic Abolitionism* (Philadelphia: University of Pennsylvania Press, 2009), 300, n. 240.

5. Kant, "Different Human Races," 17, 19.

6. Immanuel Kant, *Observations on the Feeling of the Beautiful and Sublime* (1764), in *Observations on the Feeling of the Beautiful and Sublime and Other Writings*, ed. Patrick Frierson and Paul Guyer (Cambridge: Cambridge University Press, 2011), 58–59 (2:253) [11–65]. **Compare, re "the diversity of German contact zones and entanglements with slavery in the Americas and its subsequent colonial legacy in Africa": "These different [activist] initiatives aim at opposing what German**

historian Jürgen Zimmerer has called the 'colonial amnesia of Germans,' which stands for a still widespread ignorance about the history of German colonialism in German public memory and for a continued denial of any colonial legacy in political discourse. . . . Germany's era as colonial power in a variety of African and Asian regions has been well researched during the last decades." Heike Raphael-Hernandez and Pia Wiegmink, "German Entanglements in Transatlantic Slavery: An Introduction," *Atlantic Studies* 14, no. 4 (2017): 430, 420 [419–35], https://doi.org/ 10.1080/14788810.2017.1366009.

7. See Immanuel Kant, *Toward Perpetual Peace* (1795), in *Toward Perpetual Peace and Other Writings on Politics, Peace, and History*, ed. Pauline Kleingeld, trans. David L. Colclasure (New Haven: Yale University Press, 2006), 82–84 (8:358–59) [67–109]; and Immanuel Kant, *The Metaphysics of Morals* (1797), ed. Lara Denis, trans. Mary Gregor, rev. ed. (Cambridge: Cambridge University Press, 2017), 61–62 (6:270), 72–73 (6:283). On Kant's later, "more egalitarian" racist views, see Pauline Kleingeld, "Kant's Second Thoughts on Race," *Philosophical Quarterly* 57, no. 229 (2007): 586–88 [573–92], https://doi.org/10.1111/j.1467-9213.2007.498.x. Compare, e.g., Robert Bernasconi, "Kant's Third Thoughts on Race," in *Reading Kant's Geography*, ed. Stuart Elden and Eduardo Mendieta (Albany: State University of New York Press, 2011), 291–318; and Lucy Allais, "Kant's Racism," *Philosophical Papers* 45, no. 1–2 (2016): 1–36, https://doi.org/10.1080/05568641.2016.1199170.

8. Thomas Jefferson, *Notes on the State of Virginia* (1785), in *Thomas Jefferson: Writings*, ed. Merrill D. Peterson (New York: Library of America, 1984), 270 [123–325].

9. "Jefferson's Attitudes Toward Slavery," Thomas Jefferson Foundation, accessed November 7, 2021, https://www.monticello.org/thomas-jefferson/jefferson-slavery/ jefferson-s-attitudes-toward-slavery/. Compare, e.g., Brent Staples, "The Legacy of Monticello's Black First Family," *New York Times*, July 4, 2018, https://www.nytimes. com/2018/07/04/opinion/editorials/monticello-sally-hemings-black-family.html.

10. For a sense of the economic value of human chattel to the Anglo-American world: the British government paid £20 million in compensation—£16.5 billion by an estimated 2013 value—to owners of persons freed in British colonies by the 1833 Abolition Act (after a six-year "apprenticeship"). See Nicholas Draper, *The Price of Emancipation: Slave-Ownership, Compensation and British Society at the End of Slavery* (Cambridge: Cambridge University Press, 2013), 106–7, 138–39. See also Sanchez Manning, "Britain's Colonial Shame: Slave-Owners Given Huge Payouts after Abolition," *Independent*, February 24, 2013, https://www.independent.co.uk/ news/uk/home-news/britains-colonial-shame-slave-owners-given-huge-payouts-after-abolition-8508358.html.

11. John H. Van Evrie, *Negroes and Negro "Slavery": The First an Inferior Race; the Latter Its Normal Condition* (New York: Van Evrie, Horton, 1861), 221, v.

12. See, e.g., Kenneth O'Reilly, "The Jim Crow Policies of Woodrow Wilson," *Journal of Blacks in Higher Education* 17 (1997): 117–21, https://doi.org/10.2307/2963252; and Bruce Bartlett, "Woodrow Wilson Was Even More Racist Than You Thought," *New Republic*, July 6, 2020, https://newrepublic.com/article/158356/woodrow-wilson-racism-princeton-university.

13. See Corky Siemaszko, "Sen. Mitch McConnell's Great-Great-Grandfathers Owned 14 Slaves, Bringing Reparations Issue Close to Home," NBCNews.com, July 8, 2019, https://www.nbcnews.com/politics/congress/mitch-mcconnell-ancestors-slave-owners-alabama-1800s-census-n1027511; and Bruce Schreiner, "McConnell: Black People Vote at Similar Rates to 'Americans,'" AP News, January 20, 2022, https://apnews.com/article/voting-rights-mitch-mcconnell-louisville-legislature-elections-3309dc1cd527de8032120341f6a5cb15.

14. Sven Beckert and Seth Rockman, "Introduction: Slavery's Capitalism," in *Slavery's Capitalism: A New History of American Economic Development*, ed. Sven Beckert and Seth Rockman (Philadelphia: University of Pennsylvania Press, 2016), 25 [1–27]. **On the "neglect" of slavery in Marxism, which "tends to view capitalism as a process that goes beyond slavery that is relegated to pre-capitalist and early-capitalist history,"** see David Neilson and Michael A. Peters, "Capitalism's Slavery," *Educational Philosophy and Theory* 52, no. 5 (2020): 476 [475–84], https://doi.org/10.1080/00131857.2019.1595323. See also Calvin Schermerhorn, *The Business of Slavery and the Rise of American Capitalism, 1815–1860* (New Haven: Yale University Press, 2015); and David Waldstreicher, "The Vexed Story of Human Commodification Told by Benjamin Franklin and Venture Smith," *Journal of the Early Republic* 24, no. 2 (2004): 272–73 [268–78], https://www.jstor.org/stable/4141506.

15. *An Act of April 16, 1862 [For the Release of Certain Persons Held to Service or Labor in the District of Columbia]*, "Scope and Content," National Archives and Records Administration, https://catalog.archives.gov/id/299814. See also Tera W. Hunter, "When Slaveowners Got Reparations," *New York Times*, April 16, 2019, https://www.nytimes.com/2019/04/16/opinion/when-slaveowners-got-reparations.html. **On "natural rights philosophy [that] was virtually silent on the question of how black [Americans] were to be treated when they became 'free,'"** see Winthrop D. Jordan, *The White Man's Burden: Historical Origins of Racism in the United States* (New York: Oxford University Press, 1974), 139. Compare, e.g., Robert Nozick, *Anarchy, State, and Utopia* (New York: Basic Books, 1974), 152–53.

16. Abraham Lincoln, "Fourth Joint Debate, at Charleston, September 18, 1858," in Abraham Lincoln and Stephen A. Douglas, *Political Debates Between Hon. Abraham Lincoln and Hon. Stephen A. Douglas, in the Celebrated Campaign of 1858 in Illinois* (Columbus: Follett, Foster, 1860), 136. Compare Abraham Lincoln, "Remarks of Mr. Lincoln" ("A House Divided"), in *Proceedings of the Republican State Convention, Held at Springfield, Illinois, June 16th, 1858* (Springfield, IL: Bailhache and Baker, 1858), 9 [9–12].

17. Lincoln, "Fourth Joint Debate," 136.

18. Steven Pinker, "Groups of People May Differ Genetically in Their Average Talents and Temperaments," in *What Is Your Dangerous Idea?: Today's Leading Thinkers on the Unthinkable*, ed. John Brockman (New York: Harper Perennial, 2007), 13–14 [13–15]. See also Steven Pinker, "The Fear of Inequality," in *The Blank Slate: The Modern Denial of Human Nature* (New York: Viking Penguin, 2002), 141–58.

19. Pinker, "Talents and Temperaments," 14.

20. Samuel A. Cartwright, "Report on the Diseases and Physical Peculiarities of the Negro Race," *New Orleans Medical and Surgical Journal* 7, no. 6 (1851): 707–9 [691–715].

21. Cartwright, "Peculiarities of the Negro Race," 709–13.

22. J. T., "Why Pacific-Island Nations Are So Good at Rugby," *Economist*, August 12, 2016, https://www.economist.com/game-theory/2016/08/12/why-pacific-island-nati ons-are-so-good-at-rugby.

23. See, e.g., Ned Block, "How Heritability Misleads about Race," *Cognition* 56, no. 2 (1995): 99–128, https://doi.org/10.1016/0010-0277(95)00678-R.

24. See, e.g., Robyn Autry, "How Racial Data Gets 'Cleaned' in the U.S. Census," *Atlantic*, November 5, 2017, https://www.theatlantic.com/technology/archive/2017/11/how-racial-data-gets-cleaned/541575/.

25. In social psychology, the tendency to assume that innate differences, not social variables, produce differences in behavior and performance is "the fundamental attribution error." See, e.g., Claude M. Steele, *Whistling Vivaldi: And Other Clues to How Stereotypes Affect Us* (New York: W. W. Norton, 2010), 83–84.

26. See, e.g., Dorothy Roberts, "Separating Racial Science from Racism," in *Fatal Invention: How Science, Politics, and Big Business Re-Create Race in the Twenty-First Century* (New York: New Press, 2011), 26–54.

27. See, e.g., Edward E. Baptist, *The Half Has Never Been Told: Slavery and the Making of American Capitalism* (New York: Basic Books, 2014); and Ira Katznelson, *When Affirmative Action Was White: An Untold History of Racial Inequality in Twentieth-Century America* (New York: W. W. Norton, 2005).

28. Stephen Jay Gould, *The Mismeasure of Man*, rev. ed. (1981; New York: W. W. Norton, 1996), 52.

29. For a detailed account of Agassiz's commitment to anti-black animus, see Cristoph Irmscher, "A Pint of Ink," in *Louis Agassiz: Creator of American Science* (New York: Houghton Mifflin Harcourt, 2013), 219–69.

30. See Richard E. Green et al. [Johannes Krause, Adrian W. Briggs, Tomislav Maricic, Udo Stenzel, Martin Kircher, Nick Patterson, et al.], "A Draft Sequence of the Neanderthal Genome," *Science* 328, no. 5979 (2010): 710–22, https://doi.org/10.1126/science.1188021. See also Ker Than, "Neanderthals, Humans Interbred—First Solid DNA Evidence," *National Geographic*, May 8, 2010, https://www.nationalgeograp hic.com/news/2010/5/100506-science-neanderthals-humans-mated-interbred-dna-gene/.

31. David Hume, "Of National Characters" (1753), in *Essays and Treatises on Several Subjects*, vol. 1 (London: A. Millar, 1753), 291n [277–300].

32. David Hume, "Of National Characters" (1777), in *Essays: Moral, Political, and Literary*, ed. and notes Eugene F. Miller, rev. ed. (Indianapolis: Liberty Fund, 1987), 208n [197–215]. Compare Jackson, *Let This Voice Be Heard*, 57–58, 104.

33. Robert Palter, "Hume and Prejudice," *Hume Studies* 21, no. 1 (1995): 6–7 [3–23], https://doi.org/10.1353/hms.2011.0095. Compare, e.g., Aaron Garrett and Silvia Sebastiani, "David Hume on Race," in *The Oxford Handbook of Philosophy and Race*, ed. Naomi Zack (New York: Oxford University Press, 2017), 31–43.

34. Kant, *Beautiful and Sublime*, 60–61 (2:254–55).

35. See, e.g., Kant's notes for his *Lectures on Anthropology*, 15:878. See also Kleingeld, "Kant's Second Thoughts on Race," 576–77. Compare, e.g., Andrea Weindl, "The Slave Trade of Northern Germany from the Seventeenth to the Nineteenth Centuries," in

Extending the Frontiers: Essays on the New Transatlantic Slave Trade Database, ed. David Eltis and David Richardson (New Haven: Yale University Press, 2008).

36. See, e.g., Jason T. Sharples, *The World That Fear Made: Slave Revolts and Conspiracy Scares in Early America* (Philadelphia: University of Pennsylvania Press, 2020), 5–7; and Vincent Brown, *Tacky's Revolt: The Story of an Atlantic Slave War* (Cambridge, MA: Harvard University Press, 2020), 3–6.

37. See, e.g., Gregory Maddox, ed. and intro., *Conquest and Resistance to Colonialism in Africa* (1993; repr., New York: Routledge, 2019); Shashi Tharoor, *Inglorious Empire: What the British Did to India* (London: C. Hurst, 2017); and V. G. Kiernan, *The Lords of Human Kind: European Attitudes to the Outside World in the Imperial Age* (London: Weidenfeld and Nicolson, 1969).

38. Compare Harry G. Frankfurt, *On Bullshit* (Princeton: Princeton University Press, 2005).

39. Hume, "Of National Characters" (1753), 291n.

40. See Vincent Carretta, "Who Was Francis Williams?," *Early American Literature* 38, no. 2 (2003): 214 [213–37], https://doi.org/10.1353/eal.2003.0025.

41. On Western moral theory in historical context of worldly politics, here are two examples worth considering: Kant's central emphasis on the "categorical" moral bindingness of "law," in vague abstraction; and Hume's central emphasis on "natural sympathy," via "proximity" of vague kind. Also consider Hegel's rehumanizing "master-slave" dialectic in relation to real-time horrors of European transatlantic slavery.

42. *Dred Scott v. Sandford*, 60 U.S. 393, 403, 407 (1857). Compare, e.g., Paul Finkelman, *Supreme Injustice: Slavery in the Nation's Highest Court* (Cambridge, MA: Harvard University Press, 2018).

43. The *Dred Scott* Court enacted "originalist" interpretation of the Constitution. Compare, e.g., Jamal Greene, "Selling Originalism," *Georgetown Law Journal* 97 (2009): 657–721.

44. See, e.g., Lawrence B. Solum, "Originalism Versus Living Constitutionalism: The Conceptual Structure of the Great Debate," *Northwestern University Law Review* 113, no. 6 (2019): 1244 [1243–1296], https://scholarlycommons.law.northwestern.edu/nulr/vol113/iss6/1. Compare, e.g., Justin Driver, "Reactionary Rhetoric and Liberal Legal Academia," *Yale Law Journal* 123, no. 8 (2014): 2632–34 [2616–42], http://www.jstor.org/stable/43617002.

45. On the "neglected history of Nazi efforts to mine American race law for inspiration during the making of the Nuremberg Laws [and] what it tells us about Nazi Germany, about the modern history of racism, and especially about America," see James Q. Whitman, *Hitler's American Model: The United States and the Making of Nazi Race Law* (Princeton: Princeton University Press, 2017), 2.

46. Ira Katznelson, "What America Taught the Nazis," *Atlantic*, November 2017, https://www.theatlantic.com/magazine/archive/2017/11/what-america-taught-the-nazis/540630/. See also David Pilgrim, "Who Was Jim Crow?," Jim Crow Museum (online), September 2000 (rev. 2012), https://www.ferris.edu/HTMLS/news/jimcrow/who/index.htm.

47. Richard Rothstein, *The Color of Law: A Forgotten History of How Our Government Segregated America* (New York: Liveright, 2017), 183. See also Deborah N. Archer, "The New Housing Segregation: The Jim Crow Effects of Crime-Free Housing Ordinances," *Michigan Law Review* 118, no. 2 (2019): 184 [173–232], https://doi.org/10.36644/mlr.118.2.new.

48. *Plessy*, 163 U.S. at 537.

49. *Dred Scott*, 60 U.S. at 410. **On the neglect of the case in White America and its legal culture,** see Cass R. Sunstein, "Constitutional Myth-Making: Lessons from the *Dred Scott* Case," University of Chicago Law Occasional Paper, No. 37 (1996): 1–24, http://chicagounbound.uchicago.edu/occasional_papers/7/.

50. Compare Jim Sidanius and Felicia Pratto, "Sex and Power: The Intersecting Political Psychologies of Patriarchy and Arbitrary-Set Hierarchy," in *Social Dominance: An Intergroup Theory of Social Hierarchy and Oppression* (Cambridge: Cambridge University Press, 1999), 263–98.

51. See, e.g., Brenda E. Stevenson, "What's Love Got to Do with It? Concubinage and Enslaved Women and Girls in the Antebellum South," in *Sexuality and Slavery: Reclaiming Intimate Histories in the Americas*, ed. Daina Ramey Berry and Leslie M. Harris (Athens, GA: University of Georgia Press, 2018), 159–88; and Thomas A. Foster, *Rethinking Rufus: Sexual Violations of Enslaved Men* (Athens, GA: University of Georgia Press, 2019).

52. See again Lincoln, "Fourth Joint Debate," 136.

53. See, e.g., Nina G. Jablonski, *Living Color: The Biological and Social Meaning of Skin Color* (Berkeley: University of California Press, 2012), 10–14; Kiera Butler, "One Disease Hits Mostly People of Color. One Mostly Whites. Which One Gets Billions in Funding?," *Mother Jones*, May 4, 2015, https://www.motherjones.com/environment/2015/05/sickle-cell-cystic-fibrosis-funding-race/; and David Tan, "The Asian Flush Syndrome: Red-Faced over Governmental Inaction," *Asian Scientist*, October 11, 2012, https://www.asianscientist.com/2012/10/features/red-faced-over-inaction-asian-flush-syndrome-2012/.

54. Julian Baggini, "Is the University of Edinburgh Right to 'Cancel' David Hume?," *Prospect*, September 15, 2020, https://www.prospectmagazine.co.uk/philosophy/edinburgh-university-cancel-david-hume-rename-building. See also James Farr, "Locke, Natural Law, and New World Slavery," *Political Theory* 36, no. 4 (2008): 495–522, https://doi.org/10.1177/0090591708317899. **On the 1770s indirect exchange between Hume and his antislavery critic James Beattie, who defended a " 'one race' theory,"** compare Jackson, *Let This Voice Be Heard*, 104.

55. Toni Morrison, "A Humanist View" (speech, Portland State University, May 30, 1975), Portland State Library Special Collections, 33:06–33:40 [06:50–43:01], accessed February 14, 2021, https://soundcloud.com/portland-state-library/portland-state-black-studies-1.

56. **On "an approach to propaganda . . . with an emphasis on obfuscation and on getting targets to act in the interest of the propagandist without realizing that they have done so,"** compare Christopher Paul and Miriam Matthews, *The Russian "Firehose of*

Falsehood" Propaganda Model: Why It Might Work and Options to Counter It (Santa Monica, CA: RAND Corporation, 2016), https://www.rand.org/pubs/perspectives/PE198.html. The RAND report is "not optimistic about the effectiveness of traditional counterpropaganda efforts," particularly given the "persuasive benefits that . . . propagandists gain from presenting the first version of events (which then must be dislodged by true accounts at much greater effort)."

57. American Association of Physical Anthropologists, "AAPA Statement on Biological Aspects of Race," *American Journal of Physical Anthropology* 101, no. 4 (1996): 569–70, https://doi.org/10.1002/ajpa.1331010408.

58. American Association of Physical Anthropologists, "AAPA Statement on Race and Racism," March 8, 2019, http://www.physanth.org/about/position-statements/aapa-statement-race-and-racism-2019/.

Section 5

1. See Emmanuel Chukwudi Eze, "The Color of Reason: The Idea of 'Race' in Kant's Anthropology," in *Postcolonial African Philosophy: A Critical Reader*, ed. Emmanuel Chukwudi Eze (Cambridge, MA: Blackwell, 1997), 115–17 [103–31]. On the race concept's formative philosophical history, see Justin E. H. Smith, *Nature, Human Nature, and Human Difference: Race in Early Modern Philosophy* (Princeton: Princeton University Press, 2015).

2. For a case study skeptical of the notion of cultural (as compared to biological) racism, see Hans Siebers and Marjolein H. J. Dennissen, "Is It Cultural Racism? Discursive Exclusion and Oppression of Migrants in the Netherlands," *Current Sociology* 63, no. 3 (2015): 470–89, https://doi.org/10.1177/0011392114552504.

3. Johann Gottfried von Herder, "Ideas on the Philosophy of the History of Humankind" (1784), in *The Idea of Race*, ed. Robert Bernasconi and Tommy L. Lott (Indianapolis: Hackett, 2000), 26 [23–26].

4. Johann Friedrich Blumenbach, *On the Natural Variety of Mankind* (1775), in *The Anthropological Treatises of Johann Friedrich Blumenbach*, ed. and trans. Thomas Bendyshe (London: Anthropological Society, 1865), 99 [69–141].

5. Johann Friedrich Blumenbach, *On the Natural Variety of Mankind* (1795), in *Idea of Race*, 27–29 [27–37].

6. Blumenbach, *Natural Variety* (1795), 32–33.

7. See M. F. Ashley Montagu, "The Origin of the Concept of 'Race,'" in *Man's Most Dangerous Myth: The Fallacy of Race*, 2nd ed. (New York: Columbia University Press, 1945), 12–13 [1–26]; Stephen Jay Gould, *The Mismeasure of Man*, rev. ed. (1981; New York: W. W. Norton, 1996), 405; and Smith, *Nature, Human Nature*, 259.

8. Blumenbach, *Natural Variety* (1795), 31, 28.

9. Peter J. Kitson, introduction to Johann Friedrich Blumenbach, *On the Natural Variety of Mankind* (1795), in *Slavery, Abolition and Emancipation: Writings in the British Romantic Period*, vol. 8, *Theories of Race*, ed. Peter J. Kitson (London: Pickering and Chatto, 1999), 142 [141–42].

10. G. W. F. Hegel, "Anthropology," from *Encyclopedia of the Philosophical Sciences* (1830), in *Idea of Race*, 39 [38–44].

11. Hegel, "Anthropology," 40, 42. **On "the Enlightenment's 'racial' discourse" whereby Hegel's "writings in political philosophy changed European historical perspectives into concrete projects of international politics and economics (imperialism, colonialism, and the trans-national corporation),"** compare Emmanuel Chukwudi Eze, ed., *Race and the Enlightenment: A Reader* (Cambridge, MA: Blackwell, 1997), 7–8.

12. **On the "causal role [Hegel] gives to a biological category, namely race,"** see Darrel Moellendorf, "Racism and Rationality in Hegel's Philosophy of Subjective Spirit," *History of Political Thought* 13, no. 2 (1992): 243 [243–55], https://www.jstor.org/stable/26214088.

13. **On Hegel's "pro-colonialist interpretation" of world history in light of his view that "freedom can be recognized and practiced only in classical, Christian and modern Europe,"** see Alison Stone, "Hegel and Colonialism," *Hegel Bulletin* 41, no. 2 (2020): 247 [247–70], https://doi.org/10.1017/hgl.2017.17.

14. Hegel, "Anthropology," 39.

15. See, e.g., Michael Le Page, "New World Map Is a More Accurate Earth and Shows Africa's Full Size," *New Scientist*, August 22, 2018, https://www.newscientist.com/article/2177132-new-world-map-is-a-more-accurate-earth-and-shows-africas-full-size/; and Michael C. Campbell and Sarah A. Tishkoff, "African Genetic Diversity: Implications for Human Demographic History, Modern Human Origins, and Complex Disease Mapping," *Annual Review of Genomics and Human Genetics* 9 (2008): 403–33, https://doi.org/10.1146/annurev.genom.9.081307.164258.

16. See Charles Darwin, "On the Races of Men," from *The Descent of Man* (1871), in *Idea of Race*, 56, 55, 62 [54–78].

17. Darwin, "Races of Men," 66.

18. Darwin, "Races of Men," 77.

19. See Charles Darwin, *The Descent of Man, and Selection in Relation to Sex*, vol. 1 (London: John Murray, 1871), 201.

20. **On "skin reflectance [as] strongly correlated with absolute latitude and UV radiation levels,"** see Nina G. Jablonski and George Chaplin, "The Evolution of Human Skin Coloration," *Journal of Human Evolution* 39, no. 1 (2000): 57 [57–106], https://doi.org/10.1006/jhev.2000.0403.

21. Nicholas Wade, "What Science Says About Race and Genetics," *Time*, May 9, 2014, https://time.com/91081/what-science-says-about-race-and-genetics/.

22. Wade, "Race and Genetics."

23. See Alan Derickson, "'A Widespread Superstition': The Purported Invulnerability of Workers of Color to Occupational Heat Stress," *American Journal of Public Health* 109, no. 10 (2019): 1330 [1329–35], https://doi.org/10.2105/AJPH.2019.305246. Compare, e.g., John H. Van Evrie, *Negroes and Negro "Slavery": The First an Inferior Race; the Latter Its Normal Condition* (New York: Van Evrie, Horton, 1861), 251.

24. **For "intuitions" that would reduce the race concept's "logical core" to "skin color, shape, ancestry, and aboriginal habitat [*sic*],"** compare Michael O. Hardimon, "The Ordinary Concept of Race," *Journal of Philosophy* 100, no. 9 (2003): 441, 451–52 [437–55], https://doi.org/10.5840/jphil2003100932.

25. Compare, e.g., Cécile Vidal, "Violence, Slavery and Race in Early English and French America," *The Cambridge World History of Violence*, ed. Robert Antony, Stuart Carroll, and Caroline Dodds Pennock (Cambridge: Cambridge University Press, 2020), 36–54; Catherine M. Cameron, Paul Kelton, and Alan C. Swedlun, eds., *Beyond Germs: Native Depopulation in North America* (Tucson: University of Arizona Press, 2015); and Patrick Wolfe, "Settler Colonialism and the Elimination of the Native," *Journal of Genocide Research* 8, no. 4 (2006): 387–409, https://doi.org/10.1080/14623520601056240.

26. For a succinct discussion of race propaganda in contemporary America, see Jason Stanley, *How Fascism Works: The Politics of Us and Them* (New York: Random House, 2018), 82–84. On race propaganda fueling (super)naturalist belief in race, see Karen E. Fields and Barbara J. Fields, *Racecraft: The Soul of Inequality in American Life* (London: Verso, 2014). See also Ibram X. Kendi, *Stamped from the Beginning: The Definitive History of Racist Ideas in America* (New York: Nation Books, 2016).

27. See, e.g., Manisha Sinha, "The Anglo-American Abolition Movement," in *The Slave's Cause: A History of Abolition* (New Haven: Yale University Press, 2016), 97–129.

28. See, e.g., Christopher L. Miller, *The French Atlantic Triangle: Literature and Culture of the Slave Trade* (Durham: Duke University Press, 2008); and Duncan Bell, "John Stuart Mill on Colonies," *Political Theory* 38, no. 1 (2011): 34–64, https://doi.org/10.1177/0090591709348186. Compare, e.g., Mercer Cook, "Jean-Jacques Rousseau and the Negro," *Journal of Negro History* 21, no. 3 (1936): 294–303, https://doi.org/10.2307/2714619.

29. On "the largest demographic catastrophe in world history," whereby "90 to 95 percent of the native population of the Americas lost their lives after European arrival," see David S. Jones, *Rationalizing Epidemics: Meanings and Uses of American Indian Mortality Since 1600* (Cambridge, MA: Harvard University Press, 2004), 9. On "the unique horrors of the African slave trade, during the course of which at least 30,000,000—and possibly as many as 40,000,000 to 60,000,000—Africans were killed, most of them . . . before they even had a chance to begin working as human chattel on plantations in the Indies and the Americas," see David E. Stannard, *American Holocaust: Columbus and the Conquest of the New World* (New York: Oxford University Press, 1992), 151. On the estimated total mortality rate of 50 percent for enslaved persons moved from Africa to the Americas, see Herbert S. Klein et al. [Stanley L. Engerman, Robin Haines, and Ralph Shlomowitz], "Transoceanic Mortality: The Slave Trade in Comparative Practice," *William and Mary Quarterly* 58, no. 1 (2001): 97 [93–118], https://doi.org/10.2307/2674420.

30. Compare, e.g., Seth D. Kaplan, "The Limits of Western Human Rights Discourse," in *Human Rights in Thick and Thin Societies: Universality without Uniformity* (Cambridge: Cambridge University Press, 2018), 103–33.

31. See, e.g., Eric Williams, *Capitalism and Slavery* (Chapel Hill: University of North Carolina Press, 1944); Walter Johnson, *River of Dark Dreams: Slavery and Empire in the Cotton Kingdom* (Cambridge, MA: Harvard University Press, 2013); Sven Beckert, *Empire of Cotton: A Global History* (New York: Knopf, 2014); and Manoranjan Mohanty, "Inequality from the Perspective of the Global South," in *The Oxford Handbook of Global Studies*, ed. Mark Juergensmeyer, Saskia Sassen, and Manfred B.

Steger (New York: Oxford University Press, 2019), 211–27, https://doi.org/10.1093/oxfordhb/9780190630577.013.42.

Section 6

1. **For "self-reported black race" of persons in Chicago,** see Rajesh Kumar et al. [Hui-Ju Tsai, Xiumei Hong, Xin Liu, Guoying Wang, Colleen Pearson, Katherin Ortiz, et al.], "Race, Ancestry, and Development of Food-Allergen Sensitization in Early Childhood," *Pediatrics* 128, no. 4 (2011): e821–e829, https://doi.org/10.1542/peds.2011-0691.

2. See, e.g., Michael F. Seldin et al. [Russell Shigeta, Pablo Villoslada, Carlo Selmi, Jaakko Tuomilehto, Gabriel Silva, John W. Belmont, et al.], "European Population Substructure: Clustering of Northern and Southern Populations," *PLOS Genetics* 2, no. 9 (2006): e143, https://doi.org/10.1371/journal.pgen.0020143. **On genetic differentiation within (sub)continental regions,** see L. Luca Cavalli-Sforza, Paolo Menozzi, and Alberto Piazza, *The History and Geography of Human Genes*, abridged paperback edition (Princeton: Princeton University Press, 1994), 268–72.

3. Cavalli-Sforza et al., *Human Genes*, 19.

4. See, e.g., Carol Anderson, *White Rage* (New York: Bloomsbury Press, 2016).

5. See, e.g., Jason Stanley, "The Ideology of Elites: A Case Study," in *How Propaganda Works* (Princeton: Princeton University Press, 2015), 269–91.

6. See for critiques, e.g., Randall Kennedy, "Persuasion and Distrust: A Comment on the Affirmative Action Debate," *Harvard Law Review* 99, no. 6 (1986): 1327–46, https://doi.org/10.2307/1341257; and Erika K. Wilson, "Why Diversity Fails: Social Dominance Theory and the Entrenchment of Racial Inequality," *National Black Law Journal* 26, no. 1 (2017): 129–53, https://escholarship.org/uc/item/2zn704q4.

7. See, e.g., Derrick Bell, *Silent Covenants: Brown v. Board of Education and the Unfulfilled Hopes for Racial Reform* (New York: Oxford University Press, 2004); and Richard Rothstein, *Brown v. Board at 60: Why Have We Been So Disappointed? What Have We Learned?* (Washington, DC: Economic Policy Institute, 2014).

8. Lawrence Bobo, James R. Kluegel, and Ryan A. Smith, "Laissez-Faire Racism: The Crystallization of a 'Kinder, Gentler' Anti-Black Ideology," in *Racial Attitudes in the 1990s: Continuity and Change*, ed. Steven A. Tuch and Jack K. Martin (Westport, CT: Praeger, 1997), 16 [15–42].

9. See, e.g., Neil Bhutta et al. [Andrew C. Chang, Lisa J. Dettling, and Joanne W. Hsu], "Disparities in Wealth by Race and Ethnicity in the 2019 Survey of Consumer Finances," FEDS Notes, Washington, DC: Board of Governors of the Federal Reserve System, September 28, 2020, https://doi.org/10.17016/2380-7172.2797; and William Darity Jr. et al. [Darrick Hamilton, Mark Paul, Alan Aja, Anne Price, Antonio Moore, and Caterina Chiopris], *What We Get Wrong About Closing the Racial Wealth Gap* (Durham, NC: Samuel DuBois Cook Center on Social Equity at Duke University and Insight Center for Community Economic Development, 2018), 2–4 [2–67], https://socialequity.duke.edu/wp-content/uploads/2020/01/what-we-get-wrong.pdf.

10. See, e.g., Hakeem Jefferson, "Storming the U.S. Capitol Was About Maintaining White Power in America," FiveThirtyEight, January 8, 2021, https://fivethirtyei ght.com/features/storming-the-u-s-capitol-was-about-maintaining-white-power-in-america/; and Victor Luckerson, "Living in the Age of the White Mob," *New Yorker*, January 15, 2021, https://www.newyorker.com/news/dispatch/living-in-the-age-of-the-white-mob. Compare, e.g., Ashley Jardina, *White Identity Politics* (Cambridge: Cambridge University Press, 2019), 3–4.

11. Julie Zauzmer Weil, "Help Us Identify Members of Congress Who Enslaved People," *Washington Post*, January 14, 2022, https://www.washingtonpost.com/history/inte ractive/2022/submit-congress-enslaved-database/.

12. See, e.g., Jerome McCristal Culp Jr., "Colorblind Remedies and the Intersectionality of Oppression: Policy Arguments Masquerading as Moral Claims," *New York University Law Review* 69 (1994): 180–81 [162–96]; and Michelle Wilde Anderson, "Colorblind Segregation: Equal Protection as a Bar to Neighborhood Integration," *California Law Review* 92, no. 3 (2004): 883–84 [841–84], https://www.jstor.org/stable/i277010.

13. See US Census Bureau, "History: Through the Decades: Questionnaires," accessed November 14, 2021, https://www.census.gov/history/www/through_the_decades/ questionnaires/.

14. Michael Crowley, "Trump's '1776 report' Defends America's Founding on the Basis of Slavery and Blasts Progressivism.," *New York Times*, January 18, 2021, https:// www.nytimes.com/2021/01/18/us/trump-1776-commission-report.html. See, e.g., Paul Finkelman, *Slavery and the Founders: Race and Liberty in the Age of Jefferson*, 3rd ed. (New York: Routledge, 2014). See also David Kindy, "New Research Suggests Alexander Hamilton Was a Slave Owner," *Smithsonian*, November 10, 2020, https:// www.smithsonianmag.com/history/new-research-alexander-hamilton-slave-owner-180976260/; and Erica Armstrong Dunbar, *Never Caught: The Washingtons' Relentless Pursuit of Their Runaway Slave, Ona Judge* (New York: 37 INK, 2017).

15. See, e.g., John Rankin, *Letters on American Slavery* (1826; Boston: Garrison and Knapp, 1833); and David M. Chalmers, *Hooded Americanism: The History of the Ku Klux Klan*, 3rd ed. (1965; Durham, NC: Duke University Press, 1981). Compare Patrice Douglass, Selamawit D. Terrefe, and Frank B. Wilderson, "Afro-Pessimism," Oxford Bibliographies, August 28, 2018, https://doi.org/10.1093/obo/9780190280 024-0056.

16. "History of the U.S. Capitol Building," Buildings and Grounds section, Architect of the Capitol, accessed January 18, 2021, https://www.aoc.gov/explore-capitol-cam pus/buildings-grounds/capitol-building/history.

17. "Slave Labor Commemorative Marker," Art section, Architect of the Capitol, accessed January 18, 2021, https://www.aoc.gov/explore-capitol-campus/art/slave-labor-commemorative-marker.

18. Julie Zauzmer Weil, Adrian Blanco, and Leo Dominguez, "More Than 1,700 Congressmen Once Enslaved Black People. This Is Who They Were, and How They Shaped the Nation.," *Washington Post*, January 10, 2022, https://www.washingtonp ost.com/history/interactive/2022/congress-slaveowners-names-list/. See also Gillian Brockell, "The Senate's First Woman Was Also Its Last Enslaver," *Washington Post*,

January 10, 2022, https://www.washingtonpost.com/history/2022/01/10/rebecca-fel
ton-last-enslaver/.

19. See Ta-Nehisi Coates, "Martin Luther King Makes the Case for Reparations," *Atlantic*,
June 12, 2014, https://www.theatlantic.com/business/archive/2014/06/martin-lut
her-king-makes-the-case-for-reparations/372696/. Compare, e.g., Christine Hauser,
"Judge Allows Part of Lawsuit by Tulsa Massacre Survivors Seeking Reparations,"
New York Times, May 3, 2022, https://www.nytimes.com/2022/05/03/us/tulsa-massa
cre-lawsuit-reparations.html; Andre M. Perry, Anthony Barr, and Carl Romer, "The
True Costs of the Tulsa Race Massacre, 100 Years Later," Brookings Institution, May 28,
2021, https://www.brookings.edu/research/the-true-costs-of-the-tulsa-race-massa
cre-100-years-later/; and Patricia Mazzei and Livia Albeck-Ripka, "Surfside Condo
Collapse Victims Reach $997 Million Settlement," *New York Times*, May 11, 2022,
https://www.nytimes.com/2022/05/11/us/surfside-condo-collapse-settlement-vict
ims.html. On "the phrase, 'delay, deny and hope you die' to characterize" an approach
to approving injury settlement claims, see Ken Belson, "Black Former N.F.L. Players
Say Racial Bias Skews Concussion Payouts," *New York Times*, August 25, 2020, https://
www.nytimes.com/2020/08/25/sports/football/nfl-concussion-racial-bias.html.

20. See, e.g., William A. Darity Jr. and A. Kirsten Mullen, *From Here to
Equality: Reparations for Black Americans in the Twenty-First Century* (Chapel
Hill: University of North Carolina Press, 2020). See also Brandon M. Terry, "Requiem
for a Dream: The Problem-Space of Black Power," in *To Shape a New World: Essays on
the Political Philosophy of Martin Luther King, Jr.*, ed. Brandon M. Terry and Tommie
Shelby (Cambridge, MA: Harvard University Press, 2018), 290–324; and Gary
Younge, "The Misremembering of 'I Have a Dream,'" *Nation*, August 14, 2013, https://
www.thenation.com/article/archive/misremembering-i-have-dream/.

21. For findings that "much of the variance that explains opposition to explicitly racially targeted programs is due to anti-Black antipathy blended with perceptions
of value violation, not simply anti-redistributive attitudes or attitudes toward
group-based claims and demands," see Joshua L. Rabinowitz et al. [David O. Sears,
Jim Sidanius, and Jon A. Krosnick], "Why Do White Americans Oppose Race-
Targeted Policies? Clarifying the Impact of Symbolic Racism," *Political Psychology*
30, no. 5 (2009): 825 [805–28], https://doi.org/10.1111/j.1467-9221.2009.00726.x.
See also Katherine Cramer, "Understanding the Role of Racism in Contemporary US
Public Opinion," *Annual Review of Political Science* 23 (2020): 153–69, https://doi.
org/10.1146/annurev-polisci-060418-042842.

22. For a philosophical overview of "what is involved in having FMS [full moral status],
as opposed to a lesser degree of moral status," see Agnieszka Jaworska and Julie
Tannenbaum, "The Grounds of Moral Status," *Stanford Encyclopedia of Philosophy*,
March 3, 2021 (revised), https://plato.stanford.edu/entries/grounds-moral-status/
#WhatFullMoraStatFMS.

23. For fuller elaboration of a distinction between racialism (ostensibly not evalua-
tive) and racism (which is evaluative), see Kwame Anthony Appiah, "Racisms," in
Anatomy of Racism, ed. David Theo Goldberg (Minneapolis: University of Minnesota
Press, 1990), 3–17.

24. See, e.g., Rick Perlstein, "Exclusive: Lee Atwater's Infamous 1981 Interview on the Southern Strategy," *Nation*, November 13, 2012, https://www.thenation.com/arti cle/archive/exclusive-lee-atwaters-infamous-1981-interview-southern-strategy/. Compare, e.g., Eric Levitz, "Will Black Voters Still Love Biden When They Remember Who He Was?," *Intelligencer* (New York), March 12, 2019, https://nymag.com/intell igencer/2019/03/joe-biden-record-on-busing-incarceration-racial-justice-democra tic-primary-2020-explained.html.

25. See, e.g., Charles W. Mills, "White Supremacy," in *The Routledge Companion to Philosophy of Race*, ed. Paul C. Taylor, Linda Martín Alcoff, and Luvell Anderson (New York: Routledge, 2018), 475–87. Compare, e.g., Lothrop Stoddard, *The Rising Tide of Color Against White World-Supremacy* (New York: Charles Scribner's Sons, 1920).

26. Sidanius and Pratto, *Social Dominance*, 16. See also Martin Luther King Jr., "Racism and the White Backlash," in *Where Do We Go from Here: Chaos or Community?* (New York: Harper and Row, 1967), 67–101; and Malcolm X, "God's Judgment of White America (The Chickens Are Coming Home to Roost)," December 1963, in *The End of White World Supremacy: Four Speeches*, ed. Imam Benjamin Karim (New York: Seaver Books, 1971), 121–48.

27. James Bryce, *Race Sentiment as a Factor in History* (London: London University Press, 1915), 3.

28. See, e.g., Richard Hofstadter, *Social Darwinism in American Thought, 1860–1915* (Philadelphia: University of Pennsylvania Press, 1944); and Robert C. Bannister, *Social Darwinism: Science and Myth in Anglo-American Social Thought* (Philadelphia: Temple University Press, 1979). Compare William Julius Wilson, *When Work Disappears: The World of the New Urban Poor* (New York: Knopf, 1996), xv–xvi.

29. **On the relation between Jews being "a persecuted religious minority" and "religious leadership [that] transformed Judaism from a cult based on ritual sacrifices in the temple to a religion whose main norm required every Jewish man to read and to study the Torah in Hebrew,"** see Maristella Botticini and Zvi Eckstein, *The Chosen Few: How Education Shaped Jewish History, 70–1492* (Princeton: Princeton University Press, 2012), 2. **On "an alternative Jewish supremacy (justified by high IQ) that applies primarily to Jews most likely to pass as or identify as 'white' (coded here as Ashkenazi, although not all Ashkenazi Jews are white),"** which would argue for **"(white) Jewish inclusion within white supremacy,"** see Adam Shapiro, "The Dangerous Resurgence in Race Science," *American Scientist*, January 29, 2020, https:// www.americanscientist.org/blog/macroscope/the-dangerous-resurgence-in-race-science. Compare, e.g., Steven Pinker, "Groups and Genes," *New Republic*, June 26, 2006, https://newrepublic.com/article/77727/groups-and-genes.

30. Compare, e.g., Greg Grandin, "Slavery, and American Racism, Were Born in Genocide," *Nation*, January 20, 2020, https://www.thenation.com/article/society/ slavery-american-genocide-racism/; and Allen C. Guelzo, *Reconstruction: A Concise History* (New York: Oxford University Press, 2018). **On how, in 1970—"only months ago, integration seemed an irreversible process" but "[n]ow it seems that the idea's time is waning; that, as happened 93 years ago, a racial Reconstruction may be**

collapsing"—see "Nation: End of Reconstruction," *Time*, March 2, 1970, http://cont
ent.time.com/time/subscriber/article/0,33009,904202,00.html.

31. Compare, e.g., "Destabilization of Social Systems," *Encyclopedia of World Problems
and Human Potential*, Union of International Associations (UIA), October 4, 2020,
http://encyclopedia.uia.org/en/problem/139799; Derek W. Black, "Freedom,
Democracy, and the Right to Education," *Northwestern University Law Review* 116,
no. 4 (2022): 1096–97 [1031–97], https://scholarlycommons.law.northwestern.edu/
nulr/vol116/iss4/3; German Lopez, "Mass Incarceration in America, Explained in 22
Maps and Charts," Vox, October 11, 2016, https://www.vox.com/2015/7/13/8913297/
mass-incarceration-maps-charts; Libby Nelson and Dara Lind, "The School-
to-Prison Pipeline, Explained," Vox, October 27, 2015, https://www.vox.
com/2015/2/24/8101289/school-discipline-race; and Scott Winship, Richard V.
Reeves, and Katherine Guyot, "The Inheritance of Black Poverty: It's All About the
Men," Brookings Institution, March 22, 2018, https://www.brookings.edu/research/
the-inheritance-of-black-poverty-its-all-about-the-men/.

32. Compare, e.g., Stefan M. Bradley, "The Civil Rights Era Was Supposed to Drastically
Change America. It Didn't," *Washington Post*, December 23, 2020, https://www.was
hingtonpost.com/graphics/2020/national/george-floyd-america-civil-rights-timel
ine/; and Robert C. Smith, "From Protest to Incorporation: A Framework for Analysis
of Civil Rights Movement Outcomes," in *We Have No Leaders: African Americans in
the Post-Civil Rights Era*, foreword Ronald W. Walters (Albany: State University of
New York Press, 1996), 3–26.

Section 7

1. For a Western "history of ideas [that] must be coupled with a concrete knowledge of
the events that took place at the same time" . . . but has no entries for slavery or the
transatlantic slave trade in a long index, compare J. Bronowski and Bruce Mazlish,
The Western Intellectual Tradition from Leonardo to Hegel (New York: Harper and
Brothers, 1960), xi, 520–21.

2. Edith R. Sanders, "The Hamitic Hypothesis; Its Origin and Functions in Time
Perspective," *Journal of African History* 10, no. 4 (1969): 532 [521–32], https://doi.org/
10.1017/S0021853700009683. On the biblical roots of "Hamites," see Colin Kidd,
The Forging of Races: Race and Scripture in the Protestant Atlantic World, 1600–2000
(Cambridge: Cambridge University Press, 2006), 24, 74–77.

3. OED Online, s.v. "Negro, *n.* and *adj.*," accessed June 6, 2020, http://www.oed.com/
view/Entry/125898. Compare Richard B. Moore, *The Name "Negro": Its Origin and
Evil Use*, ed. and intro. W. Burghardt Turner and Joyce Moore Turner (1960; repr.,
Baltimore: Black Classic Press, 1992), 74–76.

4. See G. W. F. Hegel, "Anthropology," from *Encyclopedia of the Philosophical
Sciences* (1830), in *The Idea of Race*, ed. Robert Bernasconi and Tommy L. Lott
(Indianapolis: Hackett, 2000), 40 [38–44].

5. G. W. F. Hegel, *Lectures on the Philosophy of History* (1837), trans. J. Sibree (London: Henry G. Bohn, 1861), 95.

6. Sanders, "Hamitic Hypothesis," 532. See also Scott Trafton, *Egypt Land: Race and Nineteenth-Century American Egyptomania* (Durham, NC: Duke University Press, 2004), 13, 31–32.

7. Constantin-François Volney, *Travels through Syria and Egypt, in the Years 1783, 1784, and 1785*, vol. 1 (London: G. G. J. Robinson and J. Robinson, 1787), 82–83.

8. Volney, *Travels*, 80.

9. **On the Muslim and European histories of the "Saharan divide,"** see Ghislaine Lydon, "Saharan Oceans and Bridges, Barriers and Divides in Africa's Historiographical Landscape," *Journal of African History* 56, no. 1 (2015): 4–10 [3–22], https://doi.org/10.1017/S002185371400070X.

10. Compare, e.g., Iman Amrani, "Why Don't We Think of North Africa as Part of Africa?," *Guardian*, September 9, 2015, https://www.theguardian.com/commentisfree/2015/sep/09/north-africa-algeria-black-africa-shared-history.

11. See Amal Kandeel, "Sub-Saharan Africa Must Respect North Africa's Water Rights," Middle East Institute, March 9, 2018, accessed July 20, 2022, https://www.mei.edu/publications/sub-saharan-africa-must-respect-north-africas-water-rights. Compare, e.g., "Egypt Discovers Oil in the Sahara," *Atalayar*, June 3, 2020, https://atalayar.com/en/content/egypt-discovers-oil-sahara.

12. **For a UN list of "Sub-Saharan Africa" and "Northern Africa" countries that purportedly "does not imply any assumption regarding political or other affiliation,"** see "Methodology: Geographic Regions," United Nations Department of Economic and Social Affairs: Statistics Division, 2019, https://unstats.un.org/unsd/methodology/m49/.

13. See, e.g., Michael Russell, *View of Ancient and Modern Egypt: With an Outline of Its Natural History* (New York: Harper and Brothers, 1835), 306. **On the British colonial construction of "the Middle East" after World War I,** see Karl E. Meyer, "How the Middle East Was Invented," *New York Times*, March 13, 1991, https://www.nytimes.com/1991/03/13/opinion/editorial-notebook-how-the-middle-east-was-invented.html.

14. James Henry Breasted, *The Conquest of Civilization* (New York: Harper and Brothers, 1926), 113.

15. Hugh Trevor-Roper, "The Rise of Christian Europe," *Listener* 70, no. 1809 (1963): 871 [871–75].

16. **For an overview of the history between ancient Nubia and Egypt,** see Boyce Rensberger, "The Grandeur That Was Nubia," *Washington Post*, May 10, 1995, https://www.washingtonpost.com/archive/1995/05/10/the-grandeur-that-was-nubia/. **For a comprehensive study of ancient Nubia,** see Marjorie M. Fisher et al. [Peter Lacovara, Salima Ikram, and Sue D'Auria], eds., *Ancient Nubia: African Kingdoms on the Nile* (Cairo: American University in Cairo Press, 2012).

17. See, e.g., Sam Nixon, "Before Timbuktu: The Great Trading Centre of Tadmakka," *Current World Archaeology* 39 (2010): 40–51; and E. W. Bovill, *Caravans of the Old Sahara: An Introduction to the History of the Western Sudan* (London: Oxford University Press, 1933).

18. Of current debate is whether persons living in the Sahara region with their herds of grazing animals induced or delayed the ecological change. See David K. Wright, "Humans as Agents in the Termination of the African Humid Period," *Frontiers in Earth Science* 5, no. 4 (2017), https://doi.org/10.3389/feart.2017.00004; and Chris Brierley, Katie Manning, and Mark Maslin, "Pastoralism May Have Delayed the End of the Green Sahara," *Nature Communications* 9, no. 4018 (2018), https://doi.org/10.1038/s41467-018-06321-y.

19. See María Cerezo et al. [Alessandro Achilli, Anna Olivieri, Ugo A. Perego, Alberto Gómez-Carballa, Francesca Brisighelli, Hovirag Lancioni et al.], "Reconstructing Ancient Mitochondrial DNA Links between Africa and Europe," *Genome Research* 22 (2012): 821 [821–26], http://www.genome.org/cgi/doi/10.1101/gr.134452.111.

20. Herodotus, *The Histories*, trans. George Rawlinson, intro. Rosalind Thomas (New York: Knopf, 1997), 2.104 (174, 174n2). Compare Herodotus, *The Histories*, trans. Robin Waterfield, intro. Carolyn Dewald (Oxford: Oxford University Press, 1998), 2.104 (134).

21. Sanders, "Hamitic Hypothesis," 521.

22. C. G. Seligman, *Races of Africa* (London: Thornton Butterworth, 1930), 18–19, 96–97. For a remembrance that acknowledges his "racial, diffusionist theory [*sic*]" and celebrates "the true founder of LSE anthropology, C. G. ('Sligs') Seligman," see Adam Kuper, "C. G. Seligman, 'Sligs'" (public talk at the Royal Anthropological Institute), May 2018, https://therai.org.uk/archives-and-manuscripts/obituaries/c-g-seligman.

23. On Leopold's atrocities in Congo, see Adam Hochschild, *King Leopold's Ghost: A Story of Greed, Terror, and Heroism in Colonial Africa* (New York: Houghton Mifflin, 1998); and "King Leopold Denies Charges Against Him," *New York Times*, December 11, 1906, https://www.nytimes.com/1906/12/11/archives/king-leopold-denies-charges-against-him-is-poorer-instead-of-richer.html.

24. Henry M. Stanley, *Through the Dark Continent*, vol. 1 (New York: Harper and Brothers, 1878), 426–27. (Stanley coined "the Dark Continent" phrase.) Compare Michael F. Robinson, *The Lost White Tribe: Explorers, Scientists, and the Theory that Changed a Continent* (New York: Oxford University Press, 2016), 3–10.

25. On "Caucasian Hamites" and the "confusion of race and language," see Asya Pereltsvaig and Martin W. Lewis, *The Indo-European Controversy: Facts and Fallacies in Historical Linguistics* (Cambridge: Cambridge University Press, 2015), 29–32.

26. Verena J. Schuenemann et al. [Alexander Peltzer, Beatrix Welte, W. Paul van Pelt, Martyna Molak, Chuan-Chao Wang, Anja Furtwängler et al.], "Ancient Egyptian Mummy Genomes Suggest an Increase of Sub-Saharan African Ancestry in Post-Roman Periods," *Nature Communications* 8, no. 15694 (2017), https://doi.org/10.1038/ncomms15694.

27. Schuenemann et al., "Egyptian Mummy Genomes."

28. Ben Guarino, "DNA from Ancient Egyptian Mummies Reveals Their Ancestry," *Washington Post*, May 30, 2017, https://www.washingtonpost.com/news/speaking-of-science/wp/2017/05/30/dna-from-ancient-egyptian-mummies-reveals-their-ancestry/; and Philip Perry, "Were the Ancient Egyptians Black or White? Scientists Now Know," Big Think, June 11, 2017, https://bigthink.com/philip-perry/were-the-ancient-egyptians-black-or-white-scientists-now-know.

29. Schuenemann et al., "Egyptian Mummy Genomes."
30. Kathryn A. Bard, "Ancient Egyptians and the Issue of Race," in *Black Athena Revisited*, ed. Mary R. Lefkowitz and Guy MacLean Rogers (Chapel Hill: University of North Carolina Press, 1996), 107, 109 [103–11]. Compare Mary R. Lefkowitz, "Ancient History, Modern Myths," in *Black Athena Revisited*, 6 [3–23]; and Martin Bernal, *Black Athena: The Afroasiatic Roots of Classical Civilization*, vol. 1 (New Brunswick: Rutgers University Press, 1987), 243–46.
31. See Frank M. Snowden Jr., "Bernal's 'Blacks' and the Afrocentrists," in *Black Athena Revisited*, 118–19 [112–28].
32. Stuart Tyson Smith, "People," in *The Oxford Encyclopedia of Ancient Egypt*, vol. 3, ed. Donald B. Redford (Oxford: Oxford University Press, 2001), 28 [27–32]. **For earlier elaboration of the view that the ancient Egyptians were black if Western criteria for "the black race" are applied consistently,** see Cheikh Anta Diop, *The African Origin of Civilization: Myth or Reality*, trans. Mercer Cook (Chicago: Lawrence Hill, 1974), 43–56.
33. **On Western "liberalism" that "almost always managed to sidestep the ultimate question of equality between white people and black people,"** see Chinua Achebe, "An Image of Africa," *Massachusetts Review* 18, no. 4 (1977): 787–88 [782–94], https://www.jstor.org/stable/25088813.

Section 8

1. Thornton Stringfellow, *A Brief Examination of Scripture Testimony on the Institution of Slavery* (Richmond: Religious Herald, 1841), 27.
2. See, e.g., W. Winwood Reade, "Efforts of Missionaries Among Savages," *Journal of the Anthropological Society of London* 3 (1865): clxiii–clxxxiii, https://doi.org/10.2307/3025323; and "Bishop Colenso on African Missions," *New York Times*, June 4, 1865, https://www.nytimes.com/1865/06/04/archives/bishop-colenso-on-african-missions.html.
3. Frederick Douglass, *Oration, Delivered in Corinthian Hall, Rochester, by Frederick Douglass, July 5th, 1852* (Rochester, NY: Lee, Mann, 1852), 29–30.
4. "African-American Catholics: Whites and Blacks in the Antebellum Catholic Church," American Catholic History Classroom (Catholic University of America), https://cuomeka.wrlc.org/exhibits/show/fcc/fcc-intro/background—whites-and-blacks-.
5. See, e.g., Monica Clark, "Black Catholics Ask Pope for Apology," *National Catholic Reporter*, September 22, 2015, https://www.ncronline.org/blogs/ncr-today/black-catholics-ask-pope-apology.
6. See, e.g., Rachel L. Swarns, "Georgetown University Plans Steps to Atone for Slave Past," *New York Times*, September 1, 2016, https://www.nytimes.com/2016/09/02/us/slaves-georgetown-university.html. Compare, e.g., Alvin Powell, "Dual Message of Slavery Probe: Harvard's Ties Inseparable from Rise, and Now University Must Act," *Harvard Gazette*, April 26, 2022, https://news.harvard.edu/gazette/story/2022/04/slavery-probe-harvards-ties-inseparable-from-rise/.

7. See Lester E. Bush Jr., "Mormonism's Negro Doctrine: An Historical Overview," *Dialogue* 8, no. 1 (1973): 11, 16 [11–68].

8. "Gospel Topics Essays: Race and the Priesthood," Church of Jesus Christ of Latter-day Saints, December 2013, https://www.lds.org/topics/race-and-the-priesthood?lang=eng.

9. "Church History Topics: Slavery and Abolition," Church of Jesus Christ of Latter-day Saints, https://www.lds.org/study/history/topics/slavery-and-abolition?lang=eng.

10. See James B. Christensen, "Negro Slavery in the Utah Territory," *Phylon Quarterly* 18, no. 3 (1957): 303–04 [298–305], https://doi.org/10.2307/272985. See also Max Perry Mueller, *Race and the Making of the Mormon People* (Chapel Hill: University of North Carolina Press, 2017).

11. See, e.g., Peggy Fletcher Stack, "39 Years Later, Priesthood Ban Is History, But Racism Within Mormon Ranks Isn't, Black Members Say," *Salt Lake Tribune*, June 9, 2017, https://archive.sltrib.com/article.php?id=5371962&itype=CMSID.

12. "SBC Resolutions: Resolution on Racial Reconciliation on the 150th Anniversary of the Southern Baptist Convention," Southern Baptist Convention, 1995, http://www.sbc.net/resolutions/899/resolution-on-racial-reconciliation-on-the-150th-anniversary-of-the-southern-baptist-convention.

13. R. Albert Mohler Jr., "A Letter from the President," in "Report on Slavery and Racism in the History of the Southern Baptist Theological Seminary," Southern Baptist Theological Seminary, December 12, 2018, 2 [1–4], http://www.sbts.edu/wp-content/uploads/2018/12/Racism-and-the-Legacy-of-Slavery-Report-v4.pdf.

14. "Report on Slavery and Racism," Southern Baptist Theological Seminary, 6 [5–71].

15. See, e.g., Emma Green, "Southern Baptists and the Sin of Racism," *Atlantic*, April 7, 2015, https://www.theatlantic.com/politics/archive/2015/04/southern-baptists-wrestle-with-the-sin-of-racism/389808/; and Randall Balmer, "The Real Origins of the Religious Right," Politico, May 27, 2014, https://www.politico.com/magazine/story/2014/05/religious-right-real-origins-107133.

16. Dwight McKissic, "Resolution for the 2017 SBC Annual Meeting–Condemning the Alt-Right and White Nationalism," *SBC Voices*, May 28, 2017, http://sbcvoices.com/resolution-for-the-2017-sbc-annual-meeting-condemning-the-alt-right-white-nationalism/. See also Emma Green, "A Resolution Condemning White Supremacy Causes Chaos at the Southern Baptist Convention," *Atlantic*, June 14, 2017, https://www.theatlantic.com/politics/archive/2017/06/the-southern-baptist-convention-alt-right-white-supremacy/530244/.

17. See, e.g., Jessica Martínez and Gregory A. Smith, "How the Faithful Voted: A Preliminary 2016 Analysis," Pew Research Center, November 9, 2016, https://www.pewresearch.org/fact-tank/2016/11/09/how-the-faithful-voted-a-preliminary-2016-analysis/; and Frank Newport, "Religious Group Voting and the 2020 Election," Gallup, November 13, 2020, https://news.gallup.com/opinion/polling-matters/324410/religious-group-voting-2020-election.aspx.

18. "The Southern Baptist Convention: A Closer Look," Southern Baptist Convention, 2020, http://www.sbc.net/aboutus/acloserlook.asp.

19. Charlotte Hunt-Grubbe, "The Elementary DNA of Dr Watson," *Sunday Times*, October 14, 2007, http://www.thesundaytimes.co.uk/sto/culture/books/article73186.ece.

20. James Watson, "To Question Genetic Intelligence Is Not Racism," *Independent*, October 19, 2007, http://www.independent.co.uk/voices/commentators/james-wat son-to-question-genetic-intelligence-is-not-racism-5328720.html.

21. See Amy Harmon, "James Watson Had a Chance to Salvage His Reputation on Race. He Made Things Worse.," *New York Times*, January 1, 2019, https://www.nytimes. com/2019/01/01/science/watson-dna-genetics-race.html.

22. See, e.g., Sally Haslanger, "What Good Are Our Intuitions?," *Proceedings of the Aristotelian Society, Supplementary Volumes* 80, no. 1 (2006): 95–97 [89–118], https:// doi.org/10.1111/j.1467-8349.2006.00139.x.

23. See, e.g., William Howard Taft, "The South and the National Government," intro. Andrew Carnegie (speech, North Carolina Society, New York, December 7, 1908), 11 [9–16].

24. *United States v. Thind*, 261 U.S. 204, 210 (1923).

25. Paul C. Taylor, *Race: A Philosophical Introduction*, 2nd ed. (Cambridge: Polity Press, 2013), 46.

26. *Thind*, 261 U.S. at 206.

27. *Thind*, 261 U.S. at 208.

28. *Thind, 261 U.S. at 211.* **Compare the "White" race category on the 2020 U.S. census form, which specifies "German, Irish, English, Italian, Lebanese, Egyptian, etc."**

29. See, e.g., Justin Driver, "Supremacies and the Southern Manifesto," *Texas Law Review* 92, no. 5 (2014): 1059–61 [1053–135]; and Ruth Bloch Rubin and Gregory Elinson, "Anatomy of Judicial Backlash: Southern Leaders, Massive Resistance, and the Supreme Court, 1954–1958," *Law and Social Inquiry* 43, no. 3 (2018): 944–80, https:// doi.org/10.1111/lsi.12316.

30. Compare David Scott FitzGerald and David Cook-Martín, *Culling the Masses: The Democratic Origins of Racist Immigration Policy in the Americas* (Cambridge, MA: Harvard University Press, 2014), 6. **The authors ask, "Why have the most egregious historical discriminators [in the Americas]—the United States and Canada—allowed the transformation of their populations by letting in large numbers of formerly excluded ['ethnic'] groups?" Liberal scholars have a frustrating habit of characterizing the United States of slavery and segregation as a flawed "liberal democracy"—recasting the American caste hierarchy in terms of hypocritical "racism" or "racist policies."** Compare, e.g., Roy L. Garis, "America's Immigration Policy," *North American Review* 220, no. 824 (1924): 63–64 [63–77], http://www.jstor. org/stable/25113348.

31. See, e.g., David Waldstreicher, *Slavery's Constitution: From Revolution to Ratification* (New York: Hill and Wang, 2009).

32. See US Census Bureau, "History: Through the Decades: Questionnaires," accessed November 14, 2021, https://www.census.gov/history/www/through_the_decades/ questionnaires/. See also D'vera Cohn, "Race and the Census: The 'Negro' Controversy," Pew Research Center, January 21, 2010, http://www.pewsocialtrends.org/ 2010/01/21/race-and-the-census-the-"negro"-controversy/.

33. **On "the construction of Latin American identity in the nineteenth century . . . based on the contrast between an idealized personality, which circulated among peoples of Latin origin, particularly Hispanics, and an imagined vision of the Anglo-Saxon**

character that was especially strong in the United States"—a contrast framed by culture and religion, not a continental "race" thing—compare Stella Maris Scatena Franco, "Latinos versus Anglo-Saxons: Identity Projections in the Accounts of Latin Americans Who Traveled to the United States in the Nineteenth Century," *Almanack*, no. 16 (2017): 80–81 [80–120], https://doi.org/10.1590/2236-463320171602.

34. United Nations Educational, Scientific and Cultural Organization (UNESCO), "Statement by Experts on Race Problems," *International Social Science Bulletin* 2, no. 3 (1950): 391–96. The UNESCO team of eight social and natural scientists was led by Ashley Montagu, a dedicated race skeptic.

35. Taylor, *Race*, 71–73. Compare, e.g., Sebastián Gil-Riaño, "Relocating Anti-Racist Science: The 1950 UNESCO Statement on Race and Economic Development in the Global South," *British Journal for the History of Science* 51, no. 2 (2018): 281–303, https://doi.org/10.1017/S0007087418000286.

36. Frantz Fanon, *Black Skin, White Masks*, trans. Charles Lam Markmann (New York: Grove, 1967), 112. See also Lewis R. Gordon, *What Fanon Said: A Philosophical Introduction to His Life and Thought* (New York: Fordham University Press, 2015), 85–86.

37. See Jilana Jaxon et al. [Ryan F. Lei, Reut Shachnai, Eleanor K. Chestnut, and Andrei Cimpian], "The Acquisition of Gender Stereotypes About Intellectual Ability: Intersections with Race," *Social Issues* 75, no. 4 (2019): 1192–1215, https://doi.org/10.1111/josi.12352; and Joshua Aronson, Carrie B. Fried, and Catherine Good, "Reducing the Effects of Stereotype Threat on African American College Students by Shaping Theories of Intelligence," *Journal of Experimental Social Psychology* 38, no. 2 (2002) 113–25, https://doi.org/10.1006/jesp.2001.1491.

38. Kelly M. Hoffman et al. [Sophie Trawalter, Jordan R. Axt, and M. Norman Oliver], "Racial Bias in Pain Assessment and Treatment Recommendations, and False Beliefs about Biological Differences between Blacks and Whites," *PNAS* 113, no. 16 (2016): 4296, 4298 [4296–4301], https://doi.org/10.1073/pnas.1516047113.

39. Adam Waytz, Kelly Marie Hoffman, and Sophie Trawalter, "A Superhumanization Bias in Whites' Perceptions of Blacks," *Social Psychological and Personality Science* 6, no. 3 (2015): 352 [352–59], https://doi.org/10.1177/1948550614553642. See also Tommy J. Curry, *The Man-Not: Race, Class, Genre, and the Dilemmas of Black Manhood* (Philadelphia: Temple University Press, 2017), 171–72.

40. See, e.g., David Brion Davis, *The Problem of Slavery in the Age of Emancipation* (New York: Knopf, 2014), 15–35; Ziad Obermeyer et al. [Brian Powers, Christine Vogeli, and Sendhil Mullainathan], "Dissecting Racial Bias in an Algorithm Used to Manage the Health of Populations," *Science* 366, no. 6464 (2019): 447–53, https://doi.org/10.1126/science.aax2342; and Jill Lepore, "The Invention of the Police," *New Yorker*, July 13, 2020, https://www.newyorker.com/magazine/2020/07/20/the-invention-of-the-police.

41. Terence Keel, *Divine Variations: How Christian Thought Became Racial Science* (Stanford: Stanford University Press, 2018), 13. Keel attributes a non-continental type of "racial consciousness to the early followers of Jesus," who "drew from ancient racial practices to imagine themselves as a unique ethnic group with a distinct origin." Terence D. Keel, "Response to My Critics: The Life of Christian Racial

Forms in Modern Science," *Zygon* 54, no. 1 (2019); 264–65 [261–79], https://doi.org/10.1111/zygo.12499. See also Dorothy Roberts, *Fatal Invention: How Science, Politics, and Big Business Re-Create Race in the Twenty-First Century* (New York: New Press, 2011), 79–80.

42. See, e.g., Gary Younge, "What Black America Means to Europe," *New York Review of Books*, June 6, 2020, https://www.nybooks.com/daily/2020/06/06/what-black-amer ica-means-to-europe/.

Section 9

1. Ernst Mayr, "The Biology of Race and the Concept of Equality," *Daedalus* 131, no. 1 (2002): 89–90 [89–94], https://www.jstor.org/stable/20027740. **For Mayr's favorable review of anthropologist Carleton S. Coon's openly racialist treatise** *The Origin of Races* (New York: Knopf, 1962), see Ernst Mayr, "Origin of the Human Races," *Science* 138, no. 3538 (1962): 420–22, https://doi.org/10.1126/scie nce.138.3538.420. Compare, e.g., Ladelle McWhorter, "Racism, Eugenics, and Ernst Mayr's Account of Species," *Philosophy Today* 54, suppl. (2010): 200–207, https://doi.org/10.5840/philtoday201054Supplement66; and Peter Sachs Collopy, "Race Relationships: Collegiality and Demarcation in Physical Anthropology," *Journal of the History of the Behavioral Sciences* 51, no. 3 (2015): 237–60, https://doi.org/10.1002/jhbs.21728.

2. E. O. Wilson and W. L. Brown Jr., "The Subspecies Concept and its Taxonomic Application," *Systematic Zoology* 2, no. 3 (1953): 109 [97–111], https://doi.org/10.2307/2411818.

3. L. Luca Cavalli-Sforza, Paolo Menozzi, and Alberto Piazza, *The History and Geography of Human Genes*, abridged paperback edition (Princeton: Princeton University Press, 1994), 19.

4. Marta Mirazón Lahr, *The Evolution of Modern Human Diversity: A Study of Cranial Variation* (Cambridge: Cambridge University Press, 1996), 339.

5. See Martin W. Lewis and Kären E. Wigen, *The Myth of Continents: A Critique of Metageography* (Berkeley: University of California Press, 1997), 22–23, 27–28. **For a nineteenth-century view that posits only two continents, "the Eastern and Western," with "continent" defined as "a great extent of land nowhere entirely separated by water," whereby Europe, Asia, and Africa are "grand divisions" of the Eastern continent,** see Sidney E. Morse, *A System of Geography, for the Use of Schools* (New York: Harper and Brothers, 1844), 6.

6. Asya Pereltsvaig and Martin W. Lewis, *The Indo-European Controversy: Facts and Fallacies in Historical Linguistics* (Cambridge: Cambridge University Press, 2015), 22; see also 23–24. **On how "Aryan" segments of the Indian subcontinent became "Caucasian" in relation to "the hardening of the Hindu caste system and the installation of more commercial ties with Europe,"** see Nina G. Jablonski, *Living Color: The Biological and Social Meaning of Skin Color* (Berkeley: University of California Press, 2012), 107–8. **On how "governing a colony involved familiarity with what had**

preceded the arrival of the colonial power on the Indian scene," see Romila Thapar, "The Theory of Aryan Race and India: History and Politics," *Social Scientist* 24, no. 1–3 (1996): 4 [3–29], https://doi.org/10.2307/3520116.

7. See, e.g., David Reich et al. [Kumarasamy Thangaraj, Nick Patterson, Alkes L. Price, and Lalji Singh], "Reconstructing Indian Population History," *Nature* 461 (2009): 489–94, https://doi.org/10.1038/nature08365; and Press Trust of India, "Indian Population Originated in 3 Migration Waves from Africa, Iran and Asia," *Business Standard*, May 11, 2017, https://www.business-standard.com/article/current-affairs/indian-populat ion-originated-in-3-migration-waves-from-africa-iran-asia-117051100378_1.html.

8. See, e.g., Neha Mishra, "India and Colorism: The Finer Nuances," *Washington University Global Studies Law Review* 14, no. 4 (2015): 735–37 [725–50], https:// openscholarship.wustl.edu/law_globalstudies/vol14/iss4/14; and Brishti Basu, "The People Fighting 'Light Skin' Bias," BBC.com, August 8, 2020, https://www.bbc.com/ future/article/20200818-colourism-in-india-the-people-fighting-light-skin-bias.

9. Mayr, "Biology of Race," 90.

10. See George B. J. Busby et al. [Gavin Band, Quang Si Le, Muminatou Jallow, Edith Bougama, Valentina D. Mangano, Lucas N. Amenga-Etego Anthony Enimil, et al.], "Admixture into and within Sub-Saharan Africa," *eLife* 5 (2016): e15266 (at 2 of 44), https://doi.org/10.7554/eLife.15266.

11. See, e.g., Edward W. Said, *Orientalism* (New York: Pantheon, 1978), 76–80; and Christopher Hutton, "Orientalism and Race: Aryans and Semites," in *Orientalism and Literature*, ed. Geoffrey P. Nash (Cambridge: Cambridge University Press, 2019), 117–32. For a highly detailed, color-conscious, and linguistic-oriented elaboration of "Aryan" racialism in its contemporaneous heyday, compare John Fiske, "Who Are the Aryans?," *Atlantic*, February 1881, https://www.theatlantic.com/magazine/ archive/1881/02/who-are-the-aryans/521367/.

12. Fernando A. Villanea and Joshua G. Schraiber, "Multiple Episodes of Interbreeding Between Neanderthals and Modern Humans," *Nature Ecology and Evolution* 3, no. 1 (2019): 39 [39–44], https://doi.org/10.1038/s41559-018-0735-8. See also Martin Petr et al. [Svante Pääbo, Janet Kelso, and Benjamin Vernot], "Limits of Long-Term Selection against Neanderthal Introgression," *PNAS* 116, no. 5 (2019): 1639–44, https://doi.org/10.1073/pnas.1814338116; Sriram Sankararaman et al. [Swapan Mallick, Michael Dannemann, Kay Prüfer, Janet Kelso, Svante Pääbo, Nick Patterson, and David Reich], "The Genomic Landscape of Neanderthal Ancestry in Present-Day Humans," *Nature* 507 (2014): 354–57, https://doi.org/10.1038/nature12961; and Benjamin Vernot and Joshua M. Akey, "Resurrecting Surviving Neandertal Lineages from Modern Human Genomes," *Science* 343, no. 6174 (2014): 1017–21, https://doi. org/10.1126/science.1245938.

13. See, e.g., Antonio Cascais, "Portugal Confronts Its Slave Trade Past," Deutsche Welle (DW), March 24, 2021, https://www.dw.com/en/portugal-commemoration-transa tlantic-slave-trade/a-56976093.

14. See, e.g., David N. Livingstone, "Cultural Politics and the Racial Cartographics of Human Origins," *Transactions of the Institute of British Geographers* 35, no. 2 (2010): 204–21, https://doi.org/10.1111/j.1475-5661.2009.00377.x. For evidence that anatomically modern humankind originated around 200,000 years ago in what is

today Botswana, see Eva K. F. Chan et al. [Axel Timmermann, Benedetta F. Baldi, Andy E. Moore, Ruth J. Lyons, Sun-Seon Lee, Anton M. F. Kalsbeek et al.], "Human Origins in a Southern African Palaeo-Wetland and First Migrations," *Nature* 575 (2019): 185–89, https://doi.org/10.1038/s41586-019-1714-1.

15. Francesc Relaño, *The Shaping of Africa: Cosmographic Discourse and Cartographic Science in Late Medieval and Early Modern Europe* (Aldershot: Ashgate, 2002), 8, 36.

16. **Compare, for contemporary quasi-polygenetic or "multiregional" accounts of major races through geographic isolation,** John Hawks and Milford H. Wolpoff, "Sixty Years of Modern Human Origins in the American Anthropological Association," *American Anthropologist* 105, no. 1 (2003): 94–95 [89–100], https://doi.org/10.1525/aa.2003.105.1.89.

17. See George B. J. Busby et al. [Gavin Band, Quang Si Le, Muminatou Jallow, Edith Bougama, Valentina D. Mangano, Lucas N. Amenga-Etego Anthony Enimil, et al.], "Admixture into and within Sub-Saharan Africa," *eLife* 5 (2016): e15266 (at 1 of 44), https://doi.org/10.7554/eLife.15266.

18. Mayr, "Biology of Race," 91. Compare, e.g., Giuseppe Lippi et al. [Giuseppe Banfi, Emmanuel J. Favaloro, Joern Rittweger, and Nicola Maffulli], "Updates on Improvement of Human Athletic Performance: Focus on World Records in Athletics," *British Medical Bulletin* 87, no. 1 (2008): 7–15, https://doi.org/10.1093/bmb/ldn029.

19. **On race, running performance, and genetics,** see Daniel Macarthur, "The Gene for Jamaican Sprinting Success? No, Not Really," *Wired*, October 4, 2008, https://www.wired.com/2008/10/the-gene-for-jamaican-sprinting-success-no-not-really/.

20. Mayr, "Biology of Race," 92. **Mayr overlooks inconvenient facts: Jews of Eastern European, not Iberian, descent are the ones particularly affected by Tay-Sachs; some non-Jewish subgroups of French Canadian descent are noticeably affected; and persons who count as Jewish often have substantial non-Jewish ancestry.** See R. Myerowitz and N. D. Hogikyan, "Different Mutations in Ashkenazi Jewish and non-Jewish French Canadians with Tay-Sachs Disease," *Science* 232, no. 4758 (1986): 1646–48, https://doi.org/10.1126/science.3754980.

21. Robin O. Andreasen, "Biological Conceptions of Race," in *Philosophy of Biology*, ed. Mohan Matthen and Christopher Stephens (Amsterdam: Elsevier, 2007), 468 [455–81]. Compare, e.g., Matthew Kopec, "Clines, Clusters, and Clades in the Race Debate," *Philosophy of Science* 81, no. 5 (2014): 1053–65, https://doi.org/10.1086/677695.

22. Philip Kitcher, "Race, Ethnicity, Biology, Culture," in *Racism*, ed. Leonard Harris (Amherst, NY: Humanity Books, 1999), 103–4 [87–117]. **On his conversion to skepticism about races as natural kinds,** see Philip Kitcher, "Does 'Race' Have a Future?," *Philosophy and Public Affairs* 35, no. 4 (2007): 317 [293–317], https://doi.org/10.1111/j.1088-4963.2007.00115.x.

23. Andreasen, "Biological Conceptions of Race," 471–73.

24. **On "ancestry informative markers (AIMs)" technology that is "designed to bring about a correspondence of familiar ideas of race and supposed socially neutral DNA,"** see Duana Fullwiley, "The Biologistical Construction of Race: 'Admixture' Technology and the New Genetic Medicine," *Social Studies of Science* 38, no. 5 (2008): 695 [695–735], https://doi.org/10.1177/0306312708090796.

25. Nicholas G. Crawford et al. [Derek E. Kelly, Matthew E. B. Hansen, Marcia H. Beltrame, Shaohua Fan, Shanna L. Bowman, Ethan Jewett et al.], "Loci Associated with Skin Pigmentation Identified in African Populations," *Science* 358, no. 6365 (2017): 867–68 [867–87], https://doi.org/10.1126/science.aan8433.

26. Andreasen, "Biological Conceptions of Race," 468.

27. Neil Risch et al. [Esteban Burchard, Elad Ziv, and Hua Tang], "Categorization of Humans in Biomedical Research: Genes, Race and Disease," *Genome Biology* 3, no. 7 (2002): comment2007, https://doi.org/10.1186/gb-2002-3-7-comment2007. Compare, e.g., Charmaine D. Royal et al. [John Novembre, Stephanie M. Fullerton, David B. Goldstein, Jeffrey C. Long, Michael J. Bamshad, and Andrew G. Clark], "Inferring Genetic Ancestry: Opportunities, Challenges, and Implications," *American Journal of Human Genetics* 86, no. 5 (2010): 670 [661–73], https://doi.org/10.1016/j.ajhg.2010.03.011.

28. Risch et al., "Categorization of Humans." **For results showing that "differences among human groups, even very distant ones and no matter whether the groups are defined on a racial or on a geographical basis, represent only a small fraction of the global genetic diversity of our species,"** compare Guido Barbujani et al. [Arianna Magagni, Eric Minch, and L. Luca Cavalli-Sforza], "An Apportionment of Human DNA Diversity," *PNAS* 94, no. 9 (1997): 4518 [4516–19], https://doi.org/10.1073/pnas.94.9.4516. Compare also R. C. Lewontin, "The Apportionment of Human Diversity," *Evolutionary Biology* 6 (1972): 396 [381–98], https://doi.org/10.1007/978-1-4684-9063-3_14.

29. **On why "from a biological systematic and evolutionary taxonomical perspective, human races/continental groups or clusters have no natural meaning or objective biological reality,"** see Koffi N. Maglo, Tesfaye B. Mersha, and Lisa J. Martin, "Population Genomics and the Statistical Values of Race: An Interdisciplinary Perspective on the Biological Classification of Human Populations and Implications for Clinical Genetic Epidemiological Research," *Frontiers in Genetics* 7, no. 22 (2016), https://doi.org/10.3389/fgene.2016.00022.

30. Michael Yudell et al. [Dorothy Roberts, Rob DeSalle, and Sarah Tishkoff], "Taking Race Out of Human Genetics," *Science* 351, no. 6273 (2016): 564–65 [564–65], https://doi.org/10.1126/science.aac4951.

31. Compare, for a metaphysical distinction between "genuine biological entity" instead of natural "race" kind, Quayshawn Spencer, "How to Be a Biological Racial Realist," in Joshua Glasgow, Sally Haslanger, Chike Jeffers, and Quayshawn Spencer, *What Is Race?: Four Philosophical Views* (New York: Oxford University Press, 2019), 95 [73–110]. On Spencer's metaphysics of race, see Adam Hochman, "Against the New Racial Naturalism," *Journal of Philosophy* 110, no. 6 (2013): 331–51, https://doi.org/10.5840/jphil2013110625.

32. See Steven J. Micheletti et al. [Kasia Bryc, Samantha G. Ancona Esselmann, William A. Freyman, Meghan E. Moreno, G. David Poznik, and Anjali J. Shastri], "Genetic Consequences of the Transatlantic Slave Trade in the Americas," *American Journal of Human Genetics* 107 (2020): 270 [265–77], https://doi.org/10.1016/j.ajhg.2020.06.012; and Katarzyna Bryc et al. [Eric Y. Durand, J. Michael Macpherson, David Reich, and Joanna L. Mountain], "The Genetic Ancestry of African Americans, Latinos, and European Americans across the United States," *American Journal of Human Genetics* 96, no. 1 (2015): 42 [37–53], https://doi.org/10.1016/j.ajhg.2014.11.010.

33. See, e.g., Lundy Braun et al. [Anne Fausto-Sterling, Duana Fullwiley, Evelynn M. Hammonds, Alondra Nelson, William Quivers, Susan M. Reverby, and Alexandra E. Shields], "Racial Categories in Medical Practice: How Useful Are They?," *PLOS Medicine* 4, no. 9 (2007): 1423–28, https://doi.org/10.1371/journal.pmed.0040271.

Section 10

1. See, e.g., Charles W. Mills, "'But What Are You *Really*?' The Metaphysics of Race," in *Blackness Visible: Essays on Philosophy and Race* (Ithaca: Cornell University Press, 1998), 41–66; and Lucius Outlaw, "Against the Grain of Modernity: The Politics of Difference and the Conservation of 'Race,'" in *On Race and Philosophy* (New York: Routledge, 1996), 135–57.

2. W. E. B. Du Bois, "The Conservation of Races" (1897), in *W. E. B. Du Bois: Writings*, ed. Nathan Huggins (New York: Library of America, 1986), 816–17 [815–26].

3. Sally Haslanger, "Language, Politics and 'The Folk': Looking for 'The Meaning' of 'Race,'" *Monist* 93, no. 2 (2010): 181 [169–87], https://doi.org/10.5840/monist201093 211. See also Sally Haslanger, "Tracing the Sociopolitical Reality of Race," in Joshua Glasgow, Sally Haslanger, Chike Jeffers, and Quayshawn Spencer, *What Is Race?: Four Philosophical Views* (New York: Oxford University Press, 2019), 4–37; Ron Mallon, "Passing, Traveling and Reality: Social Constructionism and the Metaphysics of Race," *Noûs* 38, no. 4 (2004): 644–73, https://doi.org/10.1111/j.0029-4624.2004.00487.x; and Albert Atkin, *The Philosophy of Race* (Durham, NC: Acumen, 2012), 1–5.

4. Compare, e.g., Daniel Cressey, "Cryptozoology: Beastly Fakes," *Nature* 499 (2013): 406, https://doi.org/10.1038/499406a; and George M. Fredrickson, *The Black Image in the White Mind: The Debate on Afro-American Character and Destiny, 1817–1914* (New York: Harper and Row, 1971) 53–56.

5. Haslanger, "'The Meaning' of 'Race,'" 183.

6. Sally Haslanger, "Exploring *Race* in Life, in Speech, and in Philosophy: Comments on Josh Glasgow's *A Theory of Race*," *Symposia on Gender, Race and Philosophy* 5 (2009): 5 [1–9], http://web.mit.edu/sgrp/2009/no2/Haslanger1009.pdf. Compare, e.g., Joshua Glasgow, "Is Race an Illusion of a (Very) Basic Reality?," in *What Is Race?*, 117, 133 [111–49].

7. See, e.g., J. Philippe Rushton, "Race, Brain Size, and Intelligence: A Reply to Cernovsky," *Psychological Reports* 66, no. 2 (1990): 659–66, https://doi.org/10.2466/pr0.1990.66.2.659; and "Retraction Notice," *Psychological Reports*, December 24, 2020, https://doi.org/10.1177/0033294120982774.

8. Haslanger, "'The Meaning' of 'Race,'" 181.

9. See Glasgow, Haslanger, Jeffers, and Spencer, "Introduction," in *What Is Race?*, 2 [1–3].

10. See, e.g., Haslanger, "Sociopolitical Reality of Race," 24–26. Compare, e.g., Robert Carson and Hollis Robbins, "Susan Sontag: Race, Class, and the Limits of Style," *American Interest* 15, no. 4 (2020), https://www.the-american-interest.com/2019/11/29/susan-sontag-race-class-and-the-limits-of-style/.

11. Anthony Appiah, "The Uncompleted Argument: Du Bois and the Illusion of Race," in *"Race," Writing, and Difference*, ed. Henry Louis Gates Jr. (Chicago: University of Chicago Press, 1986), 35 [21–37].

12. See, e.g., K. Anthony Appiah, "Race, Culture, Identity: Misunderstood Connections," in K. Anthony Appiah and Amy Gutmann, *Color Conscious: The Political Morality of Race* (Princeton: Princeton University Press, 1996), 73n [30–105].

13. Du Bois, "Conservation of Races," 817–18. **Scientistic consolidation of European peoples into a single white/European race was not settled until the twentieth century. For a pre-consolidation hierarchical account of "the three primary races" of Europe—"*Homo Europaeus* (. . . the Aryan race); *Homo Alpinus* (. . . the "Celtic" or "Celta-Slav" race); and the so-called Mediterranean race"—see** Carlos C. Closson, "The Hierarchy of European Races," *American Journal of Sociology* 3, no. 3 (1897): 316 [314–27], https://doi.org/10.1086/210710. **(Du Bois's "Conservation" was also published in 1897.) On the "Euro-supremacist and Euro-universalist idea of Europe . . . in which, in the later nineteenth century, the idea of *Homo Europaeus* was set against the so-called 'inferior races' of the European colonies, above all in Africa,"** compare Shane Weller, "*Homo Europaeus*: 1848–1918," in *The Idea of Europe: A Critical History* (Cambridge: Cambridge University Press, 2021), 134 [114–40].

14. Du Bois, "Conservation of Races," 818.

15. **Most persons assigned to the same race do not, from their (raced) national group perspectives, nearly look alike.** See, e.g., Kurt Hugenberg et al. [Steven G. Young, Michael J. Bernstein, and Donald F. Sacco], "The Categorization-Individuation Model: An Integrative Account of the Other-Race Recognition Deficit," *Psychological Review* 117, no. 4 (2010): 1168–87, https://doi.org/10.1037/a0020463. Compare, e.g., Ryuta Itagaki, "The Anatomy of Korea-Phobia in Japan," *Japanese Studies* 35, no. 1 (2015): 49–66, https://doi.org/10.1080/10371397.2015.1007496.

16. See, e.g., Chike Jeffers, "The Cultural Theory of Race: Yet Another Look at Du Bois's 'The Conservation of Races,'" *Ethics* 123, no. 3 (2013): 421–22 [403–26], https://doi.org/10.1086/669566; and "Cultural Constructionism," in *What Is Race?*, 38–72. Compare, e.g., Tunde Adeleke, "Black Americans and Africa: A Critique of the Pan-African and Identity Paradigms," *International Journal of African Historical Studies* 31, no. 3 (1998): 506–07 [505–36], https://doi.org/10.2307/221474.

17. Compare, e.g., Jeffers, "Cultural Theory of Race," 425–26.

18. W. E. B. Du Bois, *Dusk of Dawn: An Essay Toward an Autobiography of a Race Concept* (New York: Harcourt, Brace, 1940), 133. **On Frederick Douglass rejecting, a hundred years prior, "all arguments that African-Americans had any racial, national or spiritual connection with African peoples,"** compare Daniel Kilbride, "What Did Africa Mean to Frederick Douglass?," *Slavery and Abolition* 26, no. 1 (2015): 40 [40–62], https://doi.org/10.1080/0144039X.2014.916516.

19. See, e.g., Saidiya Hartman, *Lose Your Mother: A Journey Along the Atlantic Slave Route* (New York: Farrar, Straus and Giroux, 2007).

20. Appiah, "Race, Culture, Identity," 74. See also Naomi Zack, *Philosophy of Science and Race* (New York: Routledge, 2002), 87–88.

21. See, e.g., DeNeen L. Brown, "Slavery's Bitter Roots: In 1619, '20 and odd Negroes' Arrived in Virginia," *Washington Post*, August 24, 2018, https://www.washingtonp ost.com/news/retropolis/wp/2018/08/24/slaverys-bitter-roots-in-1619-20-and-odd-negroes-arrived-in-virginia/. **For description of the later practice of "seasoning" (adjustment to the disease environment), which lasted roughly a year and was thought to reduce the high mortality rate of enslaved Africans who did survive the Middle Passage,** see Michael Mullin, *Africa in America: Slave Acculturation and Resistance in the American South and the British Caribbean, 1736–1831* (Urbana: University of Illinois Press, 1992), 86–88.

22. **Compare—on "recent scholarship [that] has documented the extent to which local populations and institutions were actively complicit in Nazi crimes, participating in and benefiting from the persecution of Jewish citizens, not only in Germany but across Europe"**—Victoria J. Barnett, "The Changing View of the 'Bystander' in Holocaust Scholarship: Historical, Ethical, and Political Implications," *Utah Law Review* 2017, no. 4 (2017): 633 [633–47], http://dc.law.utah.edu/ulr/vol2017/iss4/1.

23. Warren M. Billings, "The Law of Servants and Slaves in Seventeenth-Century Virginia," *Virginia Magazine of History and Biography* 99, no. 1 (1991): 45–46 [45–62], https://www.jstor.org/stable/4249198.

24. See, e.g., Andy Kiersz, "The Economic and Racial Inequality Problem in Baltimore Exists in Many US Cities," Business Insider, May 3, 2015, https://www.businessinsider. com/income-and-racial-inequality-maps-2015-5.

25. See, e.g., Winthrop D. Jordan, "Historical Origins of the One-Drop Racial Rule in the United States," ed. Paul Spickard, *Journal of Critical Mixed Race Studies* 1, no. 1 (2014): 98–132, https://escholarship.org/uc/item/91g761b3.

26. See, e.g., Arnold K. Ho et al. [Jim Sidanius, Daniel T. Levin, and Mahzarin R. Banaji], "Evidence for Hypodescent and Racial Hierarchy in the Categorization and Perception of Biracial Individuals," *Journal of Personality and Social Psychology* 100, no. 3 (2011): 492–506. https://doi.org/10.1037/a0021562.

27. See, e.g., Linda Alcoff, "Mestizo Identity," in *American Mixed Race: The Culture of Microdiversity*, ed. Naomi Zack (Lanham, MD: Rowman and Littlefield, 1995), 258–59 [257–78]; Cleuci de Oliveira, "Brazil's New Problem with Blackness," *Foreign Policy*, April 5, 2017, https://foreignpolicy.com/2017/04/05/brazils-new-problem-with-blackness-affirmative-action/; and Norimitsu Onishi, "A Racial Awakening in France, Where Race Is a Taboo Topic," *New York Times*, July 14, 2020, https://www. nytimes.com/2020/07/14/world/europe/france-racism-universalism.html.

28. See, e.g., Ritu Prasad, "The Awkward Questions about Slavery from Tourists in US South," BBC.com, October 2, 2019, https://www.bbc.com/news/world-us-canada-49842601; Michael T. Luongo, "Despite Everything, People Still Have Weddings at 'Plantation' Sites," *New York Times*, updated October 20, 2020, https://www.nytimes. com/2020/10/17/style/despite-everything-people-still-have-weddings-at-plantation-sites.html; and Jennifer Schuessler, "The Long Battle Over 'Gone With the Wind,'" *New York Times*, June 14, 2020, https://www.nytimes.com/2020/06/14/movies/gone-with-the-wind-battle.html. **On unfreedom for the American Negro as mostly a state of mind—"The material (not psychological) conditions of the lives of slaves**

compared favorably with those of free industrial workers"—a point that "merely emphasizes the hard lot of all workers, free or slave, during the first half of the nineteenth century"—see Robert William Fogel and Stanley L. Engerman, *Time on the Cross: The Economics of American Negro Slavery* (Boston: Little, Brown, 1974), 5.

29. On debate over whether the race concept depends on belief in natural races, see Robert Gooding-Williams, *In the Shadow of Du Bois: Afro-Modern Political Thought in America* (Cambridge, MA: Harvard University Press, 2011), 37–53.

30. See, e.g., Lucius Outlaw, "'Conserve' Races? In Defense of W. E. B. Du Bois," in *W. E. B. Du Bois on Race and Culture*, ed. Bernard W. Bell, Emily R. Grosholz, and James B. Stewart (New York: Routledge, 1996), 15–38; and Paul C. Taylor, "Appiah's Uncompleted Argument: W. E. B. Du Bois and the Reality of Race," *Social Theory and Practice* 26, no. 1 (2000): 103–28, https://doi.org/10.5840/soctheorpract20002 616. See also Cornel West, *The American Evasion of Philosophy: A Genealogy of Pragmatism* (Madison: University of Wisconsin Press, 1989), 138–50.

31. Lawrence D. Bobo, "Reclaiming a Du Boisian Perspective on Racial Attitudes," *Annals of the American Academy of Political and Social Science* 568, no. 1 (2000): 187 [186–202], https://doi.org/10.1177/000271620056800114. See also W. E. B. Du Bois, *The Philadelphia Negro: A Social Study* (Philadelphia: University of Pennsylvania Press, 1899).

32. Adam Hochman, "Replacing Race: Interactive Constructionism about Racialized Groups," *Ergo* 4, no. 3 (2017): 85, 64 [61–92], https://doi.org/10.3998/ergo.12405 314.0004.003. See also Adam Hochman, "In Defense of the Metaphysics of Race," *Philosophical Studies* 174 (2017): 2709–29, https://doi.org/10.1007/s11 098-016-0806-0.

33. Hochman, "Replacing Race," 79–80.

34. Paul C. Taylor, "What Races Are: The Metaphysics of Critical Race Theory," in *Race: A Philosophical Introduction*, 2nd ed. (Cambridge: Polity Press, 2013), 89–90 [68–119]. See also Robert Gooding-Williams, "Race, Multiculturalism and Democracy," *Constellations* 5, no. 1 (1998): 18–41, https://doi.org/10.1111/1467-8675.00071.

35. Taylor, "Appiah's Uncompleted Argument," 104. See also Michael Omi and Howard Winant, *Racial Formation in the United States*, 3rd ed. (New York: Routledge, 2005), 109.

36. See Lawrence Blum, "Racialized Groups: The Sociohistorical Consensus," *Monist* 93, no. 2 (2010): 300, 304 [298–320], https://doi.org/10.5840/monist201093 217. Compare, e.g., Adam Hochman, "Racialization: A Defense of the Concept," *Ethnic and Racial Studies* 42, no. 8 (2019): 1245–62, https://doi.org/10.1080/01419 870.2018.1527937.

37. Lawrence Blum, *"I'm Not a Racist, But . . .": The Moral Quandary of Race* (Ithaca: Cornell University Press, 2002), 147.

38. See Plato, *The Republic*, trans. G. M. A. Grube, 2nd ed. (Indianapolis: Hackett, 1992), Bk. 3, 414c–415d (91–92).

39. For default American superimposition of a "race" thing on visible African ancestry (e.g., "subjective social status shapes the way the brain responds to race, which may have implications for psychopathology") when roughly alluding

to American slave lineage, compare Keely A. Muscatell, Ethan McCormick, and Eva H. Telzer, "Subjective Social Status and Neural Processing of Race in Mexican American Adolescents," *Development and Psychopathology* 30, no. 5 (2018): 1837–48, https://doi.org/10.1017/S0954579418000949. Compare also Matthew O. Hunt, "Race, Ethnicity, and Lay Explanations of Poverty in the United States: Review and Recommendations for Stratification Beliefs Research," *Sociology of Race and Ethnicity* 2, no. 4 (2016): 393–401, https://doi.org/10.1177/2332649216666544; Jacqueline M. Chen and David L. Hamilton, "Natural Ambiguities: Racial Categorization of Multiracial Individuals," *Journal of Experimental Social Psychology* 48, no. 1 (2012): 152–64, https://doi.org/10.1016/j.jesp.2011.10.005; and Kenneth J. Arrow, "What Has Economics to Say About Racial Discrimination?," *Journal of Economic Perspectives* 12, no. 2 (1998): 91–100, https://doi.org/10.1257/jep.12.2.91.

40. Blum, *Quandary of Race*, 147.

41. See, e.g., Tommie Shelby, *Dark Ghettos: Injustice, Dissent, and Reform* (Cambridge, MA: Harvard University Press, 2016), 39–42; Ben L. Martin, "From Negro to Black to African American: The Power of Names and Naming," *Political Science Quarterly* 106, no. 1 (1991): 88–90 [83–107], https://doi.org/10.2307/2152175; and Courtney L. McCluney et al. [Kathrina Robotham, Serenity Lee, Richard Smith, and Myles Durkee], "The Costs of Code-Switching," HBR.org (Harvard Business Review), November 15, 2019, https://hbr.org/2019/11/the-costs-of-codeswitching.

42. Compare, e.g., B. Keith Payne, Heidi A. Vuletich, and Jazmin L. Brown-Iannuzzi, "Historical Roots of Implicit Bias in Slavery," *PNAS* 116, no. 24 (2019): 11693–98, https://doi.org/10.1073/pnas.1818816116.

43. See, e.g., Taylor, "What Races Are," 116–18.

44. Blum, *Quandary of Race*, 148. See also Blum, "Racialized Groups," 299–300.

45. Compare, e.g., Barbara J. Fields, "Whiteness, Racism, and Identity," *International Labor and Working-Class History*, no. 60 (2001): 48 [48–56], https://www.jstor.org/stable/27672735.

46. See, e.g., Nicholas D. Smith, "Aristotle's Theory of Natural Slavery," *Phoenix* 37, no. 2 (1983): 121–22 [109–22], https://doi.org/10.2307/1087451.

47. See, e.g., William Goodell, *The American Slave Code in Theory and Practice: Its Distinctive Features Shown by Its Statutes, Judicial Decisions, and Illustrative Facts* (New York: American and Foreign Anti-Slavery Society, 1853); and Theodore Dwight Weld, *American Slavery as It Is: Testimony of A Thousand Witnesses* (New York: American Anti-Slavery Society, 1839). See also, e.g., Stephanie E. Jones-Rogers, *They Were Her Property: White Women as Slave Owners in the American South* (New Haven: Yale University Press, 2019), 142–43; and Kia Shant'e Breaux, "Sally Hemings Family Rights Unclear," Associated Press, February 12, 2000, https://apnews.com/article/433d8378c6d48403169296dc1d17c29d.

48. Max Weber, *The Protestant Ethic and the Spirit of Capitalism*, trans. Talcott Parsons, foreword R. H. Tawney (New York: Charles Scribner's Sons, 1930), 162. See also Caitlin Rosenthal, *Accounting for Slavery: Masters and Management* (Cambridge, MA: Harvard University Press, 2018), 179–80.

49. See, e.g., Blum, *Quandary of Race*, 1–3; and Sophia Moreau, *Faces of Inequality: A Theory of Wrongful Discrimination* (New York: Oxford University Press, 2020), 14–18.

My emphasis on race ideology would imply a significant distinction between instrumental racism and emotional responses to perceptions of race (e.g., hatred). Compare J. L. A. Garcia, "The Heart of Racism," *Journal of Social Philosophy* 27, no. 1 (1996): 5–46, https://doi.org/10.1111/j.1467-9833.1996.tb00225.x. For criticism that "the 'heart' does not have to be involved in order for an action or institution to be racist, and unjust because racist," see Tommie Shelby, "Is Racism in the 'Heart'?," *Journal of Social Philosophy* 33, no. 3 (2002): 418–19 [411–20], https://doi.org/10.1111/0047-2786.00150.

50. See, e.g., Manning Marable, *How Capitalism Underdeveloped Black America: Problems in Race, Political Economy and Society* (Boston: South End Press, 1983); and Cornel West, "Book Review: *How Capitalism Underdeveloped Black America: Problems in Race, Political Economy and Society,* by Manning Marable, Boston: South End Press, 1983," *Insurgent Sociologist* 13, no. 3 (1986): 114–15, https://doi.org/10.1177/089692058601300315. My alternative title for Marable's book would be: *How Capitalism through Slavery Underdeveloped Black America.*

51. An example of studious naivete: "The state may be 88 percent white, but it cannot be easily dismissed as a nest of bigots.... It is not Alabama in the sixties.... Kansas may burn to restore the gold standard . . . but one thing it doesn't do is racism." Thomas Frank, *What's the Matter with Kansas? How Conservatives Won the Heart of America* (New York: Metropolitan Books, 2004): 179–80. For reports on color-conscious reality in Kansas, see David Condos, "What the History of 'Noose Road' Tells Us about Kansas, Race and the Lynchings of Black Men," High Plains Public Radio, March 2, 2021, https://www.hppr.org/hppr-news/2021-03-02/what-the-history-of-noose-road-tells-us-about-kansas-race-and-the-lynchings-of-black-men; and Mary L. Dudziak, "The Limits of Good Faith: Desegregation in Topeka, Kansas, 1950–1956," *Law and History Review* 5, no. 2 (1987): 390–91 [351–91], https://doi-org.ezproxy.library.tufts.edu/743891.

52. See W. E. B. Du Bois, *Black Reconstruction in America, 1860–1880* (New York: Russell and Russell, 1935), 700–701.

53. David R. Roediger, *The Wages of Whiteness: Race and the Making of the American Working Class,* rev. ed. (London: Verso, 1999), 13. Compare, e.g., Noel Ignatiev, "The Paradox of the White Worker: Studies in Race Formation," *Labour/Le Travail* 30 (1992): 233–34 [233–40], https://doi.org/10.2307/25143630; and Joe William Trotter Jr., *Workers on Arrival: Black Labor in the Making of America* (Oakland: University of California Press, 2019), 108–9.

54. See, e.g., *United States v. Thind,* 261 U.S. 204, 207 (1923).

55. See US Census Bureau, "History: Through the Decades: Questionnaires," 1790 to 1840 forms, accessed November 14, 2021, https://www.census.gov/history/www/through_the_decades/questionnaires/. See also Janell Ross, "Joe Biden Didn't Just Compromise with Segregationists. He Fought for Their Cause in Schools, Experts Say," NBCNews.com, June 25, 2019, https://www.nbcnews.com/news/nbcblk/joe-biden-didn-t-just-compromise-segregationists-he-fought-their-n1021626; and Richard Rothstein, "Private Agreements, Government Enforcement," in *Color of Law: A Forgotten History of How Our Government Segregated America* (New York: Liveright, 2017), 77–92.

56. See, e.g., Anne Case and Angus Deaton, "Rising Morbidity and Mortality in Midlife among White Non-Hispanic Americans in the 21st Century," *PNAS* 112, no. 49 (2015): 15078–83, https://doi.org/10.1073/pnas.1518393112; Leonardo Baccini and Stephen Weymouth, "Gone for Good: Deindustrialization, White Voter Backlash, and US Presidential Voting," *American Political Science Review* 115, no. 2 (2021): 550-567, https://doi.org/10.1017/S0003055421000022; and Joseph Darda, *How White Men Won the Culture Wars: A History of Veteran America* (Oakland: University of California Press, 2021).

57. See, e.g., Eric Alterman, "Class, Not Race," *Guardian*, July 3, 2007, https://www.theguardian.com/commentisfree/2007/jul/03/classnotrace. On "the fundamental issue [of] the historical development of working-class identity as racial identity," compare Herbert Hill, "The Problem of Race in American Labor History," *Reviews in American History* 24, no. 2 (1996): 189 [189–208], https://doi.org/10.1353/rah.1996.0037. Compare also Juan F. Perea, "The Echoes of Slavery: Recognizing the Racist Origins of the Agricultural and Domestic Worker Exclusion from the National Labor Relations Act," *Ohio State Law Journal* 72, no. 1 (2011): 98–99 [95–138]; Pete Daniel, "African American Farmers and Civil Rights," *Journal of Southern History* 73, no. 1 (2007): 8–12 [3–38], http://www.jstor.org/stable/27649315; Sydney H. Schanberg, "Plumbers Agree to a Wagner Plan to End Walkout," *New York Times*, May 16, 1964, https://www.nytimes.com/1964/05/16/archives/plumbers-agree-to-a-wagner-plan-to-end-walkout-but-puerto-rican.html; and Fenaba R. Addo and William A. Darity Jr., "Disparate Recoveries: Wealth, Race, and the Working Class after the Great Recession," *Annals of the American Academy of Political and Social Science* 695, no. 1 (2021): 189 [173–92], https://doi.org/10.1177/00027162211028822.

58. Fredrickson, *Black Image*, 63.

59. Stephen A. Douglas, "First Joint Debate, Ottawa, IL, August 21, 1858," in Abraham Lincoln and Stephen A. Douglas, *Political Debates between Hon. Abraham Lincoln and Hon. Stephen A. Douglas, in the Celebrated Campaign of 1858 in Illinois* (Columbus: Follett, Foster, 1860), 71 [66–73].

60. Taylor, "What Races Are," 73.

61. See, e.g., Dane Kennedy, *Decolonization: A Very Short Introduction* (New York: Oxford University Press, 2016), 2–3; and Isabel Wilkerson, *The Warmth of Other Suns: The Epic Story of America's Great Migration* (New York: Random House, 2010), 273–75.

62. Mukoma Wa Ngugi, "African in America or African American?," *Guardian*, January 14, 2011, https://www.theguardian.com/commentisfree/cifamerica/2011/jan/13/race-kenya. Compare, e.g., Mary C. Waters, Philip Kasinitz, and Asad L. Asad, "Immigrants and African Americans," *Annual Review of Sociology* 40 (2014): 369–90, https://doi.org/10.1146/annurev-soc-071811-145449.

63. See, e.g., Charles V. Hamilton, "Pan-Africanism and the Black Struggle in the U.S.," *Black Scholar* 2, no. 7 (1971): 10 [10–15], https://doi.org/10.1080/00064246.1971.11760845. Compare, e.g., Valerie Russ, "Who Is Black in America? Ethnic Tensions Flare between Black Americans and Black Immigrants," *Philadelphia Inquirer*, October 19, 2018, https://www.inquirer.com/philly/news/cynthia-erivo-harriet-tubman-movie-luvvie-ajayi-american-descendants-of-slaves-20181018.html.

64. American artist Winfred Rembert recounts his "black" experience with a "young African guy" in the 1970s: "He sat the table, just looking at the food—collard greens, pork chops, stuff like that. . . . Then when we went back to work, he said to me, 'You eat garbage. Food no good. And you messed up the bloodline.' What he meant was that we came here through slavery, and the White people messed with our women and we got children that were not pure African. He was going on and on about that." Winfred Rembert, *Chasing Me to My Grave: An Artist's Memoir of the Jim Crow South*, as told to Erin I. Kelly, foreword Bryan Stevenson (New York: Bloomsbury Press, 2021), 181.

65. See, e.g., American Anti-Slavery Society, *Caste* (New York: R. G. Williams, ~1839], 1–2, 7; W. E. B. Du Bois, "Caste in America: That Is the Root of the Trouble," *Des Moines Register and Leader*, October 19, 1904, 5 (W. E. B. Du Bois Papers, MS 312, Special Collections and University Archives, University of Massachusetts Amherst Libraries); and Gunnar Myrdal, *An American Dilemma: The Negro Problem and Modern Democracy* (New York: Harper and Brothers, 1944), 54.

66. See, e.g., David Jenkins, *Black Zion: Africa, Imagined and Real, as Seen by Today's Blacks* (New York: Harcourt Brace Jovanovich, 1975); and Beverly Lindsay and Playthell Benjamin, "Book Review: *Black Zion: The Return of Afro-Americans and West Indians to Africa*, by David Jenkins (London: Wildwood House, 1975)," *Journal of Black Studies* 8, no. 2 (1977): 251–55, https://doi.org/10.1177/002193477700800 208. Note the national specificity of the UK edition's subtitle.

67. See, e.g., Nico Slate, *Colored Cosmopolitanism: The Shared Struggle for Freedom in the United States and India* (Cambridge, MA: Harvard University Press, 2012); Ruth Ben-Ghiat, "When Fascist Aggression in Ethiopia Sparked a Movement of Black Solidarity," *Washington Post*, August 3, 2020, https://www.washingtonpost.com/outl ook/2020/08/03/when-fascist-aggression-ethiopia-sparked-movement-black-sol idarity/; Daniel S. Lucks, *Selma to Saigon: The Civil Rights Movement and the Vietnam War* (Lexington: University Press of Kentucky, 2014); and Yvonne D. Newsome, "International Issues and Domestic Ethnic Relations: African Americans, American Jews, and the Israel-South Africa Debate," *International Journal of Politics, Culture, and Society* 5, no. 1 (1991): 19–48, https://www.jstor.org/stable/20007027.

68. See, e.g., Kwame Anthony Appiah, "2017 'Great Immigrant': In His Own Words," Carnegie Corporation of New York, accessed November 24, 2021, https://www.carne gie.org/multimedia/great-immigrants-kwame-anthony-appiah/.

69. Appiah, "Uncompleted Argument," 34–35. Du Bois, after expressing solidarity with "yellow Asia" (117) in the *Dusk of Dawn* passage that Appiah targets, demystifies his own concern with race. Again: "Perhaps it is wrong to speak of it at all as 'a concept'" (133).

70. W. E. B. Du Bois, *The Negro*, afterword Robert Gregg (1915; Philadelphia: University of Pennsylvania Press, 2001), 242. See also Anindya Sekhar Purakayastha, "W. E. B. Du Bois, B. R. Ambedkar and the History of Afro-Dalit Solidarity," *Sanglap* 6, no. 1 (2019): 20–36, http://sanglap-journal.in/index.php/sanglap/article/view/116.

71. See Toni Morrison, "A Humanist View" (speech, Portland State University, May 30, 1975), Portland State Library Special Collections, 34:43–35:34 [06:50–43:01],

accessed February 14, 2021, https://soundcloud.com/portland-state-library/portl and-state-black-studies-1. See also Gillian Brockell, "Tulsa Isn't the Only Race Massacre You Were Never Taught in School. Here Are Others," *Washington Post*, June 1, 2021, https://www.washingtonpost.com/history/2021/06/01/tulsa-race-massacres-silence-schools/; and Brentin Mock, "White Americans' Hold on Wealth Is Old, Deep, and Nearly Unshakeable," Bloomberg.com, September 3, 2019, https://www.bloomberg.com/news/articles/2019-09-03/the-amazing-resiliency-of-white-wealth.

72. Jack Ewing, "United States Is the Richest Country in the World, and It Has the Biggest Wealth Gap," *New York Times*, September 23, 2020, https://www.nytimes.com/2020/09/23/business/united-states-is-the-richest-country-in-the-world-and-it-has-the-biggest-wealth-gap.html. See also Philip Alston, "Report of the Special Rapporteur on Extreme Poverty and Human Rights on His Mission to the United States of America," UN Human Rights Council, May 4, 2018, 3–5 [3–20], https://digitallibrary.un.org/record/1629536/files/A_HRC_38_33_Add-1-EN.pdf.

73. See Neil Bhutta et al. [Andrew C. Chang, Lisa J. Dettling, and Joanne W. Hsu], "Disparities in Wealth by Race and Ethnicity in the 2019 Survey of Consumer Finances," FEDS Notes, Washington: Board of Governors of the Federal Reserve System, September 28, 2020, https://doi.org/10.17016/2380-7172.2797.

74. See, e.g., P. R. Lockhart, "How Slavery Became America's First Big Business," Vox, August 16, 2019, https://www.vox.com/identities/2019/8/16/20806069/slavery-econ omy-capitalism-violence-cotton-edward-baptist.

75. See Adam Waytz, Kelly Marie Hoffman, and Sophie Trawalter, "A Superhumanization Bias in Whites' Perceptions of Blacks," *Social Psychological and Personality Science* 6, no. 3 (2015): 352 [352–59], https://doi.org/10.1177/1948550614553642. See also Julie Netherland and Helena Hansen, "White Opioids: Pharmaceutical Race and the War on Drugs That Wasn't," *BioSocieties* 12, no. 2 (2017): 217–38, http://dx.doi.org/10.1057/biosoc.2015.46; and Frank Edwards, Hedwig Lee, and Michael Esposito, "Risk of Being Killed by Police Use of Force in the United States by Age, Race–Ethnicity, and Sex," *PNAS* 116, no. 34 (2019): 16793–98, https://doi.org/10.1073/pnas.1821204116.

76. **For findings that the "black-white ['intergenerational mobility'] gap—the largest gap among those we study—is driven entirely by sharp differences in the outcomes of black and white men who grow up in families with comparable incomes,"** see Raj Chetty et al. [Nathaniel Hendren, Maggie R. Jones, and Sonya R. Porter], "Race and Economic Opportunity in the United States: An Intergenerational Perspective," *Quarterly Journal of Economics* 135, no. 2 (2020): 778 [711–83], https://doi.org/10.1093/qje/qjz042.

77. See Russell Lynes, "Rube Goldberg," *New York Times*, December 16, 1973, https://www.nytimes.com/1973/12/16/archives/rube-goldberg-his-life-and-work-by-peter-c-marzio-illustrated-322.html.

78. Compare, e.g., Quayshawn Spencer, "A Radical Solution to the Race Problem," *Philosophy of Science* 81, no. 5 (2014): 1025–26 [1025–38], https://doi.org/10.1086/677694.

Section 11

1. See, e.g., Bernard Lewis, *Race and Slavery in the Middle East: An Historical Enquiry* (New York: Oxford University Press, 1990), 28; and George Pavlu, "Recalling Africa's Harrowing Tale of Its First Slavers—The Arabs—As UK Slave Trade Abolition Is Commemorated," *New African*, March 27, 2018 (previously published as "The First Slavers," *New African*, no. 379, 1999), https://newafricanmagazine.com/16616/.

2. W. E. B. Du Bois, *The Souls of Black Folk* (Chicago: A. C. McClurg, 1903), 1–2; and W. E. B. Du Bois, "The Study of the Negro Problems," *Annals of the American Academy of Political and Social Science* 11 (1898): 7 [1–23], https://doi.org/10.1177/0002716 29801100101. See also Robert-Gooding Williams, *In the Shadow of Du Bois: Afro-Modern Political Thought in America* (Cambridge, MA: Harvard University Press, 2009), 58–65.

3. "What Is to Be Done with the Negro Element" (letter to the editor), *New York Times*, September 21, 1866, https://www.nytimes.com/1866/09/21/archives/what-is-to-be-done-with-the-negro-element-mr-beechers.html. Many commentators have misattributed authorship of the unsigned "Negro Element" letter to abolitionist Henry Ward Beecher. See, e.g., "Epilogue: 'What Is to Be Done with the Negro?' The Era of Reconstruction," in *The New York Times: Complete Civil War, 1861–1865*, ed. Harold Holzer and Craig L. Symonds (New York: Black Dog and Leventhal, 2010), 440 [440–57]. For the actual letter by Beecher that prompted the unsigned "Negro Element" response, see "National Restoration; Interesting and Important Letter from Henry Ward Beecher," *New York Times*, September 1, 1866, https://www.nytimes.com/1866/09/01/archives/national-restoration-interesting-and-important-letter-from-henry.html.

4. Compare, on "whitening" policies of assimilation in Latin American societies, Thomas E. Skidmore, "'Whitening,' the Brazilian Solution," in *Black into White: Race and Nationality in Brazilian Thought*, rev. ed. (Durham, NC: Duke University Press, 1993) 64–69; and Herbert S. Klein and Ben Vinson III, *African Slavery in Latin America and the Caribbean*, 2nd ed. (New York: Oxford University Press, 2007), 243–46.

5. K. Anthony Appiah, "Race, Culture, Identity: Misunderstood Connections," in K. Anthony Appiah and Amy Gutmann, *Color Conscious: The Political Morality of Race* (Princeton: Princeton University Press, 1996), 83 [30–105]. Compare, e.g., Stephanie M. H. Camp, "Black Is Beautiful: An American History," *Journal of Southern History* 81, no. 3 (2015): 675–90, http://www.jstor.org/stable/43918403.

6. See, e.g., Orlando Patterson, "The Social and Cultural Matrix of Black Youth," in *The Cultural Matrix: Understanding Black Youth*, ed. Orlando Patterson (Cambridge, MA: Harvard University Press, 2015), 85–86, 92–96 [45–136]; and Karolyn Tyson, William Darity Jr., and Domini R. Castellino, "It's Not 'A Black Thing': Understanding the Burden of Acting White and Other Dilemmas of High Achievement," *American Sociological Review* 70, no. 4 (2005): 600–601, [582–605], https://doi.org/10.1177/000312240507000403. See also Nikole-Hannah Jones, "School Segregation, the

Continuing Tragedy of Ferguson," ProPublica, December 19, 2014, https://www.pro
publica.org/article/ferguson-school-segregation.

7. See Appiah, "Race, Culture, Identity," 97–99; and K. Anthony Appiah, "The State and
the Shaping of Identity," in *The Tanner Lectures on Human Values*, vol. 23, ed. Grethe
B. Peterson (Salt Lake City: University of Utah Press, 2002), 286 [234–99].

8. Compare Johann Friedrich Blumenbach, *On the Natural Variety of Mankind* (1795),
in *The Idea of Race*, ed. Robert Bernasconi and Tommy L. Lott (Indianapolis: Hackett,
2000), 27–29 [27–37]; and Johann Friedrich Blumenbach, *On the Natural Variety of
Mankind* (1775), in *The Anthropological Treatises of Johann Friedrich Blumenbach*, ed.
and trans. Thomas Bendyshe (London: Anthropological Society, 1865), 99 [69–141].

9. **Strictly speaking, not all descendants of American slavery are Black American. For
about fifty years starting in 1820, several thousand migrated to Liberia (founded
by the American Colonization Society) in West Africa.** See Phillip W. Magness and
Sebastian N. Page, *Colonization after Emancipation: Lincoln and the Movement for
Black Resettlement* (Columbia: University of Missouri Press, 2011).

10. **On "historical memory of slavery and inferiority" that "attaches to the Negro,"** see
Gunnar Myrdal, *An American Dilemma: The Negro Problem and Modern Democracy*
(New York: Harper and Brothers, 1944), 54. **For an overview of the Black American
mentality of resistance,** see Joanne Grant, ed., *Black Protest: History, Documents, and
Analyses, 1619 to the Present* (Greenwich, CT: Fawcett, 1968).

11. See, e.g., Edward E. Baptist, *The Half Has Never Been Told: Slavery and the Making
of American Capitalism* (New York: Basic Books, 2014); and Carol Anderson, *White
Rage* (New York: Bloomsbury Press, 2016). See also Mehrsa Baradaran, *The Color of
Money: Black Banks and the Racial Wealth Gap* (Cambridge, MA: Harvard University
Press, 2017).

12. See, e.g., Christina M. Greer, *Black Ethnics: Race, Immigration, and the Pursuit of the
American Dream* (New York: Oxford University Press, 2013); and Tod G. Hamilton,
Immigration and the Remaking of Black America, fore. Douglas S. Massey
(New York: Russell Sage Foundation, 2019).

13. **On "the peculiar sensation" of "double-consciousness"—namely, "One ever feels
his two-ness—an American, a Negro; two souls, two thoughts, two unreconciled
strivings"—see Du Bois,** *Souls of Black Folk*, 3. See also Clint Smith, *How the Word
Is Passed: A Reckoning with the History of Slavery across America* (New York: Little,
Brown, 2021). Compare, e.g., Nisha Chittal, "The Kamala Harris Identity Debate
Shows How America Still Struggles to Talk about Multiracial People," Vox, January
20, 2021, https://www.vox.com/identities/2020/8/14/21366307/kamala-harris-black-
south-asian-indian-identity; and Hope Reese, "Chimamanda Ngozi Adichie: I Became
Black in America," JSTOR Daily, August 29, 2018, https://daily.jstor.org/chimamanda-
ngozi-adichie-i-became-black-in-america/.

14. Orlando Patterson, *Slavery and Social Death: A Comparative Study* (Cambridge,
MA: Harvard University Press, 1982), 79. See also David Brion Davis, *Inhuman
Bondage: The Rise and Fall of Slavery in the New World* (New York: Oxford University
Press, 2006), 30–32.

15. For a case study of slavery-based dishonor within a purported major race group, see Adaobi Tricia Nwaubani, "My Great-Grandfather, the Nigerian Slave-Trader," *New Yorker*, July 15, 2018, https://www.newyorker.com/culture/personal-hist ory/my-great-grandfather-the-nigerian-slave-trader. On "between 10 and 20% of Igbos—amounting to many millions of people—[who] are descendants of slaves [in Nigeria] and still face significant discrimination" there, see Adaobi Tricia Nwaubani, "Nigeria's Slave Descendants Hope Race Protests Help End Discrimination," Thomson Reuters Foundation, June 29, 2020, https://www.reuters. com/article/us-nigeria-race-descendants-feature-trfn/nigerias-slave-descendants-hope-race-protests-help-end-discrimination-idUSKBN2410BN.

16. See, e.g., Douglas S. Massey and Nancy A. Denton, *American Apartheid: Segregation and the Making of the Underclass* (Cambridge, MA: Harvard University Press, 1993); and Ellora Derenoncourt et al. [Chi Hyun Kim, Moritz Kuhn, and Moritz Schularick], "Wealth of Two Nations: The U.S. Racial Wealth Gap, 1860–2020," National Bureau of Economic Research, Working Paper 30101, June 2022, 1–40, http://dx.doi.org/ 10.3386/w30101. Compare Eleanor Marie Lawrence Brown, "An Alternative View of Immigrant Exceptionalism, Particularly as It Relates to Blacks: A Response to Chua and Rubenfeld," *California Law Review* 103, no. 4 (2015): 1013–16 [989–1017], http:// www.jstor.org/stable/24758493.

17. See, e.g., Dylan Matthews, "Wearing Blackface Is a Big Part of Dutch Christmas," Vox, December 21, 2015, https://www.vox.com/2014/12/24/7446745/netherlands-black-pete-racist; and Soraya Nadia McDonald, "American in Paris: Saul Williams Critiques His Home Country from the Outside Looking In," *Washington Post*, September 23, 2015, https://www.washingtonpost.com/news/arts-and-entertainment/wp/2015/09/ 23/american-in-paris-saul-williams-critiques-his-country-from-the-outside-looking-in/.

18. Richard Moran, "2015 Mark Sacks Lecture: Williams, History, and 'the Impurity of Philosophy,'" *European Journal of Philosophy* 24, no. 2 (2016): 322, 329 n. 7 [315–30], https://doi.org/10.1111/ejop.12149. See also George Chaplin, "An Informational Taxonomy of Race-Ideation," in *The Effects of Race*, ed. Nina G. Jablonski and Gerhard Maré (Stellenbosch: African Sun Media, 2018), 117–18 [109–38].

19. Linda Alcoff, "Mestizo Identity," in *American Mixed Race: The Culture of Microdiversity*, ed. Naomi Zack (Lanham, MD: Rowman and Littlefield, 1995), 277 [257–78]. Compare G. Cristina Mora, *Making Hispanics: How Activists, Bureaucrats, and Media Constructed a New American* (Chicago: University of Chicago Press, 2014), 4–6.

20. Michael Hanchard, "Black Cinderella? Race and the Public Sphere in Brazil," *Public Culture* 7, no. 1 (1994): 177 [165–85], https://doi.org/10.1215/08992363-7-1-165. See also Mara Loveman, Jeronimo O. Muniz, and Stanley R. Bailey, "Brazil in Black and White? Race Categories, the Census, and the Study of Inequality," *Ethnic and Racial Studies* 35, no. 8 (2012): 1468–69 [1466–83], https://doi.org/10.1080/01419 870.2011.607503; and Mary C. Karasch, *Slave Life in Rio de Janeiro, 1808–1850* (Princeton: Princeton University Press, 1987), 4–6.

21. See, e.g., Martin W. Lewis and Kären E. Wigen, *The Myth of Continents: A Critique of Metageography* (Berkeley: University of California Press, 1997), 33–35.

22. **For nonracialist presentation of those migration patterns,** compare Christopher I. Beckwith, *Empires of the Silk Road: A History of Central Eurasia from the Bronze Age to the Present* (Princeton: Princeton University Press, 2009), 30–32; and J. P. Mallory, *In Search of the Indo-Europeans: Language, Archaeology and Myth* (London: Thames and Hudson, 1989), 266–70.

23. **On "disjunctures in racial assignment" for "Asian Americans" in the United States,** compare Jennifer Lee and Karthick Ramakrishnan, "Who Counts as Asian," *Ethnic and Racial Studies* 43, no. 10 (2020): 1733–56, https://doi.org/10.1080/01419 870.2019.1671600.

24. See US Census Bureau, "About the Topic of Race," accessed August 28, 2021, https:// www.census.gov/topics/population/race/about.html. See also, e.g., Carleton S. Coon, *The Origin of Races* (New York: Knopf, 1962), 3–7. **For "Negroid," "Khoisanoid," and "Capoid" groupings of non-northern African peoples,** compare "Human Races," in *Encyclopedia of Genetics, Genomics, Proteomics and Informatics* (Dordrecht: Springer, 2008), https://doi.org/10.1007/978-1-4020-6754-9_7931.

25. US Census Bureau, "What We Do," accessed January 24, 2022, https://www.census. gov/about/what.html.

26. See Anna Brown, "The Changing Categories the U.S. Census Has Used to Measure Race," Pew Research Center, February 25, 2020, https://www.pewresearch.org/fact-tank/2020/02/25/the-changing-categories-the-u-s-has-used-to-measure-race/. **As a point of clarification, "racial categories" for Africa-identified persons have not been "included on every U.S. census since the first one in 1790":** *Slaves* **is not a race category.**

27. See US Census Bureau, "History: Through the Decades: Questionnaires," 1990 to 2020 forms, accessed November 14, 2021, https://www.census.gov/history/www/ through_the_decades/questionnaires/.

28. See US Census Bureau, "Measuring Race and Ethnicity Across the Decades: 1790–2010," by Beverly M. Pratt, Lindsay Hixson, and Nicholas A. Jones, accessed October 28, 2021, https://www.census.gov/data-tools/demo/race/MREAD_1790_2010.html. (**Census forms have never asked for "ethnicity."**) The Census Bureau authors write: "In 1900, for the first time, 'Negro' was used, in conjunction with 'Black,' to describe the population of African origin. . . . The term 'Negro' was used to refer to full-blooded individuals and the term 'of Negro descent' was used to refer to 'Mulattos.'" By those convoluted criteria, most descendants of American slavery would have counted as "mulatto."

29. See, e.g., Paul Finkelman, "How the Proslavery Constitution Led to the Civil War," *Rutgers Law Journal* 43, no. 3 (2013): 407–8 [405–38]. Finkelman writes: "The provisions for amending the Constitution gave the slave states a perpetual veto over any alteration of the Constitution by requiring that three-fourths of the states ratify any amendment." He adds: "In 1787 the value of all the slaves in the United States exceeded that of any other form of property except real estate. . . . Almost all the leaders in southern states were slaveowners."

30. See again Census Bureau, "Questionnaires," https://www.census.gov/history/www/through_the_decades/questionnaires/. Flat "black or African American" raciality made its first appearance on the 2020 census, which specifies examples of global Africa-identified peoples, namely, "African American, Jamaican, Haitian, Nigerian, Ethiopian, Somali, etc."

31. Compare, e.g., Laris Karklis and Emily Badger, "Every Term the Census Has Used to Describe America's Racial and Ethnic Groups Since 1790," *Washington Post*, November 4, 2015, https://www.washingtonpost.com/news/wonk/wp/2015/11/04/every-term-the-census-has-used-to-describe-americas-racial-groups-since-1790/; and Lawrence Blum, *"I'm Not a Racist, But . . . ": The Moral Quandary of Race* (Ithaca: Cornell University Press, 2002), 148.

32. See D'vera Cohn, "Race and the Census: The 'Negro' Controversy," Pew Research Center, January 21, 2010, http://www.pewsocialtrends.org/2010/01/21/race-and-the-census-the-"negro"-controversy/.

33. See Census Bureau, "Questionnaires," 1890 form, https://www.census.gov/history/www/through_the_decades/questionnaires/.

34. See Census Bureau, "Topic of Race," https://www.census.gov/topics/population/race/about.html.

35. See Census Bureau, "Measuring Race," under "1890," https://www.census.gov/data-tools/demo/race/MREAD_1790_2010.html.

36. See Census Bureau, "Topic of Race," https://www.census.gov/topics/population/race/about.html.

37. On research sparked by the finding that "about 2 percent of the DNA in the genomes of modern-day people with Eurasian ancestry is Neanderthal in origin," see Jef Akst, "Neanderthal DNA in Modern Human Genomes Is Not Silent," *Scientist*, September 1, 2019, https://www.the-scientist.com/features/neanderthal-dna-in-modern-human-genomes-is-not-silent-66299.

38. On "Arab Americans and others [who] have argued that being classified as white does not reflect the self-identity of Americans from the Middle East or North Africa," see "Demographic Portrait of Muslim Americans," Pew Research Center, July 26, 2017, https://www.pewforum.org/2017/07/26/demographic-portrait-of-muslim-americans/. Compare Sarah M. A. Gualtieri, *Between Arab and White: Race and Ethnicity in the Early Syrian American Diaspora* (Berkeley: University of California Press, 2009), 1–2; and Neda Maghbouleh, *The Limits of Whiteness: Iranian Americans and the Everyday Politics of Race* (Stanford: Stanford University Press, 2017), 4–5. For "Middle East"-origin identities in the United States, the "Caucasian" subcaste hierarchy no longer officially discriminates between true- and off- "white."

39. See *United States v. Thind*, 261 U.S. 204, 208 (1923).

40. See, e.g., Kiara Alfonseca, "20 Years after 9/11, Islamophobia Continues to Haunt Muslims," ABCNews.com, September 11, 2021, https://abcnews.go.com/US/20-years-911-islamophobia-continues-haunt-muslims/story?id=79732049.

Section 12

1. For example, for the very flexible (or unstable) notion of "race" as "a *cluster* concept which draws together under a single word references to biological, cultural, and geographical factors thought characteristic of a population," see Lucius Outlaw, " 'Conserve' Races?: In Defense of W. E. B. Du Bois," in *W. E. B. Du Bois on Race and Culture*, ed. Bernard W. Bell, Emily R. Grosholz, and James B. Stewart (New York: Routledge, 1996), 20 [15–38].
2. See Philip Gourevitch, "The Optimist," *New Yorker*, February 23, 2003, https://www.newyorker.com/magazine/2003/03/03/the-optimist-3.

Part II. Section 1

1. See, e.g., Steven Pinker, "Groups of People May Differ Genetically in Their Average Talents and Temperaments," in *What Is Your Dangerous Idea?: Today's Leading Thinkers on the Unthinkable*, ed. John Brockman (New York: Harper Perennial, 2007), 13–15; and Ernst Mayr, "The Biology of Race and the Concept of Equality," *Daedalus* 131, no. 1 (2002): 89–94, https://www.jstor.org/stable/20027740.
2. See, e.g., Immanuel Kant, *Observations on the Feeling of the Beautiful and Sublime* (1764), in *Observations on the Feeling of the Beautiful and Sublime and Other Writings*, ed. Patrick Frierson and Paul Guyer (Cambridge: Cambridge University Press, 2011), 58–61 (2:253–2.255) [11–65]; and David Hume, "Of National Characters" (1753), in *Essays and Treatises on Several Subjects*, vol. 1 (London: A. Millar, 1753), 291n [277–300].
3. Joshua Glasgow, Sally Haslanger, Chike Jeffers, and Quayshawn Spencer, "Introduction," in *What Is Race? Four Philosophical Views* (New York: Oxford University Press, 2019), 2 [1–3].
4. See, e.g., Neil Risch et al. [Esteban Burchard, Elad Ziv, and Hua Tang], "Categorization of Humans in Biomedical Research: Genes, Race and Disease," *Genome Biology* 3, no. 7 (2002): comment2007, https://doi.org/10.1186/gb-2002-3-7-comment2007.
5. See, e.g., Katarzyna Bryc et al. [Eric Y. Durand, J. Michael Macpherson, David Reich, and Joanna L. Mountain], "The Genetic Ancestry of African Americans, Latinos, and European Americans across the United States," *American Journal of Human Genetics* 96, no. 1 (2015): 42 [37–53], https://doi.org/10.1016/j.ajhg.2014.11.010.
6. See, e.g., Christine Tamir, "Key Findings about Black Immigrants in the U.S.," Pew Research Center, January 27, 2022, https://www.pewresearch.org/fact-tank/2022/01/27/key-findings-about-black-immigrants-in-the-u-s/.
7. "Black and White," *New York Times*, September 26, 1982, https://www.nytimes.com/1982/09/26/opinion/black-and-white.html.

Section 2

1. Chris Jones, "The Passion of Tiger Woods," *Esquire*, October 2003, http://www.esqu ire.com/sports/a644/esq1003-oct-tiger/.
2. See, e.g., Eamonn Callan, "The Ethics of Assimilation," *Ethics* 115, no. 3 (2005): 490–91 [471–500], https://doi.org/10.1086/428460.
3. Ronald R. Sundstrom, *The Browning of America and the Evasion of Social Justice* (New York: State University of New York Press, 2008), 117.
4. Sundstrom, *Browning of America*, 110. Compare, e.g., Lewis R. Gordon, "Race, Biraciality, and Mixed Race—In Theory," in *Her Majesty's Other Children: Sketches of Racism from a Neocolonial Age* (Lanham, MD: Rowman and Littlefield, 1997), 51–71; and Jon Michael Spencer, *The New Colored People: The Mixed-Race Movement in America*, foreword Richard E. van der Ross (New York: New York University Press, 1997), 1–14.
5. **For early analysis of the "tragic mulatto" archetype,** see Sterling A. Brown, "Negro Character as Seen by White Authors," *Journal of Negro Education* 2, no. 2 (1933): 180, 192–96 [179–203], http://dx.doi.org/10.2307/2292236. See also David Pilgrim, "The Tragic Mulatto Myth," Jim Crow Museum of Racist Memorabilia (online), November 2000 (rev. 2012), https://www.ferris.edu/HTMLS/news/jimcrow/mulatto/homepage. htm; and Caroline A. Streeter, *Tragic No More: Mixed-Race Women and the Nexus of Sex and Celebrity* (Amherst: University of Massachusetts Press, 2012), 2–3. In film, see also *Pinky*, dir. Elia Kazan (1949; 20th Century Fox); *Imitation of Life*, dir. John M. Stahl (1934; Universal Pictures); and *Imitation of Life*, dir. Douglas Sirk (1959; Universal Pictures).
6. Margaret Hunter, "The Persistent Problem of Colorism: Skin Tone, Status, and Inequality," *Sociology Compass* 1, no. 1 (2007): 237 [237–54], https://doi.org/10.1111/ j.1751-9020.2007.00006.x. See also Verna M. Keith and Cedric Herring, "Skin Tone and Stratification in the Black Community," *American Journal of Sociology* 97, no. 3 (1991): 760–78, https://doi.org/10.1086/229819. **On American "disparities in po- licing and punishment within the black population along the colour continuum,"** see Ellis P. Monk, "The Color of Punishment: African Americans, Skin Tone, and the Criminal Justice System," *Ethnic and Racial Studies* 42, no. 10 (2019): 1593 [1593–1612], https://doi.org/10.1080/01419870.2018.1508736. **On "Western ideals of white skin" driving a "multibillion-dollar skin-whitening market" in Asian countries,** see Crystal Tai and Tashny Sukumaran, "Asia's Addiction to Whiter Skin Runs Deep— But the Backlash Has Begun," *South China Morning Post*, February 3, 2019, https:// www.scmp.com/week-asia/society/article/2184747/asias-addiction-whiter-skin- runs-deep-backlash-has-begun.
7. Robert L. Reece, "What Are You Mixed With: The Effect of Multiracial Identification on Perceived Attractiveness," *Review of Black Political Economy* 43, no. 2 (2016): 140, 145 [139–47], https://doi.org/10.1007/s12114-015-9218-1.

8. Glenn C. Loury, *The Anatomy of Racial Inequality* (Cambridge, MA: Harvard University Press, 2002), 59–60. See also Erving Goffman, *Stigma: Notes on the Management of Spoiled Identity* (New York: Simon and Schuster, 1963), 4–5; and J. Lorand Matory, *Stigma and Culture: Last-Place Anxiety in Black America* (Chicago: University of Chicago Press, 2015), 337–38.

9. See, e.g., Tommie Shelby, *We Who Are Dark: The Philosophical Foundations of Black Solidarity* (Cambridge, MA: Harvard University Press, 2005), 241–42. **On "a continuing stigma of inferiority and dishonor" attaching to slaves freed through manumission, which "generated resentments that could lead backward to an identification with slaves,"** see David Brion Davis, *The Problem of Slavery in the Age of Emancipation* (New York: Knopf, 2014), 55.

10. See, e.g., Randall Kennedy, *Sellout: The Politics of Racial Betrayal* (New York: Pantheon, 2008), 176–78.

11. See, e.g., Robert L. Reece, "Genesis of U.S. Colorism and Skin Tone Stratification: Slavery, Freedom, and Mulatto-Black Occupational Inequality in the Late 19th Century," *Review of Black Political Economy* 45, no. 1 (2018): 3–21, https://doi.org/10.1177/0034644618770761.

12. John Dollard, *Caste and Class in a Southern Town*, 3rd ed. (Garden City, NY: Doubleday, 1949), 69.

13. Sundstrom, *Browning of America*, 120.

14. See Edward E. Telles, *Race in Another America: The Significance of Skin Color in Brazil* (Princeton: Princeton University Press, 2004), 1–3; and Mohamed Adhikari, *Not White Enough, Not Black Enough: Racial Identity in the South African Coloured Community* (Athens: Ohio University Press, 2005), 2–5.

15. **For "results documenting biases in the social categorization and perception of biracials [that] have implications for resistance to change in the American racial hierarchy,"** see Arnold K. Ho et al. [Jim Sidanius, Daniel T. Levin, and Mahzarin R. Banaji], "Evidence for Hypodescent and Racial Hierarchy in the Categorization and Perception of Biracial Individuals," *Journal of Personality and Social Psychology* 100, no. 3 (2011): 492 [492–506], https://doi.org/10.1037/a0021562. See also Arnold K. Ho, Nour S. Kteily, and Jacqueline M. Chen, "'You're One of Us': Black Americans' Use of Hypodescent and Its Association with Egalitarianism," *Journal of Personality and Social Psychology* 113, no. 5 (2017): 753–68, http://dx.doi.org/10.1037/pspi0000107.

16. See, e.g., Linda Martín Alcoff, "Latino/as, Asian Americans, and the Black-White Binary," *Journal of Ethics* 7, no. 1 (2003): 5–27, https://doi.org/10.1023/A:1022870628484.

17. Tina Fernandes Botts, "Editor's Introduction: Toward a Mixed Race Theory," in *Philosophy and the Mixed Race Experience*, ed. and intro. Tina Fernandes Botts (London: Lexington Books, 2016), 3 [1–17]. See also Sundstrom, *Browning of America*, 66–75. Compare, e.g., Richard Rodriguez, *Brown: The Last Discovery of America* (New York: Viking, 2002), 30.

18. See, e.g., Anthony G. Greenwald et al. [T. Andrew Poehlman, Eric Luis Uhlmann, and Mahzarin R. Banaji], "Understanding and Using the Implicit Association Test: III.

Meta-Analysis of Predictive Validity," *Journal of Personality and Social Psychology* 97, no. 1 (2009): 17–41, https://doi.org/10.1037/a0015575. There is controversy about the Implicit Association Test's reliability and usefulness. See, e.g., Keith Payne, Laura Niemi, and John M. Doris, "How to Think About 'Implicit Bias,'" *Scientific American*, March 27, 2018, https://www.scientificamerican.com/article/how-to-think-about-implicit-bias/.

19. Biographies of Anatole Broyard and Phillipa Schuyler portray sad cases of "passing" out of blackness. See Bliss Broyard, *One Drop: My Father's Hidden Life—A Story of Race and Family Secrets* (New York: Little, Brown, 2007); and Kathryn Talalay, *Composition in Black and White: The Life of Philippa Schuyler* (New York: Oxford University Press, 1995).

20. Compare, e.g., Cox News Service, "'Multiracial' Category Stirs Rights Debate," *Chicago Tribune*, July 21, 1996, http://articles.chicagotribune.com/1996-07-21/news/9607210214_1_multiracial-category-census-bureau-la-raza.

21. There has been a renewed push to racially recast Douglass. See, e.g., Tanya Katerí Hernández, "Frederick Douglass: A Multi-Racial Trailblazer," *Baltimore Sun*, February 8, 2018, https://www.baltimoresun.com/opinion/op-ed/bs-ed-op-0209-douglass-race-20180208-story.html. For a nineteenth-century perspective—"Whether Mr. Douglass, who is of mixed race, inclines in his intelligence more to his white father than to his colored mother need not be nicely balanced, yet in some respects he differs materially from the African stock. He never has been imitative in his manner of thought"—compare "Our First Man of Color," *New York Times*, May 3, 1891, https://www.nytimes.com/1891/05/03/archives/our-first-man-of-color-frederick-douglass-the-colored-orator-by.html. Douglass himself endorsed a black (American) social identity: "I now understood what had been to me a most perplexing difficulty—to wit, the white man's power to enslave the black man." Frederick Douglass, *Narrative of the Life of Frederick Douglass, an American Slave. Written by Himself.* (Boston: Anti-Slavery Office, 1845), 33.

Section 3

1. John F. Kennedy, "Radio and Television Report to the American People on Civil Rights," June 11, 1963, John F. Kennedy Presidential Library, https://www.jfklibrary.org/asset-viewer/archives/JFKWHA/1963/JFKWHA-194-001/JFKWHA-194-001. For a compilation of state and local segregation laws that were active in 1950, see Pauli Murray, ed., *States' Laws on Race and Color*, foreword Davison M. Douglas (1951; repr., Athens: University of Georgia Press, 1997).

2. See Amy Traub et al. [Laura Sullivan, Tatjana Meschede, and Tom Shapiro], "The Asset Value of Whiteness: Understanding the Racial Wealth Gap," Demos, February 6, 2017, https://www.demos.org/research/asset-value-whiteness-understanding-racial-wealth-gap.

3. See, e.g., Ian F. Haney López, "Post-Racial Racism: Racial Stratification and Mass Incarceration in the Age of Obama," *California Law Review* 98, no. 3 (2010): 1023–74, https://doi.org/10.15779/Z38H696.

4. See, e.g., Jennifer L. Hochschild, "You Win Some, You Lose Some: Explaining the Pattern of Success and Failure in the Second Reconstruction," in *Taking Stock: American Government in the Twentieth Century*, ed. Morton Keller and R. Shep Melnick (New York: Cambridge University Press, 1999), 219–48.

5. *Regents of the Univ. of California v. Bakke*, 438 U.S. 265, 266–67 (1978). Compare, e.g., Ian F. Haney López, "'A Nation of Minorities': Race, Ethnicity, and Reactionary Colorblindness," *Stanford Law Review* 59, no. 4 (2007): 985–1063, https://www.jstor.org/stable/40040347.

6. Compare John F. Kennedy, Executive Order 10925—Establishing the President's Committee on Equal Employment Opportunity, March 6, 1961, John F. Kennedy Presidential Library, https://www.jfklibrary.org/asset-viewer/archives/JFKPOF/081/JFKPOF-081-005. See also Louis Menand, "The Changing Meaning of Affirmative Action, *New Yorker*, January 13, 2020, https://www.newyorker.com/magazine/2020/01/20/have-we-outgrown-the-need-for-affirmative-action.

7. See *Adarand Constructors, Inc. v. Peña*, 515 U.S. 200, 226–28 (1995). Compare, e.g., Adam Winkler, "Fatal in Theory and Strict in Fact: An Empirical Analysis of Strict Scrutiny in the Federal Courts," *Vanderbilt Law Review* 59, no. 3 (2006): 823–25 [793–871], https://scholarship.law.vanderbilt.edu/vlr/vol59/iss3/3.

8. Michelle Alexander, *The New Jim Crow: Mass Incarceration in the Age of Colorblindness* (New York: New Press, 2010), 100–101. See also Bruce Western, "The Scope and Causes of the Prison Boom," in *Punishment and Inequality in America* (New York: Russell Sage Foundation, 2006), 9–82.

9. **See, e.g., *Dred Scott v. Sandford* (1857), which held that Black Americans could not be American citizens, had no legal standing in federal court, and thus could not sue for freedom from slavery; *Elk v. Wilkins* (1884), which denied birthright citizenship to Native Americans; and *Korematsu v. United States* (1944), which permitted internment of Japanese Americans during World War II.** Compare Reva B. Siegel, "Foreword: Equality Divided," *Harvard Law Review* 127, no. 1 (2013): 2–5 [1–94], https://www.jstor.org/stable/i23741021.

10. William F. Russell and Taylor Branch, *Second Wind: The Memoirs of an Opinionated Man* (New York: Random House, 1979), 187.

11. Daniel Cox, Juhem Navarro-Rivera, and Robert P. Jones, "Race, Religion, and Political Affiliation of Americans' Core Social Networks," PRRI (Public Religion Research Institute), August 28, 2014, https://www.prri.org/research/poll-race-religion-politics-americans-social-networks/.

12. US Census Bureau, *Overview of Race and Hispanic Origin: 2010*, by Karen R. Humes, Nicholas A. Jones, and Roberto R. Ramirez (Washington, DC: US Department of Commerce, 2011), 4 [1–23], https://www.census.gov/content/dam/Census/library/publications/2011/dec/c2010br-02.pdf.

13. Lincoln Quillian et al. [Devah Pager, Ole Hexel, and Arnfinn H. Midtbøen], "Meta-analysis of Field Experiments Show No Change in Racial Discrimination in Hiring

over Time," *PNAS* 114, no. 41 (2017): 10871 [10870–75], https://doi.org/10.1073/pnas.1706255114.

14. US Census Bureau, *Negro Population 1790–1915* (Washington, DC: Government Printing Office, 1918), 209, https://www.census.gov/library/publications/1918/dec/negro-population-1790-1915.html. **Via a "Negro" race label, the Census Bureau retroactively superimposed a "race" thing on its *Slaves* category (1790–1860)—and today continues to do so via a "[b]lack or African" label. But "race" did not appear on the census until 1900, as part of a *color or race* category, which was previously *color* alone.** See this volume, Part I, Section 11, "Enter 'Geoancestry.'"

15. See US Census Bureau, "Measuring Race and Ethnicity Across the Decades: 1790–2010," by Beverly M. Pratt, Lindsay Hixson, and Nicholas A. Jones, last revised September 4, 2015, accessed October 28, 2021, https://www.census.gov/data-tools/demo/race/MREAD_1790_2010.html; and Anna Brown, "The Changing Categories the U.S. Census Has Used to Measure Race," Pew Research Center, February 25, 2020, https://www.pewresearch.org/fact-tank/2020/02/25/the-changing-categories-the-u-s-has-used-to-measure-race/. **As clarified in Part I (this volume), *Slaves* is not a race category.**

16. See, e.g., Aliya Saperstein and Andrew M. Penner, "Racial Fluidity and Inequality in the United States," *American Journal of Sociology* 118, no. 3 (2012): 676–727, https://doi.org/10.1086/667722.

17. **On segregation of blood by American color caste,** see Thomas A. Guglielmo, "'Red Cross, Double Cross': Race and America's World War II-Era Blood Donor Service," *Journal of American History* 97, no. 1 (2010): 63–90, https://doi.org/10.2307/jahist/97.1.63.

18. See, e.g., Associated Press, "Obama's True Colors: Black, White . . . or Neither?," NBCNews.com, December 14, 2008, https://www.nbcnews.com/id/wbna28216005.

19. **For a detailed account of purposes the one-drop rule has served,** see Winthrop D. Jordan, "Historical Origins of the One-Drop Racial Rule in the United States," ed. Paul Spickard, *Journal of Critical Mixed Race Studies* 1, no. 1 (2014): 98–132, https://escholarship.org/uc/item/91g761b3.

20. See Orlando Patterson, *Slavery and Social Death: A Comparative Study* (Cambridge, MA: Harvard University Press, 1982), 79. **I have not used the term "homegrown" lightly. On the breeding of enslaved Americans,** see Ned Sublette and Constance Sublette, *The American Slave Coast: A History of the Slave-Breeding Industry* (Chicago: Lawrence Hill Books, 2016); and Manning Marable, *How Capitalism Underdeveloped Black America: Problems in Race, Political Economy and Society* (Boston: South End Press, 1983), 72–76.

21. **For comparison of color caste stratification in America with the Hindu *varna* ("color") caste system in India,** see Gerald D. Berreman, "Caste in India and the United States," *American Journal of Sociology* 66, no. 2 (1960): 120–27, https://doi.org/10.1086/222839.

22. Virginia General Assembly, *Preservation of Racial Integrity* (1924), https://www.encyclopediavirginia.org/Preservation_of_Racial_Integrity_1924. **Previously, Virginia law less restrictively specified one-fourth (1705) and one-eighth (1785) African**

ancestry as amounts dividing white and black. See Joshua D. Rothman, *Notorious in the Neighborhood: Sex and Families Across the Color Line in Virginia, 1787–1861* (Chapel Hill: University of North Carolina Press, 2003), 208–10.

23. Walter A. Plecker, "The New Virginia Law to Preserve Racial Integrity," *Virginia Health Bulletin*, March 1924, http://edu.lva.virginia.gov/dbva/items/show/226. Compare John Eligon, e.g., "Quadroon? Moor? Virginia Sued for Making Those Who Wed Say What They Are," *New York Times*, September 8, 2019, https://www.nytimes.com/2019/09/08/us/virginia-marriage-race.html.

24. See James O'Byrne, "Many Feared Naomi Drake and Powerful Racial Whim," *Times-Picayune*, August 16, 1993, https://www.nola.com/news/politics/article_5f54a83c-6981-518e-8396-5ba24a07fc60.html.

Section 4

1. David Theo Goldberg, "Made in the USA: Racial Mixing 'n Matching," in *American Mixed Race: The Culture of Microdiversity*, ed. Naomi Zack (Lanham, MD: Rowman and Littlefield, 1995), 237 [237–56].

2. *Dred Scott v. Sandford*, 60 U.S. 393, 407 (1857). Compare, e.g., Frederick Douglass, *My Bondage and My Freedom* (New York: Miller, Orton and Mulligan, 1855), 51–52, 58–59; Kathleen Brown, *Good Wives, Nasty Wenches, and Anxious Patriarchs: Gender, Race, and Power in Colonial Virginia* (Chapel Hill: University of North Carolina Press, 1996), 1–2, 197–98; and Thavolia Glymph, *Out of the House of Bondage: The Transformation of the Plantation Household* (New York: Cambridge University Press, 2008), 4–5.

3. Linda Alcoff, "Mestizo Identity," in *American Mixed Race: The Culture of Microdiversity*, ed. Naomi Zack (Lanham, MD: Rowman and Littlefield, 1995), 259 [257–78].

4. Naomi Zack, "Life after Race," in *American Mixed Race*, 300, 305 [297–307].

5. E. Franklin Frazier, *Black Bourgeoisie* (New York: Free Press, 1957), 20. See also John Dollard, *Caste and Class in a Southern Town*, 3rd ed. (Garden City, New York: Doubleday, 1949), 186–87.

6. James Weldon Johnson, *The Autobiography of an Ex-Colored Man* (Boston: Sherman, French, 1912).

7. See, e.g., Allyson Hobbs, *A Chosen Exile: A History of Racial Passing in American Life* (Cambridge, MA: Harvard University Press, 2014), 167–69. **On how " 'black' passing narratives cast doubt on passing as a form of racial 'liberation,' drawing on metaphors of concealment and disguise to highlight the compromised agency of the subject who 'crosses over,' "** see Gayle Wald, *Crossing the Line: Racial Passing in Twentieth-Century U.S. Literature and Culture* (Durham: Duke University Press, 2000), 16. See also Martha J. Cutter, "'As White as Most White Women': Racial Passing in Advertisements for Runaway Slaves and the Origins of a Multivalent Term," *American Studies* 54, no. 4 (2016): 73–97, http://www.jstor.org/stable/44982355.

8. Molly McElroy, "Secrets of Famous 1930s 'Blonde Bombshell of Rhythm' Revealed with Help from UW Library," *UW News*, March 27, 2012, https://www.washington.edu/news/2012/03/27/secrets-of-famous-1930s-blonde-bombshell-of-rhythm-revealed-with-help-from-uw-library/. For the source story, see Phyllis Fletcher, "Secrets of a Blonde Bombshell," *Studio 360* (radio program), PRI, September 30, 2011, https://www.pri.org/stories/2011-09-30/secrets-blonde-bombshell.

9. Hobbs, *Chosen Exile*, 164.

10. "All Niggers, More or Less!," *News and Courier* (Charleston), October 17, 1895, 5.

11. For historical illustration of the difference between formally versus functionally abolishing slavery, see Douglas A. Blackmon, *Slavery by Another Name: The Re-Enslavement of Black People in America from the Civil War to World War II* (New York: Doubleday, 2008). For the classic depiction of White American hysteria about Africa-identified men in proximity to Europe-identified women, see *The Birth of a Nation*, dir. D. W. Griffith (1915; David W. Griffith).

12. Barbara Young Welke, "Miscegenation and the Racial State," *Contemporary Sociology* 41, no. 3 (2012): 284 [283–87], https://doi.org/10.1177/0094306112443511a.

13. Equal Justice Initiative, *Lynching in America: Targeting Black Veterans* (Montgomery, AL: Equal Justice Initiative, 2017), 45 [1–52], https://eji.org/wp-content/uploads/2019/10/lynching-in-america-targeting-black-veterans-web.pdf. See also Jamiles Larty and Sam Morris, "How White Americans Used Lynchings to Terrorize and Control Black People," *Guardian*, April 26, 2018, https://www.theguardian.com/us-news/2018/apr/26/lynchings-memorial-us-south-montgomery-alabama.

14. See, e.g., Khary Oronde Polk, "Communicable Subjects: African American Soldiers Trip the Global Color Line," in *Contagions of Empire: Scientific Racism, Sexuality, and Black Military Workers Abroad, 1898–1948* (Chapel Hill: University of North Carolina Press, 2020), 166–212; Gary I. Schulman, "Race, Sex, and Violence: A Laboratory Test of the Sexual Threat of the Black Male Hypothesis," *American Journal of Sociology* 79, no. 5 (1974): 1260–77, https://doi.org/10.1086/225677; and Faye Crosby, Stephanie Bromley, and Leonard Saxe, "Recent Unobtrusive Studies of Black and White Discrimination and Prejudice: A Literature Review," *Psychological Bulletin* 87, no. 3 (1980): 546–63, https://doi.org/10.1037/0033-2909.87.3.546.

15. William Waller Hening, ed., Act XII, "Negro Womens Children to Serve According to the Condition of the Mother," *Laws of Virginia*, December 1662, in *The Statutes at Large; Being a Collection of All the Laws of Virginia, from the First Session of the Legislature, in the Year 1619*, vol. 2 (Richmond: Samuel Pleasants, 1809–23), 170. Compare Jennifer L. Morgan, *Laboring Women: Reproduction and Gender in New World Slavery* (Philadelphia: University of Pennsylvania Press, 2004), 71–72; and Brown, *Good Wives, Nasty Wenches*, 196.

16. See, e.g., Jennifer Lisa Vest, "Being and Not Being, Knowing and Not Knowing," in *Philosophy and the Mixed Race Experience*, ed. and intro. Tina Fernandes Botts (London: Lexington Books, 2016), 113n3 [93–116]; and Joshua Glasgow, *A Theory of Race* (New York: Routledge, 2009), 63–64.

17. See *Doe v. State, Department of Health and Human Resources*, 470 So. 2d 369 (La. Ct. App. 1985). Compare Michael Omi, "Racial Identity and the State: The Dilemmas of

Classification," *Law and Inequality* 15, no. 1 (1997): 7–23, http://scholarship.law.umn. edu/lawineq/vol15/iss1/2.

18. Art Harris, "Louisiana Court Sees No Shades of Gray in Woman's Request," *Washington Post*, May 21, 1983, https://www.washingtonpost.com/archive/politics/ 1983/05/21/louisiana-court-sees-no-shades-of-gray-in-womans-request/ddb0f 1df-ba5d-4141-9aa0-6347e60ce52d/. For an overview of the Phipps case and the Supreme Court's approach to race classification since *Korematsu* (1944), see Virginia R. Domínguez, *White by Definition: Social Classification in Creole Louisiana* (New Brunswick: Rutgers University Press, 1986), 1–5.

19. Harris, "Louisiana Court Sees No Shades of Gray."

20. Frances Frank Marcus, "Louisiana Repeals Black Blood Law," *New York Times*, July 6, 1983, https://www.nytimes.com/1983/07/06/us/louisiana-repeals-black-blood-law.html.

21. "Black and White," *New York Times*. Similar nonacknowledgment of slave rape has been typical in reference to a sexual "relationship" or "affair" between Thomas Jefferson and his slave Sally Hemings, through whom he fathered at least four children, who were born into slavery. See, e.g., "Thomas Jefferson and Sally Hemings: A Brief Account," Thomas Jefferson Foundation, accessed November 7, 2021, https:// www.monticello.org/thomas-jefferson/jefferson-slavery/thomas-jefferson-and-sally-hemings-a-brief-account/; and Annette Gordon-Reed, *Thomas Jefferson and Sally Hemings: An American Controversy* (Charlottesville: University of Virginia Press, 1997), 23. Compare, e.g., Fawn M. Brodie, *Thomas Jefferson: An Intimate History* (New York: W. W. Norton, 1974), 292–93. Under chattel slavery, there is no conceptual room for owners to engage in noncoercive sexual acts with slaves: (human) property has no autonomous legal standing.

22. Times Wire Services, "Birth Certificate Race Data Upheld in Louisiana," *Los Angeles Times*, December 21, 1985, https://www.latimes.com/archives/la-xpm-1985-12-21-mn-2927-story.html.

23. See US Census Bureau, "History: Through the Decades: Questionnaires," accessed November 14, 2021, https://www.census.gov/history/www/through_the_decades/ questionnaires/; and US Census Bureau, "Measuring Race and Ethnicity across the Decades: 1790–2010," by Beverly M. Pratt, Lindsay Hixson, and Nicholas A. Jones, accessed October 28, 2021, https://www.census.gov/data-tools/demo/race/MREAD_ 1790_2010.html.

24. US Census Bureau, "Introduction," in *Population of the United States in 1860* (Washington, DC: Government Printing Office, 1864), ix–xi [iii–cvii], https://www2. census.gov/library/publications/decennial/1860/population/1860a-02.pdf; see also https://www.census.gov/library/publications/1864/dec/1860a.html.

25. See, e.g., Michael O. Emerson and Christian Smith, *Divided by Faith: Evangelical Religion and the Problem of Race in America* (New York: Oxford University Press, 2000). Compare, e.g., Tom Gjelten, "To Understand How Religion Shapes America, Look to its Early Days," NPR, *All Things Considered*, June 28, 2017, https://www.npr. org/2017/06/28/534765046/smithsonian-exhibit-explores-religious-diversitys-role-in-u-s-history.

26. Calvin Schermerhorn, "Why the Racial Wealth Gap Persists, More Than 150 Years After Emancipation," *Washington Post*, June 19, 2019, https://www.washingtonpost.com/outlook/2019/06/19/why-racial-wealth-gap-persists-more-than-years-after-emancipation/. On **"the racial wealth gap as the most robust indicator of the cumulative economic effects of white supremacy in the United States,"** see William A. Darity Jr. and A. Kirsten Mullen, "A Program of Black Reparations," in *From Here to Equality: Reparations for Black Americans in the Twenty-First Century* (Chapel Hill: University of North Carolina Press, 2020), 263 [256–70]. See also Thomas Craemer et al. [Trevor Smith, Brianna Harrison, Trevon Logan, Wesley Bellamy, and William Darity Jr.], "Wealth Implications of Slavery and Racial Discrimination for African American Descendants of the Enslaved," *Review of Black Political Economy* 47, no. 3 (2020): 218–54, https://doi.org/10.1177/0034644620926516.

27. See Howard Schuman et al. [Charlotte Steeh, Lawrence Bobo, and Maria Krysan], *Racial Attitudes in America: Trends and Interpretations*, rev. ed. (Cambridge, MA: Harvard University Press, 1997), 191–95.

28. Martin Luther King Jr., "Where Are We?," in *Where Do We Go from Here: Chaos or Community?* (New York: Harper and Row, 1967), 8 [1–22].

29. See, e.g., Ashley Lee, "Ava DuVernay's Prison Doc '13th' Spotlights Donald Trump's Violent Rally Language (video)," *Hollywood Reporter*, October 13, 2016, https://www.hollywoodreporter.com/news/politics-news/donald-trump-rallies-ava-duvernays-938117/.

Section 5

1. **For the claim that "there is no American ancient soil, no founding race, but there is a common ancestral experience of moving from 'there' to 'here'"**—which **disappears Native Americans and descendants of American slavery, who obviously were not immigrants**—see Bernard A. Weisberger, "A Nation of Immigrants," *American Heritage* 45, no. 1 (1994), https://www.americanheritage.com/nation-immigrants. Compare, e.g., William Booth, "The Myth of the Melting Pot: One Nation, Indivisible: Is It History?," *Washington Post*, February 22, 1998, https://www.washingtonpost.com/wp-srv/national/longterm/meltingpot/melt0222.htm; and US Commission on Civil Rights, *Broken Promises: Continuing Federal Funding Shortfall for Native Americans* (Washington, DC: US Commission on Civil Rights, 2018), 2.

2. See Katarzyna Bryc et al. [Eric Y. Durand, J. Michael Macpherson, David Reich, and Joanna L. Mountain], "The Genetic Ancestry of African Americans, Latinos, and European Americans across the United States," *American Journal of Human Genetics* 96, no. 1 (2015): 44–45 [37–53], https://doi.org/10.1016/j.ajhg.2014.11.010. The **study relies on genetic testing customers; White Americans anxious about black family skeletons would be reluctant customers, depressing the count of "mixed" White Americans.**

3. See, e.g., William H. Frey, *Diversity Explosion: How New Racial Demographics Are Remaking America* (Washington, DC: Brookings Institution Press, 2014); and Anthony Daniel Perez and Charles Hirschman, "The Changing Racial and Ethnic Composition of the U.S. Population: Emerging American Identities," *Population and Development Review* 35, no. 1 (2009): 1–51, https://doi.org/10.1111/j.1728-4457.2009.00260.x.

4. Ronald R. Sundstrom, *The Browning of America and the Evasion of Social Justice* (New York: State University of New York Press, 2008), 51.

5. Jennifer Lee and Frank D. Bean, "Reinventing the Color Line: Immigration and America's New Racial/Ethnic Divide," *Social Forces* 86, no. 2 (2007): 561 [561–86], https://doi.org/10.1093/sf/86.2.561. See also E. Tammy Kim, "The Perils of 'People of Color,'" *New Yorker*, July 29, 2020, https://www.newyorker.com/news/annals-of-activism/the-perils-of-people-of-color. Compare, e.g., David Zahniser et al. [Julia Wick, Benjamin Oreskes, Dakota Smith, and Gustavo Arellano], "Racist Remarks in Leaked Audio of L.A. Council Members Spark Outrage, Disgust," *Los Angeles Times*, October 9, 2022, https://www.latimes.com/california/story/2022-10-09/city-council-leaked-audio-nury-martinez-kevin-de-leon-gil-cedillo.

6. See, e.g., Farah Stockman, "'We're Self-Interested': The Growing Identity Debate in Black America," *New York Times*, November 8, 2019, https://www.nytimes.com/2019/11/08/us/slavery-black-immigrants-ados.html. **Featured in the article is the ADOS (American Descendants of Slavery) grassroots movement that "prioritizes reparations for descendants of chattel slavery in the United States."** See ADOS Advocacy Foundation, "About Us," accessed September 30, 2021, https://adosfoundation.org/mission-statement.

7. **A version of "interest convergence" is suggested here—with Black Americans on one side and nonblack "people of color" on the other.** See Derrick A. Bell Jr., "*Brown v. Board of Education* and the Interest-Convergence Dilemma," *Harvard Law Review* 93, no. 3 (1980): 518–33, https://www.jstor.org/stable/1340546.

8. Claire Jean Kim, "The Trial of Peter Liang and Confronting the Reality of Asian American Privilege," *Los Angeles Times*, April 21, 2016, http://www.latimes.com/opinion/opinion-la/la-ol-peter-liang-asian-american-privilege-20160421-snap-story.html.

9. See, e.g., Claire Jean Kim, "For Chinese American Conservatives, Race Is a Weapon," *Nation*, October 8, 2019, https://www.thenation.com/article/archive/asian-conservatives-affirmative-action/.

10. See, e.g., Michael Omi and Howard Winant, *Racial Formation in the United States*, 3rd ed. (New York: Routledge, 2005), 21–23.

11. See US Census Bureau, "Measuring Race and Ethnicity Across the Decades: 1790–2010," under "1930," by Beverly M. Pratt, Lindsay Hixson, and Nicholas A. Jones, accessed October 28, 2021, https://www.census.gov/data-tools/demo/race/MREAD_1790_2010.html.

12. US Census Bureau, *Overview of Race and Hispanic Origin: 2010*, by Karen R. Humes, Nicholas A. Jones, and Roberto R. Ramirez (Washington, DC: U.S. Department of Commerce, 2011), 6 [1–23]. See also Nate Cohn, "More Hispanics Declaring

Themselves White," *New York Times*, May 21, 2014, https://www.nytimes.com/2014/05/22/upshot/more-hispanics-declaring-themselves-white.html. **On "Latin" in contrast to, by culture and religion, "Anglo-Saxon" European peoples in the Americas,** see Stella Maris Scatena Franco, "Latinos versus Anglo-Saxons: Identity Projections in the Accounts of Latin Americans Who Traveled to the United States in the Nineteenth Century," *Almanack*, no. 16 (2017): 80–81 [80–120], https://doi.org/10.1590/2236-463320171602. Compare, e.g., Jorge J. E. Gracia, *Surviving Race, Ethnicity, and Nationality: A Challenge for the 21st Century* (Lanham, MD: Rowman and Littlefield, 2005), 66–67.

13. See, e.g., G. Cristina Mora, *Making Hispanics: How Activists, Bureaucrats, and Media Constructed a New American* (Chicago: University of Chicago Press, 2014), xiii–xv, 98–101. See also Miguel Salazar, "The Problem with Latinidad," *Nation*, September 16, 2019, https://www.thenation.com/article/archive/hispanic-heritage-month-latinidad/.

14. Compare, e.g., Monica McDermott and Frank L. Samson, "White Racial and Ethnic Identity in the United States," *Annual Review of Sociology* 31 (2005): 245–61, https://doi.org/10.1146/annurev.soc.31.041304.122322. **On "nonwhite to white race change" that corresponds to switching from non-Republican to Republican voting, driven by persons "originally identifying as Mixed and Hispanic,"** see Alexander Agadjanian and Dean Lacy, "Changing Votes, Changing Identities? Racial Fluidity and Vote Switching in the 2012–2016 US Presidential Elections," *Public Opinion Quarterly* 85, no. 3 (2021): 737–52, https://doi.org/10.1093/poq/nfab045.

15. Noel Ignatiev, *How the Irish Became White* (New York: Routledge, 1995), 2. See also Kerby A. Miller, *Emigrants and Exiles: Ireland and the Irish Exodus to North America* (New York: Oxford University Press, 1985), 318.

16. Ignatiev, *How the Irish Became White*, 59. **Similarly on some "race" thing re the "unwhite" status of certain European immigrant peoples (whom the US government nevertheless admitted to the country and legally counted as "white persons"),** see Josh Zeitz, "The 1965 Law That Gave the Republican Party Its Race Problem," Politico, August 20, 2016, https://www.politico.com/magazine/story/2016/08/immigration-1965-law-donald-trump-gop-214179/. Compare, e.g., Jennifer Median, "How Immigration Politics Drives Some Hispanic Voters to the G.O.P. in Texas," *New York Times*, February 28, 2022, https://www.nytimes.com/2022/02/28/us/politics/border-grievance-politics.html.

17. See, e.g., Peter Vellon, "'Between White Men and Negroes': The Perception of Southern Italian Immigrants Through the Lens of Italian Lynchings," in *Anti-Italianism: Essays on a Prejudice*, ed. William J. Connell and Fred Gardaphé (New York: Palgrave Macmillan, 2010), 23–32.

18. See, e.g., Cheryl Lynn Greenberg, *Troubling the Waters: Black-Jewish Relations in the American Century* (Princeton: Princeton University Press, 2006), 15–16.

19. *The National Jewish Population Survey 2000-01: Strength, Challenge and Diversity in the American Jewish Population*, United Jewish Communities, September 2003, 16–19, 27 [1–32], https://www.jewishdatabank.org/content/upload/bjdb/NJPS2000_Strength_Challenge_and_Diversity_in_the_American_Jewish_Population.pdf.

20. On Jewish "ethnic revival" forged alongside a more active role in American white caste politics starting in the mid-1960s, see Eric L. Goldstein, *The Price of Whiteness: Jews, Race, and American Identity* (Princeton: Princeton University Press, 2006), 212–17.

21. See, e.g., Orlando Patterson, "Race by the Numbers," *New York Times*, May 8, 2001, https://www.nytimes.com/2001/05/08/opinion/race-by-the-numbers.html; Richard Alba, "The Likely Persistence of a White Majority," *American Prospect*, January 11, 2016, https://prospect.org/civil-rights/likely-persistence-white-majority/; and José Jorge Mendoza, "Latinx and the Future of Whiteness in American Democracy," *APA Newsletter on Hispanic/Latino Issues in Philosophy* 16, no. 2 (2017): 6–10, https://www.apaonline.org/resource/collection/60044C96-F3E0-4049-BC5A-271C673FA 1E5/HispanicV16n2.pdf. Compare Lee and Bean, "Reinventing the Color Line," 579–80.

22. US Census Bureau, "Explanation of Race and Hispanic Origin Categories," September 15, 1999, https://www2.census.gov/programs-surveys/popest/technical-documentat ion/file-layouts/1990-2000/90s-rh-doc.txt. Compare US Office of Management and Budget, "Revisions to the Standards for the Classification of Federal Data on Race and Ethnicity," *Federal Register* 62, no. 210 (October 30, 1997): 58782, https://www.govi nfo.gov/content/pkg/FR-1997-10-30/pdf/97-28653.pdf.

23. See, e.g., Lisa Deaderick, "Latina Professors Discuss Use of 'Negrito' and 'Negrita' in Latin Culture, after J.Lo Controversy," *San Diego Union-Tribune*, November 1, 2020, https://www.sandiegouniontribune.com/columnists/story/2020-11-01/latina-pro fessors-discuss-use-of-negrito-negrita-in-latin-culture-after-j-lo-controversy.

24. By contrast, Brazil (like Argentina re its "black" population) employed the "whitening thesis," whereby "an influx of European immigrants and their mating with Afro-Brazilians would eventually whiten the population"—which Afro-Brazilian critics decried as "'policies of forced miscegenation [to] make the Negro race disappear'" locally. See George Reid Andrews, "Why the Afro-Argentines Disappeared," in *The Afro-Argentines of Buenos Aires, 1800–1900* (Madison: University of Wisconsin Press, 1980), 111 [93–112].

25. At the level of individuals, comparable or higher socioeconomic status can partially offset being of lower color caste. See, e.g., Lewis F. Carter, "Racial-Caste Hypogamy: A Sociological Myth?," *Phylon* 29, no. 4 (1968): 350 [347–50], https://doi.org/10.2307/274016; and Florencia Torche and Peter Rich, "Declining Racial Stratification in Marriage Choices? Trends in Black/White Status Exchange in the United States, 1980–2010," *Sociology of Race and Ethnicity* 3, no. 1 (2016): 31–49, https://doi.org/10.1177/2332649216648464.

Section 6

1. Naomi Zack, *Thinking about Race*, 2nd ed. (Belmont, CA: Thomson Wadsworth, 2006), 25.

2. Compare, on the American judiciary's "two common pitfalls" of "asymmetric racial recognition and gratuitous racial recognition," Justin Driver, "Recognizing Race," *Columbia Law Review* 112 (2012): 404 [404–57], https://www.jstor.org/stable/41354775.

3. Zack, *Thinking about Race*, 27.

4. Virginia Domínguez, *White by Definition: Social Classification in Creole Louisiana* (New Brunswick: Rutgers University Press, 1986), 5. **For a rare official dispute about an individual's alleged black identity,** see Susan Diesenhouse, "Boston Case Raises Questions on Misuse of Affirmative Action," *New York Times*, October 9, 1988, http://www.nytimes.com/1988/10/09/us/boston-case-raises-questions-on-misuse-of-affirmative-action.html.

5. *Loving v. Virginia* 388 U.S. 1, 11–12 (1967). Compare, e.g., Dorothy E. Roberts, "*Loving v. Virginia* as a Civil Rights Decision," *New York Law School Law Review* 59 (2014–15): 176–78 [175–209], https://digitalcommons.nyls.edu/nyls_law_review/vol59/iss1/8/.

6. Cheryl I. Harris, "Whiteness as Property," *Harvard Law Review* 106, no. 8 (1993): 1734 [1707–91], https://www.jstor.org/stable/1341787. Compare, e.g., "Judges Rule Rights Laws Safeguard Whites Equally," *New York Times*, June 26, 1976, https://www.nytimes.com/1976/06/26/archives/judges-rule-rights-laws-safeguard-whites-equally-cite-2-statutes-as.html.

7. Harris, "Whiteness as Property," 1735–36. See also Charles Mills, "White Right: The Idea of a Herrenvolk Ethics," in *Blackness Visible: Essays on Philosophy and Race* (Ithaca, NY: Cornell University Press, 1998), 154–55, 164–65 [139–66].

8. See, e.g., August H. Nimtz, "Violence and/or Nonviolence in the Success of the Civil Rights Movement: The Malcolm X–Martin Luther King, Jr. Nexus," *New Political Science* 38, no. 1 (2016): 1–22, https://doi.org/10.1080/07393148.2015.1125116; and Peniel E. Joseph, *The Sword and the Shield: The Revolutionary Lives of Malcolm X and Martin Luther King Jr.* (New York: Basic Books, 2020), 312–13.

9. See, e.g., Vann R. Newkirk II, "The Whitewashing of King's Assassination," *Atlantic*, February 2018, https://www.theatlantic.com/magazine/archive/2018/02/how-to-kill-a-revolution/552518/.

10. See, e.g., Richard Rothstein, *The Color of Law: A Forgotten History of How Our Government Segregated America* (New York: Liveright, 2017), 81–83; and Michael Jones-Correa, "The Origins and Diffusion of Racial Restrictive Covenants," *Political Science Quarterly* 115, no. 4 (2000–2001): 544, 559 [541–68], https://doi.org/10.2307/2657609.

11. **The law at issue was the Massachusetts Racial Imbalance Act of 1965.** See, e.g., Matthew F. Delmont, *Why Busing Failed: Race, Media, and the National Resistance to School Desegregation* (Oakland: University of California Press, 2016), 78–83; and Ronald P. Formisano, *Boston Against Busing: Race, Class, and Ethnicity in the 1960s and 1970s* (Chapel Hill: University of North Carolina Press, 1991), 44–47.

12. On the **"inaccurate conflation of [Martin Luther] King's activism with the ideology of colorblindness [that] began in earnest during the Reagan administration,"** see Justin Gomer and Christopher Petrella, "Reagan Used MLK Day to Undermine

Racial Justice," *Boston Review*, January 15, 2017, http://bostonreview.net/race-polit ics/justin-gomer-christopher-petrella-reagan-used-mlk-day-undermine-racial-just ice. See also Tim Naftali, "Ronald Reagan's Long-Hidden Racist Conversation with Richard Nixon," *Atlantic*, July 30, 2019, https://www.theatlantic.com/ideas/archive/ 2019/07/ronald-reagans-racist-conversation-richard-nixon/595102/.

13. Compare, e.g., *Regents of the Univ. of California v. Bakke*, 438 U.S. 265 (1978); and *Grutter v. Bollinger*, 539 U.S. 306 (2003).

14. See, e.g., Frank Newport, "Reparations and Black Americans' Attitudes about Race," Gallup.com, March 1, 2019, https://news.gallup.com/opinion/polling-matters/247 178/reparations-black-americans-attitudes-race.aspx.

15. Steven A. Holmes, "People Can Claim One or More Races on Federal Forms," *New York Times*, October 30, 1997, http://www.nytimes.com/1997/10/30/us/people-can-claim-one-or-more-races-on-federal-forms.html.

16. See, e.g., John Rawls, *A Theory of Justice*, rev. ed. (Cambridge, MA: Harvard University Press, 1999), 3–19. See also Iris Marion Young, "The Ideal of Impartiality and the Civic Public," in *Justice and the Politics of Difference* (Princeton: Princeton, NJ 1990), 112 [96–121].

17. Naomi Zack, "The Fluid Symbol of Mixed Race," *Hypatia* 25, no. 4 (2010): 886 [875–90], https://doi.org/10.1111/j.1527-2001.2010.01121.x.

18. Zack, *Thinking about Race*, 22.

19. See, e.g., Rothstein, *Color of Law*, 107–9.

20. See, e.g., "Realness," Ali G interviews Andy Rooney, 02:25–03:02, *Da Ali G Show* (TV series), HBO, August 22, 2004.

21. **On "modes of blackness" (e.g., racial, cultural, political) characteristic of Black American social identity,** see Lionel K. McPherson and Tommie Shelby, "Blackness and Blood: Interpreting African American Identity," *Philosophy and Public Affairs* 32, no. 2 (2004): 176–77 [171–92], https://doi.org/10.1111/j.1088-4963.2004.00010.x.

22. John Brown, *Address of John Brown to the Virginia Court* (Boston: C. C. Mead, 1859), https://www.loc.gov/item/rbpe.06500500/.

23. W. E. B. Du Bois, *John Brown* (Philadelphia: George W. Jacobs, 1909), 7–8.

24. See Barbara Foley, *Spectres of 1919: Class and Nation in the Making of the New Negro* (Champaign: University of Illinois Press, 2003), 32.

25. Frederick Douglass, "A Lecture on John Brown, Delivered at Harper's Ferry and Sundry Other Places," 1860, 9 [1–34], http://hdl.loc.gov/loc.mss/ms000009.mss11 879.00392.

26. **For an unexplained criterion of "primary ancestry in sub-Saharan Africa" for racial "Africans" (as in blacks/Negroes),** see Neil Risch et al. [Esteban Burchard, Elad Ziv, and Hua Tang], "Categorization of Humans in Biomedical Research: Genes, Race and Disease," *Genome Biology* 3, no. 7 (2002): comment2007, https://doi.org/10.1186/gb-2002-3-7-comment2007.

27. See, e.g., Peggy Pascoe, *What Comes Naturally: Miscegenation Law and the Making of Race in America* (New York: Oxford University Press, 2009), 27; and US Census Bureau, *Negro Population 1790–1915* (Washington, DC: Government Printing Office, 1918), 209, https://www.census.gov/library/publications/1918/dec/negro-populat ion-1790-1915.html.

Section 7

1. See, e.g., Kim Parker et al. [Juliana Menasce Horowitz, Rich Morin, and Mark Hugo Lopez], "Multiracial in America: Proud, Diverse and Growing in Numbers," Pew Research Center, June 11, 2015, https://www.pewsocialtrends.org/2015/06/11/mult iracial-in-america/.

2. Ronald R. Sundstrom, *The Browning of America and the Evasion of Social Justice* (New York: State University of New York Press, 2008), 113.

3. Compare *Lawrence v. Texas*, 539 U.S. 558 (2003).

4. See, e.g., Susan Saulny and Jacques Steinberg, "On College Forms, a Question of Race, or Races, Can Perplex," *New York Times*, June 13, 2011, https://www.nytimes.com/2011/06/14/us/14admissions.html. Compare Adrian Piper, "Passing for White, Passing for Black," *Transition* 58 (1992): 8–9 [4–32], https://doi.org/10.2307/2934966. **Piper dismisses the prospect that persons of mixed African and European descent would choose a black racial identity in order to gain a possible "affirmative action" advantage—a tactical choice that multiracials in the *Times* article admit to making.**

5. See, e.g., Jon Michael Spencer, *The New Colored People: The Mixed-Race Movement in America*, foreword Richard E. van der Ross (New York: New York University Press, 1997), 4–5.

6. **For comparison, on how amount of African ancestry shapes racial identity nationally and regionally in Brazil,** see Dóra Chor, Alexandre Pereira, and Antonio G. Pacheco et al., "Context-Dependence of Race Self-Classification: Results from a Highly Mixed and Unequal Middle-Income Country," *PLOS One* 14, no. 5 (2019): e0216653, https://doi.org/10.1371/journal.pone.0216653.

7. See, e.g., Ismail K. White and Chryl N. Laird, *Steadfast Democrats: How Social Forces Shape Black Political Behavior* (Princeton: Princeton University Press, 2020); and Michael C. Dawson, *Behind the Mule: Race and Class in African-American Politics* (Princeton: Princeton University Press, 1994). **For the "Southern Strategy" of white caste politics,** see Kevin P. Phillips, *The Emerging Republican Majority* (New Rochelle, NY: Arlington House, 1969); and Rick Perlstein, "Exclusive: Lee Atwater's Infamous 1981 Interview on the Southern Strategy," *Nation*, November 13, 2012, https://www.thenation.com/article/archive/exclusive-lee-atwaters-infamous-1981-interview-southern-strategy/.

8. Compare, e.g., Spencer, *New Colored People*, xi–xii.

9. Sundstrom, *Browning of America*, 128.

10. Naomi Zack, *Thinking about Race*, 2nd ed. (Belmont, CA: Thomson Wadsworth, 2006), 27.

11. Zack, *Thinking about Race*, 27.

12. Compare, e.g., Hope Reese, "Chimamanda Ngozi Adichie: I Became Black in America," *JSTOR Daily*, August 29, 2018, https://daily.jstor.org/chimamanda-ngozi-adichie-i-became-black-in-america/. **Adichie would have us imagine that Africans who immigrate to the United States might not grasp that "black" is supposed to be a race designation categorically meant to include Black Americans, Afro-Caribbeans, and non-northern Africans. Yet African immigrants are keenly aware of the bottom caste status of descendants of American slavery.** See, e.g.,

Adaobi Tricia Nwaubani, "When the Slave Traders Were African," *Wall Street Journal*, September 20, 2019, https://www.wsj.com/articles/when-the-slave-traders-were-afri can-11568991595.

13. Sundstrom, *Browning of America*, 129.

14. Compare, e.g., Carrie B. Dohe, *Jung's Wandering Archetype: Race and Religion in Analytical Psychology* (London: Routledge, 2016), 75–76.

15. See, e.g., Linda Martín Alcoff, *Visible Identities: Race, Gender, and the Self* (New York: Oxford University Press, 2006), 278–79.

16. Naomi Zack, *Race and Mixed Race* (Philadelphia: Temple University Press, 1993), 4.

17. Linda Alcoff, "Mestizo Identity," in *American Mixed Race: The Culture of Microdiversity*, ed. Naomi Zack (Lanham, MD: Rowman and Littlefield, 1995), 258–59 [257–78].

18. Michael Hanchard, "Black Cinderella? Race and the Public Sphere in Brazil," *Public Culture* 7, no. 1 (1994): 177 [165–85], https://doi.org/10.1215/08992363-7-1-165. See also Edward E. Telles, "Racial Classification," in *Race in Another America: The Significance of Skin Color in Brazil* (Princeton: Princeton University Press, 2004), 78–106.

19. Alcoff, "Mestizo Identity," 277. **For contemporary controversy about visible black-ness under Brazil's multiracial classification scheme (e.g., "These [affirmative ac-tion] spots are for people who are phenotypically black"),** see Cleuci de Oliveira, "Brazil's New Problem with Blackness," *Foreign Policy*, April 5, 2017, https://foreig npolicy.com/2017/04/05/brazils-new-problem-with-blackness-affirmative-action/.

20. Hanchard, "Black Cinderella," 177.

21. Alcoff, "Mestizo Identity," 277.

22. W. E. B. Du Bois, *The Souls of Black Folk* (Chicago: A. C. McClurg, 1903), 3.

23. See, e.g., Jennifer L. Morgan, "*Partus sequitur ventrem*: Law, Race, and Reproduction in Colonial Slavery," *Small Axe* 22, no. 1 (2018): 1–17, https://doi.org/10.1215/07990 537-4378888.

24. Alcoff, "Mestizo Identity," 277–78.

25. **On "the Hispanic portion of the White population" as equivalent to "the White portion of the Hispanic population,"** see US Census Bureau, "Explanation of Race and Hispanic Origin Categories," September 15, 1999, https://www2.census.gov/ programs-surveys/popest/technical-documentation/file-layouts/1990-2000/90s-rh-doc.txt.

26. L. Z. Granderson, "The Next Great Black Tennis Player Isn't Black or White—She's Madison Keys," *Undefeated*, July 4, 2016, https://theundefeated.com/features/the-next-great-black-tennis-player-isnt-black-or-white-shes-madison-keys/.

27. **For psychoanalytic and field study of unstable racial psychologies in reaction to anti-black stereotypes and prejudices,** see Frantz Fanon, *Black Skin, White Masks*, trans. Charles Lam Markmann (New York: Grove Press, 1967).

28. Naomi Zack, "Life after Race," in *American Mixed Race: The Culture of Microdiversity*, ed. Naomi Zack (Lanham, MD: Rowman and Littlefield, 1995), 301 [297–307].

Section 8

1. See, e.g., Winfred Rembert, *Chasing Me to My Grave: An Artist's Memoir of the Jim Crow South*, as told to Erin I. Kelly, foreword Bryan Stevenson (New York: Bloomsbury Press, 2021), 181.

2. On speculation that "most or all African-Americans suffer from a debilitating form of alienation that causes them to be estranged and divided" from themselves, see Howard McGary, *Race and Social Justice* (Malden, MA: Blackwell, 1999), 19. See also Sandra Graham, "Motivation in African Americans," *Review of Educational Research* 64, no. 1 (1994): 103–4 [55–117], https://doi.org/10.3102/00346543064001055.

3. Toni Morrison, "A Humanist View" (speech, Portland State University, May 30, 1975), Portland State Library Special Collections, 33:40–34:08 [06:50–43:01], accessed February 14, 2021, https://soundcloud.com/portland-state-library/portland-state-black-studies-1.

4. See US Census Bureau, "History: Through the Decades: Questionnaires," 1790 to 1880 forms, November 14, 2021, https://www.census.gov/history/www/through_the_decades/questionnaires/.

5. See, e.g., Lionel K. McPherson and Tommie Shelby, "Blackness and Blood: Interpreting African American Identity," *Philosophy and Public Affairs* 32, no. 2 (2004): 183 [171–92], https://doi.org/10.1111/j.1088-4963.2004.00010.x.

6. Naomi Zack, *Race and Mixed Race* (Philadelphia: Temple University Press, 1993), 141. Compare, e.g., E. B. Reuter, "The Superiority of the Mulatto," *American Journal of Sociology* 23, no. 1 (1917): 83–106, https://doi.org/10.1086/212720; and Willard B. Gatewood Jr., "Aristocrats of Color: South and North The Black Elite, 1880–1920," *Journal of Southern History* 54, no. 1 (1988): 3–20, https://doi.org/10.2307/2208518.

7. See Larry E. Tise, *The American Counterrevolution: A Retreat from Liberty, 1783–1800* (Mechanicsburg, PA: Stackpole Books, 1998), 220.

8. On a distinctively American "sense of racial pride and identity [that] would not prevent ['most black leaders'] from seeking cultural assimilation and economic and political integration in the United States" in the late nineteenth century, see Wilson Jeremiah Moses, *The Golden Age of Black Nationalism, 1850–1925* (New York: Oxford University Press, 1978), 55. See also Ben L. Martin, "From Negro to Black to African American: The Power of Names and Naming," *Political Science Quarterly* 106, no. 1 (1991): 90–92 [83–107], https://doi.org/10.2307/2152175.

9. Zack, *Race and Mixed Race*, 143.

10. Self-reported "race" on census forms is a recent phenomenon: "It was not until 1960 that people could select their own race. Prior to that, an individual's race [or 'color'] was determined by census takers, known as enumerators." Anna Brown, "The Changing Categories the U.S. Census Has Used to Measure Race," Pew Research Center, February 25, 2020, https://www.pewresearch.org/fact-tank/2020/02/25/the-changing-categories-the-u-s-has-used-to-measure-race/.

11. Compare, on "the profound loss of self and the erasure of 'true history' that permanent passing entailed," Allyson Hobbs, *A Chosen Exile: A History of Racial Passing in American Life* (Cambridge, MA: Harvard University Press, 2014), 156.

12. Post-WWII America appears to be an outlier for Western race science intrigue. Compare: "There was an emerging consensus among both liberals and Afrikaner nationalist intellectuals on the centrality of culture rather than race in South African debate.... 'Scientific' arguments of race difference had a diminishing impact on the nature of this debate.... [N]ational socialism in Germany made scientific racism increasingly unacceptable intellectually." Paul Rich, "Race, Science, and the Legitimization of White Supremacy in South Africa, 1902–1940," *International Journal of African Historical Studies* 23, no. 4 (1990): 686 [665–86], https://doi.org/10.2307/219503.

13. Naomi Zack, "Life after Race," in *American Mixed Race: The Culture of Microdiversity*, ed. Naomi Zack (Lanham, MD: Rowman and Littlefield, 1995), 299 [297–307].

14. See, e.g., Emma Green, "The Trouble with Wearing Turbans in America," *Atlantic*, January 27, 2015, https://www.theatlantic.com/politics/archive/2015/01/the-trouble-with-wearing-turbans-in-america/384832/.

15. Ronald R. Sundstrom, *The Browning of America and the Evasion of Social Justice* (New York: State University of New York Press, 2008), 114.

16. Sundstrom, *Browning of America*, 125.

17. Zack, *Race and Mixed Race*, 75.

18. Sundstrom, *Browning of America*, 116. Compare, e.g., Trina Jones, "Shades of Brown: The Law of Skin Color," *Duke Law Journal* 49, no. 6 (2000): 1487–1557, https://doi.org/10.2307/1373052.

19. See, e.g., Megan E. Williams, "The *Crisis* Cover Girl: Lena Horne, the NAACP, and Representations of African American Femininity, 1941–1945," *American Periodicals* 16, no. 2 (2006): 202–3, [200–218], https://doi.org/10.1353/amp.2006.0019.

20. Adrian Piper, "Passing for White, Passing for Black," *Transition* 58 (1992): 6 [4–32], https://doi.org/10.2307/2934966.

21. Adrian Piper, "Thwarted Projects, Dashed Hopes, A Moment of Embarrassment," 2012, APRA Foundation, Berlin, accessed July 10, 2020, http://www.adrianpiper.com/news_sep_2012.shtml.

Section 9

1. See US Census Bureau, "Measuring Race and Ethnicity across the Decades: 1790–2010," by Beverly M. Pratt, Lindsay Hixson, and Nicholas A. Jones, accessed October 28, 2021, https://www.census.gov/data-tools/demo/race/MREAD_1790_2010.html. Compare, e.g., Edward Telles, "Mixed and Unequal: New Perspectives on Brazilian Ethnoracial Relations," in *Pigmentocracies: Ethnicity, Race, and Color in Latin America* (Chapel Hill: University of North Carolina Press, 2014), 172–217.

2. See US Census Bureau, "History: Through the Decades: Questionnaires," 1900 form, accessed November 14, 2021, https://www.census.gov/history/www/through_the_deca des/questionnaires/.

3. See, e.g., Angela R. Dixon and Edward E. Telles, "Skin Color and Colorism: Global Research, Concepts, and Measurement," *Annual Review of Sociology* 43 (2017): 408–09 [405–24], https://doi.org/10.1146/annurev-soc-060116-053315; and E. Franklin Frazier, *Black Bourgeoisie* (New York: Free Press, 1957), 20.

4. See Plato, *The Republic*, trans. G. M. A. Grube, 2nd ed. (Indianapolis: Hackett, 1992), Bk. 3, 414c–415d (91–92).

5. Edward E. Telles, *Race in Another America: The Significance of Skin Color in Brazil* (Princeton: Princeton University Press, 2004), 79, 171.

6. See Orlando Patterson, *Slavery and Social Death: A Comparative Study* (Cambridge, MA: Harvard University Press, 1982), 13.

7. See, e.g., Naomi Zack, *Race and Mixed Race* (Philadelphia: Temple University Press, 1993), 34–35.

8. Zack, *Race and Mixed Race*, 65.

9. Compare, e.g., Adrienne D. Davis, "The Private Law of Race and Sex: An Antebellum Perspective," *Stanford Law Review* 51, no. 2 (1999): 221–88, https://doi.org/10.2307/1229269; and Kia Shant'e Breaux, "Sally Hemings Family Rights Unclear," Associated Press, February 12, 2000, https://apnews.com/article/433d8378c6d48403169296dc1 d17c29d.

10. Compare, e.g., Harriet Jacobs, *Incidents in the Life of a Slave Girl. Written by Herself*, ed. L. Maria Child (Boston: 1861), 55–57; and Michael Kilian, "Jefferson's Heirs Reject Hemings' Kin," *Chicago Tribune*, May 6, 2002, https://www.chicagotribune.com/news/ct-xpm-2002-05-06-0205060232-story.html.

11. See, e.g., Lionel K. McPherson, "The Moral Insignificance of 'Bare' Personal Reasons," *Philosophical Studies* 10, no. 1 (2002): 29–47, https://doi.org/10.1023/A:1019864618 762. Compare, e.g., Kimberly Leighton, "Addressing the Harms of Not Knowing One's Heredity: Lessons from Genealogical Bewilderment," *Adoption and Culture* 3 (2012), 63–107, https://www.jstor.org/stable/10.26818/adoptionculture.3.2012.0063.

12. On the role of "fictive, or quasi, kin [in] *binding unrelated adults to one another and thereby infusing enlarged slave communities with conceptions of obligation that had flowed initially from kin obligations rooted in blood and marriage*," see Herbert G. Gutman, *The Black Family in Slavery and Freedom, 1750–1925* (New York: Pantheon Books, 1976), 220. See also Libra R. Hilde, *Slavery, Fatherhood, and Paternal Duty in African American Communities over the Long Nineteenth Century* (Chapel Hill: University of North Carolina Press, 2020), 74–75.

13. See Naomi Zack, "Life after Race," in *American Mixed Race: The Culture of Microdiversity*, ed. Naomi Zack (Lanham, MD: Rowman and Littlefield, 1995), 299 [297–307].

14. Ronald R. Sundstrom, *The Browning of America and the Evasion of Social Justice* (New York: State University of New York Press, 2008), 128.

15. Sundstrom, *Browning of America*, 129. Compare, e.g., Kim M. Williams, *Mark One or More: Civil Rights in Multiracial America* (Ann Arbor: University of Michigan Press, 2006), 85–86.

16. Sundstrom, *Browning of America*, 130.

Section 10

1. See Thomas D. Morris, *Southern Slavery and the Law, 1619–1860* (Chapel Hill: University of North Carolina Press, 1996), 44–45. See also Taunya Lovell Banks, "Dangerous Woman: Elizabeth Key's Freedom Suit—Subjecthood and Racialized Identity in Seventeenth Century Colonial Virginia," *Akron Law Review* 41, no. 3 (2008): 812–15 [799–837], http://dx.doi.org/10.2139/ssrn.672121.

2. See US Census Bureau, "Measuring Race and Ethnicity Across the Decades: 1790–2010," under "1900," by Beverly M. Pratt, Lindsay Hixson, and Nicholas A. Jones, accessed October 28, 2021, https://www.census.gov/data-tools/demo/race/MREA D_1790_2010.html.

3. See, e.g., "Jefferson's Attitudes Toward Slavery," Thomas Jefferson Foundation, accessed November 7, 2021, https://www.monticello.org/thomas-jefferson/jefferson-slavery/jefferson-s-attitudes-toward-slavery/; and Paul Finkelman, Slavery and the Founders: Race and Liberty in the Age of Jefferson, 3rd ed. (New York: Routledge, 2014).

4. See, e.g., W. E. B. Du Bois, *Black Reconstruction in America, 1860–1880* (New York: Russell and Russell, 1935); and Eric Foner, *Reconstruction: America's Unfinished Revolution, 1863–1877* (New York: Harper and Row, 1988).

Part III. Section 1

* Martin Luther King Jr., "Racism and the White Backlash," in *Where Do We Go from Here: Chaos or Community?* (New York: Harper and Row, 1967), 95 [67–101].

1. See, e.g., Steven J. Micheletti et al. [Kasia Bryc, Samantha G. Ancona Esselmann, William A. Freyman, Meghan E. Moreno, G. David Poznik, and Anjali J. Shastri], "Genetic Consequences of the Transatlantic Slave Trade in the Americas," *American Journal of Human Genetics* 107 (2020): 270 [265–77], https://doi.org/10.1016/j.ajhg.2020.06.012; and Christine Tamir, "Key Findings about Black Immigrants in the U.S.," Pew Research Center, January 27, 2022, https://www.pewresearch.org/fact-tank/2022/01/27/key-findings-about-black-immigrants-in-the-u-s/.

2. See US Census Bureau, "Measuring Race and Ethnicity Across the Decades: 1790–2010," by Beverly M. Pratt, Lindsay Hixson, and Nicholas A. Jones, accessed October 28, 2021, https://www.census.gov/data-tools/demo/race/MREAD_1790_2010.html; and US Census Bureau, "History: Through the Decades: Questionnaires," accessed November 14, 2021, https://www.census.gov/history/www/through_the_decades/questionnaires/.

3. See, e.g., Calvin Schermerhorn, "Why the Racial Wealth Gap Persists, More Than 150 Years After Emancipation," *Washington Post*, June 19, 2019, https://www.washing

tonpost.com/outlook/2019/06/19/why-racial-wealth-gap-persists-more-than-years-after-emancipation/; and Mehrsa Baradaran, *The Color of Money: Black Banks and the Racial Wealth Gap* (Cambridge, MA: Harvard University Press, 2017), 8–9. See also Eleanor Brown and June Carbone, "Race, Property, and Citizenship," *Northwestern University Law Review* 116 (2021): 120–47, https://scholarlycommons.law.northwestern.edu/nulr_online/314; and Dania V. Francis et al. [Darrick Hamilton, Thomas W. Mitchell, Nathan A. Rosenberg, and Bryce Wilson Stucki], "Black Land Loss: 1920–1997," *AEA Papers and Proceedings* 112 (2022): 38–42, https://doi.org/10.1257/pandp.20221015.

4. See James McCune Smith, "Introduction," in Frederick Douglass, *My Bondage and My Freedom*, intro. James McCune Smith (New York: Miller, Orton and Mulligan, 1855), xxiii [xvii–xxxi]. For earlier comparison between American caste and Hindu caste, see American Anti-Slavery Society, *Caste* (New York: R. G. Williams, ~1839].

5. Daniel Kilbride, "What Did Africa Mean to Frederick Douglass?," *Slavery and Abolition* 26, no. 1 (2015): 40 [40–62], https://doi.org/10.1080/0144039X.2014.916516.

6. Compare, e.g., Tunde Adeleke, "Black Americans and Africa: A Critique of the Pan-African and Identity Paradigms," *International Journal of African Historical Studies* 31, no. 3 (1998): 506–7 [505–36], https://doi.org/10.2307/221474; "The History of Pan-Africanism," *New Internationalist*, August 5, 2000, https://newint.org/features/2000/08/05/simply; and Hakim Adi, *Pan-Africanism: A History* (London: Bloomsbury Academic, 2018), 19.

7. **On the importance of disaggregating "blacks" in America by national origin re social and economic outcomes,** see Tod G. Hamilton, *Immigration and the Remaking of Black America*, forewword Douglas S. Massey (New York: Russell Sage Foundation, 2019), 16–17. See also Eleanor Marie Lawrence Brown, "An Alternative View of Immigrant Exceptionalism, Particularly as It Relates to Blacks: A Response to Chua and Rubenfeld," *California Law Review* 103, no. 4 (2015): 1013–16 [989–1017], http://www.jstor.org/stable/24758493. Compare, e.g., Onoso Imoagene, *Beyond Expectations: Second-Generation Nigerians in the United States and Britain* (Oakland: University of California Press, 2017), 3.

8. See, e.g., Juan Williams, "Obama's Color Line," *New York Times*, November 30, 2007, https://www.nytimes.com/2007/11/30/opinion/30williams.html.

9. Compare, e.g., Gene A. Fisher and Suzanne Model, "Cape Verdean Identity in a Land of Black and White," *Ethnicities* 12, no. 3 (2012): 354–79, https://doi.org/10.1177/1468796811419599. **On the finding that "white people [in America] tend to prefer and give better opportunities to Afro-Caribbeans over African-Americans,"** see Lauretta Charlton, "Study Examines Why Black Americans Remain Scarce in Executive Suites," *New York Times*, December 9, 2019, https://www.nytimes.com/2019/12/09/us/black-in-corporate-america-report.html. **On "African foreigner privilege,"** see Mukoma Wa Ngugi, "African in America or African American?," *Guardian*, January 14, 2011, https://www.theguardian.com/commentisfree/cifamerica/2011/jan/13/race-kenya.

10. See, e.g., Carleton S. Coon, *The Origin of Races* (New York: Knopf, 1962), 3–7; and Ernst Mayr, "Origin of the Human Races," *Science* 138, no. 3538 (1962): 420–22, https://doi.org/10.1126/science.138.3538.420. Compare, e.g., Terence Keel, *Divine Variations: How Christian Thought Became Racial Science* (Stanford: Stanford University Press, 2018), 13.

11. See, e.g., Quayshawn Spencer, "How to Be a Biological Racial Realist," in Joshua Glasgow, Sally Haslanger, Chike Jeffers, and Quayshawn Spencer, *What Is Race?: Four Philosophical Views* (New York: Oxford University Press, 2019), 73–110. **The US Census Bureau, which "must adhere to the 1997 Office of Management and Budget (OMB) standards on race," claims that the categories "reflect a social definition of race recognized in this country and not an attempt to define race biologically." Yet the OMB guideline for "White" suggests otherwise: "A person having origins in any of the original peoples of Europe, the Middle East, or North Africa." (Of course, "the Middle East" is in Asia, and "North Africa" is in Africa.)** See US Census Bureau, "About the Topic of Race," accessed August 28, 2021, https://www.census.gov/topics/population/race/about.html.

12. See, e.g., Raj Chetty et al. [Nathaniel Hendren, Maggie R. Jones, and Sonya R. Porter], "Race and Economic Opportunity in the United States: An Intergenerational Perspective," *Quarterly Journal of Economics* 135, no. 2 (2020): 711–83, https://doi.org/10.1093/qje/qjz042; Ashley Nellis, *The Color of Justice: Racial and Ethnic Disparity in State Prisons*, Sentencing Project, October 13, 2021, https://www.sentencingproject.org/wp-content/uploads/2016/06/The-Color-of-Justice-Racial-and-Ethnic-Disparity-in-State-Prisons.pdf; and Emily Widra and Tiana Herring, *States of Incarceration: The Global Context 2021*, Prison Policy Initiative, September 2021, https://www.prisonpolicy.org/global/2021.html.

13. Compare, e.g., Jesse Washington (Associated Press), "Some Blacks Insist: 'I'm Not African-American,'" NBCNews.com, February 5, 2012, http://www.nbcnews.com/id/46264191/ns/us_news-life/t/some-blacks-insist-im-not-african-american/; and Keishel Williams, "For Some Children of Immigrants, 'African American' Doesn't Fit Their Unique, Black Experiences in the US," Insider, July 16, 2021, https://www.insider.com/children-of-immigrants-detail-struggles-with-blackness-in-america-2021-7.

14. See Jim Crow Museum of Racist Memorabilia (also online), Ferris State University, Big Rapids, MI, https://www.ferris.edu/HTMLS/news/jimcrow/index.htm.

15. Compare, e.g., Michael O'Donnell, "How LBJ Saved the Civil Rights Act," *Atlantic*, April 2014, https://www.theatlantic.com/magazine/archive/2014/04/what-the-hells-the-presidency-for/358630/.

16. See, e.g., Equal Justice Initiative, "How Segregation Survived," in *Segregation in America* (Montgomery, AL: Equal Justice Initiative, 2018), https://segregationinamerica.eji.org/report/how-segregation-survived.html; Richard Rothstein, *The Color of Law: A Forgotten History of How Our Government Segregated America* (New York: Liveright, 2017); Matthew Bloch, Amanda Cox, and Tom Giratikanon, "Mapping Segregation," *New York Times*, July 8, 2015, https://www.nytimes.com/interactive/2015/07/08/us/census-race-map.html; Eduardo Bonilla-Silva and David G. Embrick, "'Every Place Has a Ghetto . . .': The Significance of Whites' Social and Residential Segregation," *Symbolic Interaction* 30, no. 3 (2007): 323–45, https://doi.org/10.1525/si.2007.30.3.323; and Henry W. McGee Jr., "Afro-American Resistance to Gentrification and the Demise of Integrationist Ideology in the United States," *Urban Lawyer* 23, no. 1 (1991): 25–44, http://www.jstor.org/stable/27894698.

17. Christine Percheski and Christina Gibson-Davis, "A Penny on the Dollar: Racial Inequalities in Wealth among Households with Children," *Socius* 6 (2020): 1 [1–17], https://doi.org/10.1177/2378023120916616.

18. Emily Badger et al. [Claire Cain Miller, Adam Pearce, and Kevin Quealy], "Extensive Data Shows Punishing Reach of Racism for Black Boys," *New York Times*, March 19, 2018, https://www.nytimes.com/interactive/2018/03/19/upshot/race-class-white-and-black-men.html. See also Raj Chetty et al. [Nathaniel Hendren, Frina Lin, Jeremy Majerovitz, and Benjamin Scuderi], "Childhood Environment and Gender Gaps in Adulthood," *American Economic Review* 106, no. 5 (2016): 282–88, http://dx.doi.org/10.1257/aer.p20161073; Elizabeth Hinton, LeShae Henderson, and Cindy Reed, *An Unjust Burden: The Disparate Treatment of Black Americans in the Criminal Justice System*, Vera Institute of Justice, May 2018, 1–20, https://www.vera.org/downloads/publications/for-the-record-unjust-burden-racial-disparities.pdf; and Jugal K. Patel et al. [Tim Arango, Anjali Singhvi and Jon Huang], "Black, Homeless and Burdened by L.A.'s Legacy of Racism," *New York Times*, December 22, 2019, https://www.nytimes.com/interactive/2019/12/22/us/los-angeles-homeless-black-residents.html.

19. **For radical feminist critique of how the American women's movement has targeted Black males,** see Alison Edwards, *Rape, Racism, and the White Women's Movement* (Chicago: Sojourner Truth Organization, 1979). See also Tommy J. Curry, "Decolonizing the Intersection: Black Male Studies as a Critique of Intersectionality's Indebtedness to Subculture of Violence Theory," in *Critical Psychology Praxis: Psychosocial Non-Alignment to Modernity/Coloniality*, ed. Robert K. Beshara (New York: Routledge, 2021), 132–54; Stephanie E. Jones-Rogers, *They Were Her Property: White Women as Slave Owners in the American South* (New Haven: Yale University Press, 2019), 130–31; and Bharti Khurana et al. [Denise A. Hines, Benjamin A. Johnson, Elizabeth A. Bates, Nicola Graham-Kevan, and Randall T. Loder], "Injury Patterns and Associated Demographics of Intimate Partner Violence in Men Presenting to U.S. Emergency Departments," *Aggressive Behavior* 48, no. 3 (2022): 298–308, https://doi.org/10.1002/ab.22007. Compare, e.g., Kimberle Crenshaw, "Demarginalizing the Intersection of Race and Sex: A Black Feminist Critique of Antidiscrimination Doctrine, Feminist Theory and Antiracist Politics," *University of Chicago Legal Forum* (1989): 139–67, https://chicago unbound.uchicago.edu/uclf/vol1989/iss1/8; Kimberle Crenshaw, "Mapping the Margins: Intersectionality, Identity Politics, and Violence against Women of Color," *Stanford Law Review* 43, no. 6 (1991): 1241–99, https://doi.org/10.2307/1229039; and Jennifer C. Nash, "Re-Thinking Intersectionality," *Feminist Review* 89 (2008): 1–15, https://doi.org/10.1057/fr.2008.4.

20. See, e.g., David Brion Davis, "A Review of the Conflicting Theories on the Slave Family," *Journal of Blacks in Higher Education* no. 16 (1997): 100–103, https://doi.org/10.2307/2962919. Compare, e.g., Jim Sidanius and Felicia Pratto, "Sex and Power: The Intersecting Political Psychologies of Patriarchy and Arbitrary-Set Hierarchy," in *Social Dominance: An Intergroup Theory of Social Hierarchy and Oppression* (Cambridge: Cambridge University Press, 1999), 263–98.

21. See, e.g., Carol Anderson, "Rolling Back Civil Rights," in *White Rage* (New York: Bloomsbury Press, 2016), 98–137; Thomas Craemer et al. [Trevor Smith, Brianna Harrison, Trevon Logan, Wesley Bellamy, and William Darity Jr.], "Wealth Implications of Slavery and Racial Discrimination for African American Descendants of the Enslaved," *Review of Black Political Economy* 47, no. 3 (2020): 218–54, https://doi.org/10.1177/0034644620926516; and Valerie Russ, "Controversial Movement Is Challenging Traditional Black Organizations with 'Project Takeover,'" *Philadelphia Inquirer*, November 15, 2019, https://www.inquirer.com/news/civil-rights-comcast-ados-naacp-takeover-chapters-new-jersey-harriet-20191115.html.

22. **For context: "The Marshall Plan, the historic U.S. aid initiative to speed western Europe's recovery after World War II, is rightly legendary for its vision and accomplishments. The $13.2 billion the United States dedicated to the Plan from 1948 to 1952 would be worth a substantial $135 billion in today's money."** Benn Steil and Benjamin Della Rocca, "It Takes More Than Money to Make a Marshall Plan," Council on Foreign Relations, April 9, 2018, https://www.cfr.org/blog/it-takes-more-money-make-marshall-plan. Compare, e.g., Bridget Read, "Doing the Work at Work," *New York*, May 24, 2021, https://www.thecut.com/article/diversity-equity-inclusion-industrial-companies.html. **On "the psychology underlying [American] collective willful ignorance" of "racial economic inequality [that] is a foundational feature of the United States,"** see Michael W. Kraus et al. [Ivuoma N. Onyeador, Natalie M. Daumeyer, Julian M. Rucker, and Jennifer A. Richeson], "The Misperception of Racial Economic Inequality," *Perspectives on Psychological Science* 14, no. 6 (2019): 899 [899–921], https://doi.org/10.1177/1745691619863049.

23. Wilson Jeremiah Moses, ed., *Classical Black Nationalism: From the American Revolution to Marcus Garvey* (New York: New York University Press, 1996), 5.

24. See Tommie Shelby, *We Who Are Dark: The Philosophical Foundations of Black Solidarity* (Cambridge, MA: Harvard University Press, 2005).

25. See, e.g., "'Absolute Chaos' in Minneapolis as Protests Grow Across U.S.," *New York Times*, May 29, 2020, https://www.nytimes.com/2020/05/29/us/floyd-protests-usa.html; and Derrick Bryson Taylor, "George Floyd Protests: A Timeline," *New York Times*, November 5, 2021, https://www.nytimes.com/article/george-floyd-protests-timeline.html. Compare "Fatal Force," *Washington Post*, updated July 28, 2022, https://www.washingtonpost.com/graphics/investigations/police-shootings-database/.

26. See, e.g., Amanda Barroso and Rachel Minkin, "Recent Protest Attendees Are More Racially and Ethnically Diverse, Younger than Americans Overall," Pew Research Center, June 24, 2020, https://www.pewresearch.org/fact-tank/2020/06/24/recent-protest-attendees-are-more-racially-and-ethnically-diverse-younger-than-americans-overall/; Liam Dillon and Jaclyn Cosgrove, "'I guess America is finally listening.' Why George Floyd Protests Have Spread to Affluent White Suburbs," *Los Angeles Times*, June 6, 2020, https://www.latimes.com/california/story/2020-06-06/george-floyd-protests-suburbs; and Adam Serwer, "The New Reconstruction," *Atlantic*, October 2020, https://www.theatlantic.com/magazine/archive/2020/10/the-next-reconstruction/615475/.

27. Compare, e.g., Charles P. Henry, "A Political and Legal History of Reparations and Race Relations," in *Long Overdue: The Politics of Racial Reparations* (New York: New York University Press, 2007), 9–32. See also Allen J. Davis, "Reparations in the United States: An Historical Timeline of Reparations Payments Made From 1783 through 2021 by the United States Government, States, Cities, Religious Institutions, Universities, Corporations, and Communities," UMass Amherst Libraries, accessed February 18, 2022, https://guides.library.umass.edu/reparations.

Section 2

1. *Plessy v. Ferguson*, 163 U.S. 537, 537–38 (1896).
2. *Plessy*, 163 U.S. at 557. See also Peter S. Canellos, *The Great Dissenter: The Story of John Marshall Harlan, America's Judicial Hero* (New York: Simon and Schuster, 2021), 347–51. Compare Paul Finkelman, *Supreme Injustice: Slavery in the Nation's Highest Court* (Cambridge, MA: Harvard University Press, 2018); David A. Bateman, Ira Katznelson, and John S. Lapinski, *Southern Nation: Congress and White Supremacy after Reconstruction* (Princeton: Princeton University Press, 2018); and Louis Menand, "The Supreme Court Case That Enshrined White Supremacy in Law," *New Yorker*, January 28, 2019, https://www.newyorker.com/magazine/2019/02/04/the-supreme-court-case-that-enshrined-white-supremacy-in-law.
3. *Plessy*, 163 U.S. at 561.
4. See, e.g., Taunya Lovell Banks, "Still Drowning in Segregation: Limits of Law in Post-Civil Rights America," *Law and Inequality* 32, no. 2 (2014): 220–23 [215–55], https://scholarship.law.umn.edu/lawineq/vol32/iss2/8/.
5. See Steve Luxenberg, *Separate: The Story of Plessy v. Ferguson, and America's Journey from Slavery to Segregation* (New York: W. W. Norton, 2019), 194.
6. See, e.g., Alana Semuels, "Segregation Has Gotten Worse, Not Better, and It's Fueling the Wealth Gap Between Black and White Americans," *Time*, June 19, 2020, https://time.com/5855900/segregation-wealth-gap/. **On stereotypes of "the Latin" (a type of Southern European) dating to the early twentieth century in the United States,** see Allen Woll, "How Hollywood Has Portrayed Hispanics," *New York Times*, March 1, 1981, https://www.nytimes.com/1981/03/01/movies/how-hollywood-has-portrayed-hispanics.html. **On the 1960s emergence of "the model minority" stereotype when "anxieties" about the Black rights movement "caused white Americans to further invest in positive portrayals of Asian Americans,"** see Jeff Guo, "The Real Reasons the U.S. Became Less Racist toward Asian Americans," *Washington Post*, November 29, 2016, https://www.washingtonpost.com/news/wonk/wp/2016/11/29/the-real-reason-americans-stopped-spitting-on-asian-americans-and-started-praising-them/. **On "vast differences—in culture, ethnicity and language—[that] exist among the 567 federally recognized Indian nations across the United States,"** see Kevin Gover, "Five Myths about American Indians," *Washington Post*, November 22, 2017, https://www.washingtonpost.com/outlook/five-myths/five-myths-about-american-indians/2017/11/21/41081cb6-ce4f-11e7-a1a3-0d1e45a6de3d_story.html.

7. See, e.g., Esther J. Cepeda, "No One Wins the Oppression Olympics," *Chicago Tribune*, October 11, 2017, https://www.chicagotribune.com/suburbs/post-trib une/opinion/ct-ptb-cepeda-oppression-st-1012-20171011-story.html; and Amy Harmon, "BIPOC or POC? Equity or Equality? The Debate over Language on the Left," *New York Times*, November 1, 2021, https://www.nytimes.com/2021/11/01/ us/terminology-language-politics.html. Compare, e.g., Audra D. S. Burch and Luke Vander Ploeg, "Buffalo Shooting Highlights Rise of Hate Crimes Against Black Americans," *New York Times*, May 16, 2022, https://www.nytimes.com/2022/05/ 16/us/hate-crimes-black-african-americans.html; Jennifer C. Lee and Samuel Kye, "Racialized Assimilation of Asian Americans," *Annual Review of Sociology* 42, no. 1 (2016): 253–73, https://doi.org/10.1146/annurev-soc-081715-074310; and Margaret A. Simons, "Racism and Feminism: A Schism in the Sisterhood," *Feminist Studies* 5, no. 2 (1979): 384–401, https://doi.org/10.2307/3177603.

8. *Plessy*, 163 U.S. at 559. See also *Plessy* at 555. Compare Jamal Greene, "The Anticanon," *Harvard Law Review* 125, no. 2 (2011): 414 [379–475], https://www. jstor.org/stable/41306728. *Pace* Greene, Harlan's "color-blind" aspirational view **of the Reconstruction Amendments did not pretend that *Plessy* was "race neutral in a formal sense." Harlan rejected the Court's "separate but equal" doctrine as a barefaced, legalistic lie that read those amendments in morally shameful (if inter- pretively plausible) light of White commitment to "caste" hierarchy re the "colored citizen" as a descendant of American slavery.**

9. *Plessy*, 163 U.S. at 559.

10. *Plessy*, 163 U.S. at 562.

11. See, e.g., Pauli Murray, ed., *States' Laws on Race and Color*, foreword Davison M. Douglas (1951; repr., Athens: University of Georgia Press, 1997).

12. George C. Wallace, "Inaugural Address of Governor George Wallace," January 14, 1963, Montgomery, Alabama, Department of Archives and History, http://digital. archives.alabama.gov/cdm/ref/collection/voices/id/2952. Compare Dan T. Carter, *The Politics of Rage: George Wallace, the Origins of the New Conservatism, and the Transformation of American Politics* (New York: Simon and Schuster, 1995), 9–11; and Reginald Horsman, *Race and Manifest Destiny: The Origins of American Racial Anglo- Saxonism* (Cambridge, MA: Harvard University Press, 1981) 44–45.

13. **For the view that Western "freedom" historically was conceived in diametrical con- trast to slavery,** see Orlando Patterson, *Freedom in the Making of Western Culture* (New York: Basic Books, 1991), 403–6. **On "white freedom as the belief (and prac- tice) that freedom is central to white racial identity, and that only white people can or should be free,"** see Tyler Stovall, *White Freedom: The Racial History of an Idea* (Princeton: Princeton University Press, 2021), 11.

14. Samuel Cardwell, "'The People Whom He Foreknew': The English as a Chosen People in Bede's Historia Ecclesiastica," *Journal of the Australian Early Medieval Association* 11 (2015): 41 [41–66].

15. See, e.g., Gerald Horne, *The Counter-Revolution of 1776: Slave Resistance and the Origins of the United States of America* (New York: New York University Press, 2014), 40–41; and David Waldstreicher, *Slavery's Constitution: From Revolution to Ratification* (New York: Hill and Wang, 2009), 15–17.

16. *Plessy*, 163 U.S. at 563.
17. For instance, for Chief Justice John Roberts's closing platitude, "The way to stop discrimination on the basis of race is to stop discriminating on the basis of race," see *Parents Involved in Community Schools v. Seattle School District No. 1*, 551 U.S. 701, 748 (2007). Compare, e.g., Michelle Alexander, "The Rebirth of Caste," in *The New Jim Crow: Mass Incarceration in the Age of Colorblindness* (New York: New Press, 2010), 20–58; and Sheryll Cashin, "More Opportunity Hoarding: Separate and Unequal Schools," in *White Space, Black Hood: Opportunity Hoarding and Segregation in the Age of Inequality* (Boston: Beacon Press, 2021), 127–44.
18. *Plessy*, 163 U.S. at 562.
19. See, e.g., Canellos, *Great Dissenter*, 18–20.

Section 3

1. Anthony Appiah, "The Uncompleted Argument: Du Bois and the Illusion of Race," in *"Race," Writing, and Difference*, ed. Henry Louis Gates Jr. (Chicago: University of Chicago Press, 1986), 35 [21–37]. See also K. Anthony Appiah, "Race, Culture, Identity: Misunderstood Connections," in K. Anthony Appiah and Amy Gutmann, *Color Conscious: The Political Morality of Race* (Princeton: Princeton University Press, 1996), 30–105.
2. W. E. B. Du Bois, "The Conservation of Races" (1897), in *W. E. B. Du Bois: Writings*, ed. Nathan Huggins (New York: Library of America, 1986), 816 [815–26].
3. Du Bois, "Conservation of Races," 816–17.
4. Du Bois, "Conservation of Races," 818.
5. W. E. B. Du Bois, *Dusk of Dawn: An Essay Toward an Autobiography of a Race Concept* (New York: Harcourt, Brace, 1940), 117.
6. Appiah, "Uncompleted Argument," 34–35.
7. See, e.g., Anindya Sekhar Purakayastha, "W. E. B. Du Bois, B. R. Ambedkar and the History of Afro-Dalit Solidarity," *Sanglap* 6, no. 1 (2019): 22–23 [20–36], http://sanglap-journal.in/index.php/sanglap/article/view/116; and Isabel Wilkerson, *Caste: The Origins of Our Discontents* (New York: Random House, 2020), 25–27.
8. Du Bois, *Dusk of Dawn*, 139. Compare, e.g., David H. Gans, "'We Do Not Want to Be Hunted': The Right to Be Secure and Our Constitutional Story of Race and Policing," *Columbia Journal of Race and Law* 11, no. 2 (2021): 270–79 [239–342], https://journals.library.columbia.edu/index.php/cjrl/article/view/8230. On "the emasculating effects of caste distinctions," see W. E. B. Du Bois, "Of Mr. Booker T. Washington and Others," in *The Souls of Black Folk* (Chicago: A. C. McClurg, 1903), 59 [41–59]. See also Robert-Gooding Williams, *In the Shadow of Du Bois: Afro-Modern Political Thought in America* (Cambridge, MA: Harvard University Press, 2009), 163–64; and Tommie Shelby, *We Who Are Dark: The Philosophical Foundations of Black Solidarity* (Cambridge, MA: Harvard University Press, 2005), 62.
9. See, e.g., Sharika Crawford, "The 'African Personality' Returns: The Controversy over Gandhi at the University of Ghana," *Perspectives on History*, March 1, 2017,

https://www.historians.org/publications-and-directories/perspectives-on-hist
ory/march-2017/the-african-personality-returns-the-controversy-over-gan
dhi-at-the-university-of-ghana; and Laura Chrisman, "Du Bois in Transnational
Perspective: The Loud Silencing of Black South Africa," *Current Writing* 16, no. 2
(2004): 18–30, https://doi.org/10.1080/1013929X.2004.9678192.

10. Compare, e.g., HKS Misinformation Review Editorial Staff, "Retraction Note
to: 'Disinformation Creep: ADOS and the Strategic Weaponization of Breaking
News,'" *Harvard Kennedy School (HKS) Misinformation Review*, December 20, 2021,
https://doi.org/10.37016/mr-2020-86.

11. Appiah, "Uncompleted Argument," 35.

12. Juliana Menasce Horowitz, Anna Brown, and Kiana Cox, "Race in America 2019,"
Pew Research Center, April 9, 2019, https://www.pewsocialtrends.org/2019/04/09/
race-in-america-2019/.

13. See, e.g., Sharon Birch-Jeffrey, "African Migrants Keen to Retain Their Cultural Values
Abroad," Africa Renewal (UN), December 2018–March 2019, https://www.un.org/
africarenewal/magazine/december-2018-march-2019/african-migrants-keen-ret
ain-their-cultural-values-abroad. **Contrast, on how "social science representations
of post-1965 black immigrants in the United States employ the concept of 'ethnicity' in ways that reinforce the racialist myth of Black (American) cultural inferiority,"** Jemima Pierre, "Black Immigrants in the United States and the 'Cultural
Narratives' of Ethnicity," *Identities* 11, no. 2 (2004): 141–70, https://doi.org/10.1080/
10702890490451929.

14. See, e.g., David A. Hollinger, "The Concept of Post-Racial: How Its Easy Dismissal
Obscures Important Questions," *Daedalus* 140, no. 1 (2011): 177–78 [174–82], http://
www.jstor.org/stable/25790452.

15. Du Bois, "Conservation of Races," 821–22. Compare, e.g., Wilson J. Moses, "The
Poetics of Ethiopianism: W. E. B. Du Bois and Literary Black Nationalism," *American
Literature* 47, no. 3 (1975): 425–26 [411–26], https://doi.org/10.2307/2925341.

16. See Du Bois, "Conservation of Races," 825.

Section 4

1. See W. E. B. Du Bois, "The Talented Tenth" (1903), in *W. E. B. Du Bois: Writings*, ed.
Nathan Huggins (New York: Library of America, 1986), 842 [842–61]. Compare, e.g.,
Khiara M. Bridges, "Excavating Race-Based Disadvantage Among Class-Privileged
People of Color," *Harvard Civil Rights-Civil Liberties Law Review* 53 (2018): 65–130.

2. See Gregory Mixon, "'Good Negro–Bad Negro': The Dynamics of Race and Class
in Atlanta During the Era of the 1906 Riot," *Georgia Historical Quarterly* 81, no. 3
(1997): 593 [593–621], https://www.jstor.org/stable/40583748. See also Guion Griffis
Johnson, "Southern Paternalism Toward Negroes After Emancipation," *Journal of
Southern History* 23, no. 4 (1957): 483–509, https://doi.org/10.2307/2954388.

3. Booker T. Washington, "The Atlanta Exposition Address" (1901), in *The Booker T. Washington Reader* (Radford, VA: Wilder, 2008), 92–93 [92–99]. Compare, e.g., Ward Connerly, "Inestimable Harm," *National Review*, April 5, 2010, https://www.nationalreview.com/corner/inestimable-harm-ward-connerly/.

4. **The intellectual tradition of Black (American) solidarity has included Africa-identified individuals who are not homegrown "blacks"—for example, Caribbean American philosophers Bernard Boxill and Charles Mills, whose personal reflections are instructive.** See, e.g., Bernard Boxill, "The Value of Memory," *Proceedings and Addresses of the American Philosophical Association* 85, no. 2 (2011): 83 [81–88], https://www.jstor.org/stable/41575751; and Charles W. Mills, "The Red and the Black," *Proceedings and Addresses of the American Philosophical Association* 90 (2016): 103 [90–113], https://www.jstor.org/stable/26622940.

5. Compare, e.g., Christopher Alan Bracey, *Saviors or Sellouts: The Promise and Peril of Black Conservatism, from Booker T. Washington to Condoleezza Rice* (Boston: Beacon Press, 2008).

6. See, e.g., David Brion Davis, *The Problem of Slavery in the Age of Emancipation* (New York: Knopf, 2014), 54–55.

7. Tommie Shelby, *We Who Are Dark: The Philosophical Foundations of Black Solidarity* (Cambridge, MA: Harvard University Press, 2005), 201.

8. See, e.g., Allyson Hobbs, *A Chosen Exile: A History of Racial Passing in American Life* (Cambridge, MA: Harvard University Press, 2014), 164.

9. See, e.g., W. E. B. Du Bois, "Of Mr. Booker T. Washington and Others," in *The Souls of Black Folk* (Chicago: A. C. McClurg, 1903), 41–59.

10. See, e.g., Elijah Anderson, "'The White Space,'" *Sociology of Race and Ethnicity* 1, no. 1 (2015): 11–16 [10–21], https://doi.org/10.1177/2332649214561306; and Lawrence Bobo, "Prejudice as Group Position: Micro-Foundations of a Sociological Approach to Racism and Race Relations," *Journal of Social Issues* 55, no. 3 (1999): 464–66 [445–72], https://doi.org/10.1111/0022-4537.00127.

11. See, e.g., Richard Rothstein, "Suppressed Incomes," in *The Color of Law: A Forgotten History of How Our Government Segregated America* (New York: Liveright, 2017), 153–75; Dick Startz, "The Achievement Gap in Education: Racial Segregation Versus Segregation by Poverty," Brookings Institution, January 20, 2020, https://www.brookings.edu/blog/brown-center-chalkboard/2020/01/20/the-achievement-gap-in-education-racial-segregation-versus-segregation-by-poverty/. Compare, e.g., James D. Anderson, *The Education of Blacks in the South, 1860–1935* (Chapel Hill: University of North Carolina Press, 1988); and Kenneth B. Clark, *Dark Ghetto: Dilemmas of Social Power* (New York: Harper and Row, 1965).

12. Kwame Anthony Appiah, *The Ethics of Identity* (Princeton: Princeton University Press, 2005), 185. Compare Lionel K. McPherson and Tommie Shelby, "Blackness and Blood: Interpreting African American Identity," *Philosophy and Public Affairs* 32 (2004): 181–83 [171–92], https://doi.org/10.1111/j.1088-4963.2004.00010.x.

13. For "criteria to determine eligibility for a Black reparations program," which would solve the potential issue of intergenerational "white passing" family members of American slave lineage, see William A. Darity Jr. and A. Kirsten Mullen, "A Program of Black Reparations," in *From Here to Equality: Reparations for Black Americans in the Twenty-First Century* (Chapel Hill: University of North Carolina Press, 2020), 258–59 [256–70].

14. See François Bernier, "A New Division of the Earth" (1684), in *The Idea of Race*, ed. Robert Bernasconi and Tommy L. Lott (Indianapolis: Hackett, 2000), [1–4]; and Immanuel Kant, "Of the Different Human Races" (1777), in *Idea of Race*, [8–22].

15. See William Waller Hening, ed., Act XII, "Negro Womens Children to Serve According to the Condition of the Mother," *Laws of Virginia*, December 1662, in *The Statutes at Large; Being a Collection of All the Laws of Virginia, From the First Session of the Legislature, in the Year 1619*, vol. 2 (Richmond: Samuel Pleasants, 1809–23), 170.

16. See Toni Morrison, "A Humanist View" (speech, Portland State University, May 30, 1975), Portland State Library Special Collections, 35:48–36:06 [6:50–43:01], accessed February 14, 2021, https://soundcloud.com/portland-state-library/portland-state-black-studies-1.

17. For an earnest example of the view that "results" from early scientistic "searching for the origin of 'blackness'" can "help us see how Enlightenment thinkers justified chattel slavery," see Henry Louis Gates Jr. and Andrew S. Curran, "Inventing the Science of Race," *New York Review of Books*, December 16, 2021, https://www.nybooks.com/articles/2021/12/16/inventing-the-science-of-race/.

18. See William Goodell, *The American Slave Code in Theory and Practice: Its Distinctive Features Shown by Its Statutes, Judicial Decisions, and Illustrative Facts* (New York: American and Foreign Anti-Slavery Society, 1853); and Theodore Dwight Weld, *American Slavery as It Is: Testimony of a Thousand Witnesses* (New York: American Anti-Slavery Society, 1839).

19. "University Race-Sensitive Admissions Programs Are Not Helping Black Students Who Most Need Assistance," *Journal of Blacks in Higher Education* 56 (2007): 6–7 [6–8], https://www.jstor.org/stable/25073686. See also Joanna Walters, "'Any Black Student Will Do,'" *Guardian*, May 29, 2007, https://www.theguardian.com/education/2007/may/29/internationaleducationnews.highereducation. Compare, e.g., Craig Steven Wilder, *Ebony and Ivy: Race, Slavery, and the Troubled History of America's Universities* (New York: Bloomsbury, 2013), 280–84.

20. See, e.g., Martha S. West, "The Historical Roots of Affirmative Action," *Berkeley La Raza Law Journal* 10, no. 2 (1998): 607–8, [607–30], https://doi.org/10.15779/Z38JW8Q; Associated Press, "The 40 Who Fell in the Turbulence of the U.S. Battles for Civil Rights," *New York Times*, November 4, 1989, https://www.nytimes.com/1989/11/04/us/the-40-who-fell-in-the-turbulence-of-the-us-battles-for-civil-rights.html; and Jess Bidgood, "Black Immigrants Have Quadrupled Since 1980, Study Says," *New York Times*, April 9, 2015, https://www.nytimes.com/2015/04/10/us/black-immigrants-have-quadrupled-since-1980-study-says.html.

21. For criticism of Supreme Court Justice Lewis F. Powell Jr.'s view that "affirmative action was not a way of righting historical—and ongoing—wrongs against Black

people" but "a way to achieve diversity, a compelling state interest because it benefited all students," see Adam Harris, "This Is the End of Affirmative Action," *Atlantic*, September 2021, https://www.theatlantic.com/magazine/archive/2021/09/the-end-of-affirmative-action/619488/. See also, e.g., Derrick Bell, "Diversity's Distractions," *Columbia Law Review* 103, no. 6 (2003): 1622–33, https://doi.org/10.2307/3593396.

22. See, e.g., William Julius Wilson, *The Declining Significance of Race: Blacks and Changing American Institutions*, 3rd ed. (Chicago: University of Chicago Press, 2012). Compare, e.g., Lawrence Bobo, James R. Kluegel, and Ryan A. Smith, "Laissez-Faire Racism: The Crystallization of a 'Kinder, Gentler' Anti-Black Ideology," in *Racial Attitudes in the 1990s: Continuity and Change*, ed. Steven A. Tuch and Jack K. Martin (Westport, CT: Praeger, 1997), 15–42.

23. See, e.g., *Plessy v. Ferguson*, 163 U.S. 537, 538 (1896); and Walter A. Plecker, "The New Virginia Law to Preserve Racial Integrity," *Virginia Health Bulletin*, March 1924, http://edu.lva.virginia.gov/dbva/items/show/226.

24. See W. E. B. Du Bois, "Criteria of Negro Art" (1926), in *Writings*, 993 [993–1002].

25. For raced confusion about a community of Americans "raised to identify as black" but who "might register to most as white by appearance" and have "hardly a trace of black ancestry left in their blood" (confusion readily resolved by my "Slaves" caste analysis of "the black race" in America), compare Khushbu Shah, "They Look White But Say They're Black: A Tiny Town in Ohio Wrestles with Race," *Guardian*, July 25, 2019, https://www.theguardian.com/us-news/2019/jul/25/race-east-jackson-ohio-appalachia-white-black.

26. See, e.g., Kwame Anthony Appiah, "2017 'Great Immigrant': In His Own Words," Carnegie Corporation of New York, accessed November 24, 2021, https://www.carnegie.org/multimedia/great-immigrants-kwame-anthony-appiah/; and Alejandro Portes and Rubén G. Rumbaut, *Immigrant America: A Portrait*, 3rd ed. (Berkeley: University of California Press, 2006). Compare, e.g., Roxanne Dunbar-Ortiz, *Not "A Nation of Immigrants": Settler Colonialism, White Supremacy, and a History of Erasure and Exclusion* (Boston: Beacon Press, 2021), xiii–xvi; Isabella C. Aslarus, "Are We in the Minority?," *Harvard Crimson*, October 15, 2020, https://www.thecrimson.com/article/2020/10/15/gaasa-scrut/; and Wesley Lowery, "Which Black Americans Should Get Reparations?," *Washington Post*, September 18, 2019, https://www.washingtonpost.com/national/which-americans-should-get-reparations/2019/09/18/271cf744-cab1-11e9-a4f3-c081a126de70_story.html.

Section 5

1. On the 1970 Louisiana legal blackness standard of more than "one thirty-second" African ancestry (aka "Negro blood"), see Art Harris, "Louisiana Court Sees No Shades of Gray in Woman's Request," *Washington Post*, May 21, 1983, https://www.washingtonpost.com/archive/politics/1983/05/21/louisiana-court-sees-no-shades-of-gray-in-womans-request/ddb0f1df-ba5d-4141-9aa0-6347e60ce52d/.

2. Skin color genetics would be a very unreliable indicator of black/Negro raciality. See, e.g., Nicholas G. Crawford et al. [Derek E. Kelly, Matthew E. B. Hansen, Marcia

H. Beltrame, Shaohua Fan, Shanna L. Bowman, Ethan Jewett et al.], "Loci Associated with Skin Pigmentation Identified in African Populations," *Science* 358, no. 6365 (2017): 867–68 [867–87], https://doi.org/10.1126/science.aan8433.

3. See, e.g., Mara Loveman, Jeronimo O. Muniz, and Stanley R. Bailey, "Brazil in Black and White? Race Categories, the Census, and the Study of Inequality," *Ethnic and Racial Studies* 35, no. 8 (2012): 1468–69 [1466–83], https://doi.org/10.1080/01419870.2011.607503. See also David I. Kertzer and Dominique Arel, eds., *Census and Identity: The Politics of Race, Ethnicity, and Language in National Censuses* (Cambridge: Cambridge University Press, 2002).

4. See US Census Bureau, "History: Through the Decades: Questionnaires," 1960 to 2020 forms, accessed November 14, 2021, https://www.census.gov/history/www/through_the_decades/questionnaires/.

5. See Sonya Rastogi et al. [Tallese D. Johnson, Elizabeth M. Hoeffel, and Malcolm P. Drewery Jr.], "The Black Population: 2010," US Census Bureau, September 2011, 3 [1–20], https://www.census.gov/content/dam/Census/library/publications/2011/dec/c2010br-06.pdf; and Lindsay Hixson, Bradford B. Hepler, and Myoung Ouk Kim, "The White Population: 2010," US Census Bureau, September 2011, 3 [1–18], https://www.census.gov/content/dam/Census/library/publications/2011/dec/c2010br-05.pdf. Compare, e.g., Steven A. Holmes, "People Can Claim One or More Races on Federal Forms," *New York Times*, October 30, 1997, http://www.nytimes.com/1997/10/30/us/people-can-claim-one-or-more-races-on-federal-forms.html.

Section 6

1. See Wilson Jeremiah Moses, ed., *Classical Black Nationalism: From the American Revolution to Marcus Garvey* (New York: New York University Press, 1996), 1.

2. Compare, e.g., James M. McPherson, "The Dimensions of Change: The First and Second Reconstructions," *Wilson Quarterly* 2, no. 2 (1978): 142–44 [135–44], http://www.jstor.org/stable/40255407; and Heather Long and Andrew Van Dam, "The Black-White Economic Divide Is as Wide as It Was in 1968," *Washington Post*, June 4, 2020, https://www.washingtonpost.com/business/2020/06/04/economic-divide-black-households/.

3. See Tommie Shelby, *We Who Are Dark: The Philosophical Foundations of Black Solidarity* (Cambridge, MA: Harvard University Press, 2005), 169, 175.

4. See Lionel K. McPherson and Tommie Shelby, "Blackness and Blood: Interpreting African American Identity," *Philosophy and Public Affairs* 32, no. 2 (2004): 179 [171–92], https://doi.org/10.1111/j.1088-4963.2004.00010.x. Compare, e.g., K. Anthony Appiah, "The State and the Shaping of Identity," in *The Tanner Lectures on Human Values*, vol. 23, ed. Grethe B. Peterson (Salt Lake City: University of Utah Press, 2002), 283–84 [234–99].

5. See Lionel K. McPherson and Tommie Shelby, "Blackness and Philosophy," *Symposia on Gender, Race and Philosophy* 1, no. 1 (2005): 4 [1–5], https://web.mit.edu/sgrp/2005/no1/McPhersonShelby0505.pdf.

6. See Martin Luther King Jr., "I Have a Dream" (speech, Lincoln Memorial, Washington, DC, August 28, 1963).

7. See Lucius T. Outlaw Jr., "Against the Grain of Modernity: The Politics of Difference and the Conservation of 'Race,'" in *On Race and Philosophy* (New York: Routledge, 1996), 140 [135–57]. See also Iris Marion Young, "Social Movements and the Politics of Difference," in *Justice and the Politics of Difference* (Princeton: Princeton University Press, 1990), 156–83.

8. See Outlaw, "Against the Grain of Modernity," 156–57, 136.

9. **For a fuller exposition of "five familiar modes of thick blackness," which includes a "kinship" mode,** see Shelby, *We Who Are Dark*, 209–11.

10. See Outlaw, "Against the Grain of Modernity," 157.

11. See Shelby, *We Who Are Dark*, 155.

12. Shelby, *We Who Are Dark*, 214.

13. See Shelby, *We Who Are Dark*, 214.

14. Shelby, *We Who Are Dark*, 214–15.

15. See Shelby, *We Who Are Dark*, 212.

Section 7

1. See, e.g., Taylor Branch, *Parting the Waters: America in the King Years, 1954–63* (New York: Simon and Schuster, 1988). See also Kate Masur, *Until Justice Be Done: America's First Civil Rights Movement, from the Revolution to Reconstruction* (New York: Norton, 2021).

2. See David A. Bositis, *Blacks and the 2004 Democratic National Convention* (Washington, DC: Joint Center for Political and Economic Studies, 2004), 9. See also Ismail K. White and Chryl N. Laird, *Steadfast Democrats: How Social Forces Shape Black Political Behavior* (Princeton: Princeton University Press, 2020), 2–3; and Daniel S. Lucks, *Reconsidering Reagan: Racism, Republicans, and the Road to Trump* (Boston: Beacon Press, 2020), 8–10.

3. Tommie Shelby, *We Who Are Dark: The Philosophical Foundations of Black Solidarity* (Cambridge, MA: Harvard University Press, 2005), 242.

4. Compare, e.g., Sam Fulwood III, "Black Male Voters Are Tired of Being Taken for Granted," *The Hill*, October 22, 2020, https://thehill.com/opinion/campaign/522234-heres-why-young-black-male-voters-favor-trump/.

5. See Martin Luther King Jr., "Racism and the White Backlash," in *Where Do We Go from Here: Chaos or Community?* (New York: Harper and Row, 1967), 95 [67–101].

6. See, e.g., Justin Gomer and Christopher Petrella, "How the Reagan Administration Stoked Fears of Anti-White Racism," *Washington Post*, October 10, 2017, https://www.washingtonpost.com/news/made-by-history/wp/2017/10/10/how-the-reagan-administration-stoked-fears-of-anti-white-racism/; and Alan Abramowitz and Jennifer McCoy, "United States: Racial Resentment, Negative Partisanship, and Polarization in Trump's America," *Annals of the American Academy of Political and Social Science* 681, no. 1 (2019): 137–56 https://doi.org/10.1177/0002716218811309.

7. See, e.g., Tommie Shelby, "Racial Realities and Corrective Justice: A Reply to Charles Mills," *Critical Philosophy of Race* 1, no. 2 (2013): 158–60 [145–62], https://doi.org/ 10.5325/critphilrace.1.2.0145. Compare, e.g., Charles W. Mills, "Retrieving Rawls for Racial Justice?: A Critique of Tommie Shelby," *Critical Philosophy of Race* 1, no. 1 (2013): 16–17 [1–27], https://doi.org/10.5325/critphilrace.1.1.0001.

8. See, e.g., Robert C. Smith, *We Have No Leaders: African Americans in the Post-Civil Rights Era*, foreword Ronald W. Walters (Albany: State University of New York Press, 1996), 22–23.

9. Compare, e.g., Devah Pager and Hana Shepherd, "The Sociology of Discrimination: Racial Discrimination in Employment, Housing, Credit, and Consumer Markets," *Annual Review of Sociology* 34 (2008): 196 [181–209], https://doi. org/10.1146/annurev.soc.33.040406.131740.

10. William Howard Taft, "The South and the National Government," intro. Andrew Carnegie (speech, North Carolina Society, New York, December 7, 1908), 11 [9–16].

11. See Erick Trickey, "Chief Justice, Not President, Was William Howard Taft's Dream Job," *Smithsonian*, December 5, 2016, https://www.smithsonianmag.com/history/ chief-justice-not-president-was-william-howard-tafts-dream-job-180961279/. **On "jurisprudence [that] reflected Marshall's investment in slaves that was probably unmatched by any other member of the Supreme Court" and "dovetailed with his lifetime commitment to slavery and his virulent hostility to the very presence of free blacks in the United States,"** compare Paul Finkelman, "Master John Marshall and the Problem of Slavery" (Part I), University of Chicago Law Review Online, August 31, 2020, accessed August 1, 2022, https://lawreviewblog.uchicago.edu/2020/08/31/ marshall-slavery-pt1/. Compare also Paul Finkelman, "John Marshall: Slave Owner and Jurist," in *Supreme Injustice: Slavery in the Nation's Highest Court* (Cambridge, MA: Harvard University Press, 2018), 26–75.

12. Compare, e.g., Jason Sokol, *There Goes My Everything: White Southerners in the Age of Civil Rights, 1945–1975* (New York: Knopf, 2006); and Ashley V. Reichelmann and Matthew O. Hunt, "How We Repair It: White Americans' Attitudes toward Reparations," Brookings Institution, December 8, 2021, https://www.brookings.edu/ blog/how-we-rise/2021/12/08/how-we-repair-it-white-americans-attitudes-toward-reparations/.

13. Compare, e.g., Christopher D. DeSante and Candis Watts Smith, *Racial Stasis: The Millennial Generation and the Stagnation of Racial Attitudes in American Politics* (Chicago: University of Chicago Press, 2020), 98; and Adolph Reed Jr., *The South: Jim Crow and Its Afterlives*, foreword Barbara J. Fields (London: Verso, 2022), 6–9.

14. See Shelby, *We Who Are Dark*, 214.

15. See, e.g., N. S. Chiteji and Darrick Hamilton, "Family Connections and the Black-White Wealth Gap among Middle-Class Families," *Review of Black Political Economy* 30, no. 1 (2002): 9–28, https://doi.org/10.1007/BF02808169.

16. See Michael C. Dawson, *Behind the Mule: Race and Class in African-American Politics* (Princeton: Princeton University Press, 1994), 61.

17. See Shelby, *We Who Are Dark*, 240–41.

18. See, e.g., Mary Jo Wiggins, "Race, Class, and Suburbia: The Modern Black Suburb as a 'Race-Making Situation,'" *University of Michigan Journal of Law Reform* 35, no. 4 (2002): 806–8 [749–808], https://repository.law.umich.edu/mjlr/vol35/iss4/3.

19. See, e.g., Bernard R. Boxill, "Self-Respect," in *Blacks and Social Justice*, rev. ed. (Lanham, MD: Rowman and Littlefield, 1992), 187 [186–204].

20. Joseph Butler, *Fifteen Sermons* (1726), in *British Moralists 1650–1800*, vol. 1, ed. D.D. Raphael (1969; repr., Indianapolis: Hackett, 1991), 373 [325–77].

21. See, e.g., Tim Bontemps, "Michael Jordan Stands Firm on 'Republicans Buy Sneakers, Too' Quote, Says It Was Made in Jest," ESPN.com, May 4, 2020, https://www.espn.com/nba/story/_/id/29130478/michael-jordan-stands-firm-republicans-buy-sneakers-too-quote-says-was-made-jest; and Jabari Young, "Michael Jordan's Brand Donates $100 Million to Organizations Fighting Racism against Black People," CNBC.com, June 5, 2020, https://www.cnbc.com/2020/06/05/michael-jordans-brand-donates-100-million-to-anti-racist-groups.html.

22. Compare, e.g., Leah Wright Rigueur, *The Loneliness of the Black Republican: Pragmatic Politics and the Pursuit of Power* (Princeton: Princeton University Press, 2014), 5.

23. See, e.g., Elliot Hannon, "Leaked Audio Captures Bloomberg Defending Racial Profiling and Stop-and-Frisk Policing," Slate, February 11, 2020, https://slate.com/news-and-politics/2020/02/leaked-audio-bloomberg-aspen-institute-racial-profiling-stop-and-frisk-policing.html; Robert Henderson and Rebecca Marchiel, "The Keys to Ensuring a New Anti-Redlining Initiative Succeeds," *Washington Post*, November 15, 2021, https://www.washingtonpost.com/outlook/2021/11/15/keys-ensuring-new-anti-redlining-initiative-succeeds/; Ben Steverman, "A Tax Code Optimized for White Wealth Leaves Black Americans Behind," Bloomberg.com, March 10, 2021, https://www.bloomberg.com/news/features/2021-03-10/america-s-tax-code-leaves-black-people-behind-dorothy-brown; and Heidi Ledford, "Millions of Black People Affected by Racial Bias in Health-Care Algorithms," *Nature* 574 (2019): 608–9, https://doi.org/10.1038/d41586-019-03228-6.

24. See, e.g., Nelson Blackstock, "A Special Hatred for Blacks," in *Cointelpro: The FBI's Secret War on Political Freedom*, 3rd ed., intro. Noam Chomsky (New York: Pathfinder, 1988), 91–108.

25. Compare Shelby, *We Who Are Dark*, 68–71; and Bernard R. Boxill, "Fear and Shame as Forms of Moral Suasion in the Thought of Frederick Douglass," *Transactions of the Charles S. Peirce Society* 31, no. 44 (1995): 740–41, [713–44], https://www.jstor.org/stable/40320570.

Section 8

1. Compare, e.g., Peniel E. Joseph, "The Black Power Movement: A State of the Field," *Journal of American History* 96, no. 3 (2009): 752–53 [751–76], https://doi.org/10.1093/jahist/96.3.751; and "Black Is Beautiful: The Emergence of Black Culture and Identity in the 60s and 70s," National Museum of African American History and

Culture, July 8, 2019, https://nmaahc.si.edu/explore/stories/black-beautiful-emerge
nce-black-culture-and-identity-60s-and-70s.

2. **Black American identity was never premised on Black male authority: slavery
   and Jim Crow intervened in Black gender politics, by circumstances and design of
   White social dominance, preempting even the guise of Black male "patriarchal"
   status.** See, e.g., Stephanie E. Jones-Rogers, "'I Belong to de Mistis,'" in *They Were Her
   Property: White Women as Slave Owners in the American South* (New Haven: Yale
   University Press, 2019), 25–56; E. Franklin Frazier, *The Negro Family in the United
   States* (Chicago: University of Chicago Press, 1939), 37–41, 57–61; Alma Carten,
   "How Racism Has Shaped Welfare Policy in America Since 1935," The Conversation,
   August 21, 2016, https://theconversation.com/how-racism-has-shaped-welfare-
   policy-in-america-since-1935-63574; Ari Shapiro and Manuela López Restrepo,
   "Emmett Till's Family Says 'White Pedestal' Theory Has Denied Them Justice for
   Decades," NPR, July 14, 2022, https://www.npr.org/2022/07/14/1111380028/emm
   ett-till-warrant-family-arrest-lynching; and Barbara Young Welke, "Miscegenation
   and the Racial State," *Contemporary Sociology* 41, no. 3 (2012): 284 [283–87], https://
   doi.org/10.1177/0094306112443511a. **Nevertheless, "intersectionality" theory and
   praxis deem gender in combination with some "race" thing principally important for women and sexual minorities ("black women," "queer black women," etc.)
   under White American patriarchy.** See, e.g., Martha S. Jones, *Vanguard: How Black
   Women Broke Barriers, Won the Vote, and Insisted on Equality for All* (New York: Basic
   Books, 2020), 9–11; and Angela Y. Davis, *Blues Legacies and Black Feminism: Gertrude
   "Ma" Rainey, Bessie Smith, and Billie Holiday* (New York: Pantheon, 1998), 36–38. **On
   feminist "intersectionality wars,"** see Jennifer C. Nash, "Intersectionality and Its
   Discontents," *American Quarterly* 69, no. 1 (2017): 117–29, https://doi.org/10.1353/
   aq.2017.0006. Compare, e.g., William A. Smith, Man Hung, and Jeremy D. Franklin,
   "Racial Battle Fatigue and the MisEducation of Black Men: Racial Microaggressions,
   Societal Problems, and Environmental Stress," *Journal of Negro Education* 80, no. 1
   (2011): 75–77 [63–82], http://www.jstor.org/stable/41341106.

3. Compare Gil Scott-Heron, "Whitey on the Moon," *Small Talk at 125th and Lenox*,
   Flying Dutchman, 1970; and Taiyler Simone Mitchell, "'Whitey on the Moon'
   Poem Garners Social Media Attention on Anniversary of Moon Landing, Bezos's
   Spaceflight," Insider, July 20, 2021, https://www.businessinsider.com/whitey-on-the-
   moon-poem-resurfaces-amid-bezos-branson-spaceflight-2021-7.

4. See, e.g., Brentin Mock, "White Americans' Hold on Wealth Is Old, Deep, and
   Nearly Unshakeable," Bloomberg.com, September 3, 2019, https://www.bloomb
   erg.com/news/articles/2019-09-03/the-amazing-resiliency-of-white-wealth; and
   Richard Rothstein, "The Myth of *De Facto* Segregation," *Phi Delta Kappan* 100, no.
   5 (2019): 35–38, https://doi.org/10.1177/0031721719827543 (https://kappanonline.
   org/myth-de-facto-segregation-rothstein-school-districts/). Compare, e.g., Mary L.
   Dudziak, "Brown as a Cold War Case," *Journal of American History* 91, no. 1 (2004):
   32–42, https://doi.org/10.2307/3659611.

5. See, e.g., Janice Francis-Smith, "A Century Later, Reparations for Greenwood Still
   Elusive," *Journal Record*, February 22, 2022, https://journalrecord.com/2022/02/22/
   a-century-later-reparations-for-greenwood-still-elusive/; and Robert Samuels, "After

Reparations," *Washington Post*, April 3, 2020, https://www.washingtonpost.com/graphics/2020/national/rosewood-reparations/.

6. Josh Getlin, "Dinkins Descending," *Los Angeles Times*, October 28, 1990, https://www.latimes.com/archives/la-xpm-1990-10-28-vw-4885-story.html Compare H. L. Mencken, "Mencken's Reply to La Monte's Third Letter," in Robert Rives La Monte and H. L. Mencken, *Men versus the Man: A Correspondence between Robert Rives La Monte, Socialist and H. L. Mencken, Individualist* (New York: Henry Holt, 1910), 115–16 [107–22].

7. Lionel K. McPherson and Tommie Shelby, "Blackness and Philosophy," *Symposia on Gender, Race and Philosophy* 1, no. 1 (2005): 3 [1–5], https://web.mit.edu/sgrp/2005/no1/McPhersonShelby0505.pdf.

8. Compare Tommie Shelby, *We Who Are Dark: The Philosophical Foundations of Black Solidarity* (Cambridge, MA: Harvard University Press, 2005), 236–40.

9. See, e.g., Helena R. M. Radke et al. [Maja Kutlaca, Birte Siem, Stephen C. Wright, and Julia C. Becker], "Beyond Allyship: Motivations for Advantaged Group Members to Engage in Action for Disadvantaged Groups," *Personality and Social Psychology Review* 24, no. 4 (2020): 291–92 [291–315], https://doi.org/10.1177/108886832 0918698.

10. Compare, e.g., Derald Wing Sue et al. [Christina M. Capodilupo, Gina C. Torino, Jennifer M. Bucceri, Aisha M. B. Holder, Kevin L. Nadal, and Marta Esquilin], "Racial Microaggressions in Everyday Life: Implications for Clinical Practice," *American Psychologist*, 62, no. 4 (2007): 271–86, https://doi.org/10.1037/0003-066X.62.4.271.

11. **For criticism that Shelby's argument conceives of thin black social identity that might be "too thin" to inspire "the masses" to join in Black political solidarity,** see Orlando Patterson, "Being and Blackness," *New York Times*, January 8, 2006, https://www.nytimes.com/2006/01/08/books/review/being-and-blackness.html. **It is unclear why Patterson believes that political solidarity "motivated solely by the goal of racial justice" for Black Americans might have "no chance of success except among the [Black] elite." Presumably, the Black non-elite have only more urgent reason to rally behind a Black politics dedicated to material repair.**

12. Shelby, *We Who Are Dark*, 240.

13. Shelby, *We Who Are Dark*, 240.

14. The term "Jim Crow" comes from a Black male slave character created by a White male entertainer in the 1830s, a persona that defines White America's "blackface" tradition. See "The Origins of Jim Crow," Jim Crow Museum, https://www.ferris.edu/HTMLS/news/jimcrow/origins.htm. Compare, e.g., Brett Murphy, "Blackface, KKK Hoods and Mock Lynchings: Review of 900 Yearbooks Finds Blatant Racism," *USA Today*, February 20, 2019, https://www.usatoday.com/in-depth/news/investigations/2019/02/20/blackface-racist-photos-yearbooks-colleges-kkk-lynching-mockery-fraternities-black-70-s-80-s/2858921002/.

15. Shelby, *We Who Are Dark*, 241.

16. Shelby, *We Who Are Dark*, 241.

17. See, e.g., Simon Hall, "The NAACP, Black Power, and the African American Freedom Struggle, 1966–1969," *Historian* 69 (2007): 65–69 [49–82], https://doi.org/10.1111/j.1540-6563.2007.00174.x. Compare, e.g., Ira Katznelson, *When Affirmative Action*

Was White: An Untold History of Racial Inequality in Twentieth-Century America
(New York: W. W. Norton, 2005), 29; and Isaac Stanley-Becker, "'We Got Things
Done': Biden Recalls 'Civility' with Segregationist Senators," *Washington Post*, June
19, 2019, https://www.washingtonpost.com/nation/2019/06/19/joe-biden-james-
eastland-herman-talmadge-segregationists-civility/. I have not formed any critical
evaluation of the exclusionary turn of Black collective struggle under the Black
Power movement's influence. My non-exclusionary account of Black (American)
political solidarity is geared to post–civil rights politics of Black social equality in
the twenty-first century.

18. Shelby, *We Who Are Dark*, 241.
19. Compare, e.g., Ariela J. Gross, *What Blood Won't Tell: A History of Race on Trial in
America* (Cambridge, MA: Harvard University Press, 2008) 5–7; Sonya G. Chen
and Christian Hosam, "Claire Jean Kim's Racial Triangulation at 20: Rethinking
Black-Asian Solidarity and Political Science," *Politics, Groups, and Identities* 10,
no. 3 (2022): 455–60, https://doi.org/10.1080/21565503.2022.2044870; Neil Foley,
"Becoming Hispanic: Mexican Americans and the Faustian Pact with Whiteness,"
in *Reflexiones 1997: New Directions in Mexican American Studies*, ed. Neil Foley
(Austin: Center for Mexican American Studies, University of Texas at Austin,
1998), 65 [53–70]; Mukoma Wa Ngugi, "African in America or African American?,"
Guardian, January 14, 2011, https://www.theguardian.com/commentisfree/cifamer
ica/2011/jan/13/race-kenya; and Eleanor Marie Lawrence Brown, "An Alternative
View of Immigrant Exceptionalism, Particularly as It Relates to Blacks: A Response
to Chua and Rubenfeld," *California Law Review* 103, no. 4 (2015): 993, 1003–06
[989–1017], http://www.jstor.org/stable/24758493. Out of moral sympathy despite
antebellum history, I do not include Native Americans among "people of color"
groups participating in self-serving rhetorical solidarity. Compare, e.g., Ryan P.
Smith, "How Native American Slaveholders Complicate the Trail of Tears Narrative,"
Smithsonian, March 6, 2018, https://www.smithsonianmag.com/smithsonian-inst
itution/how-native-american-slaveholders-complicate-trail-tears-narrative-180968
339/; and Nicole Chavez, "Native Americans Weren't Alone on the Trail of Tears.
Enslaved Africans Were, Too," CNN.com, May 9, 2021, https://www.cnn.com/2021/
05/09/us/tulsa-massacre-native-history-alaina-roberts/index.html.
20. See, e.g., Martin Luther King Jr., "The World House," in *Where Do We Go from
Here: Chaos or Community?* (New York: Harper and Row, 1967), 173–76 [167–91]; and
Yvonne D. Newsome, "International Issues and Domestic Ethnic Relations: African
Americans, American Jews, and the Israel-South Africa Debate," *International
Journal of Politics, Culture, and Society* 5, no. 1 (1991): 19–20 [19–48], https://www.
jstor.org/stable/20007027.
21. Compare, e.g., Paul Frymer, Dara Z. Strolovitch, and Dorian T. Warren, "New Orleans
Is Not The Exception: Re-Politicizing the Study of Racial Inequality," *Du Bois Review*
3, no. 1 (2006): 45–48 [37–57], https://doi.org/10.1017/S1742058X06060048.
22. John Lennon and Yoko Ono, "Woman Is the Nigger of the World," *Some Time
in New York City*, Apple, 1972. "The nigger" belongs to the realm of non-
persons. Lennon and Ono introduce the looking-glass notion of "slave to the

slaves"—implicitly contrasting "woman" and "nigger" females, with both subordinate to "nigger" males. In reality, American "Slaves" caste males could truly control nothing other than their own volition: the enslaved in common, without any protection under Anglo-American law, were "owned" by "Free White" persons of any gender. On "*nigger*" as a homegrown "term of exclusion" that "defined, limited, and mocked" Black Americans, compare David Pilgrim, "Nigger and Caricature," Jim Crow Museum (online), September 2001 (rev. 2012), https://www.ferris.edu/HTMLS/news/jimcrow/caricature/homepage.htm.

23. See again Jones-Rogers, "'I Belong to de Mistis.'" Compare, e.g., Emily Owens, "On the Use of 'Slave Mistress,'" Black Perspectives, August 21, 2015, https://www.aaihs.org/on-the-use-of-slave-mistress/.

24. **For the US Supreme Court's vacuous ideology-rhetoric on this point,** see *Dred Scott v. Sandford*, 60 U.S. 393, 403, 407 (1857).

25. Compare, e.g., Margaret Levi and Laura Stoker, "Political Trust and Trustworthiness," *Annual Review of Political Science* 3, no. 1 (2000): 476–77 [475–507], https://doi.org/10.1146/annurev.polisci.3.1.475.

26. Compare, e.g., Peniel Joseph, "Coates vs. West: Feud Has Its Roots in Long Tradition of Black Intellectuals," CNN.com, December 21, 2017, https://www.cnn.com/2017/12/21/opinions/west-coates-feud-black-intellectual-tradition-joseph-opinion.

27. **On debate surrounding religion scholar Eddie S. Glaude Jr.'s declaration that "the Black Church is dead"**—in that the "idea of this venerable institution as central to black life and as a repository for the social and moral conscience of the nation has all but disappeared"—compare Samuel G. Freedman, "Call and Response on the State of the Black Church," *New York Times*, April 16, 2010, https://www.nytimes.com/2010/04/17/us/17religion.html.

28. See, e.g., Andrea L. Dennis, "A Snitch in Time: An Historical Sketch of Black Informing During Slavery," *Marquette Law Review* 97 no. 2 (2013): 279–334, http://scholarship.law.marquette.edu/mulr/vol97/iss2/4.

29. Compare, e.g., Sean Illing, "Is There an Uncontroversial Way to Teach America's Racist History?," Vox, June 11, 2021, https://www.vox.com/policy-and-politics/22464746/critical-race-theory-anti-racism-jarvis-givens; and Bernard R. Boxill, "The Surrender to Injustice," in *Blacks and Social Justice*, rev. ed. (Lanham, MD: Rowman and Littlefield, 1992), 226–70. Compare also William Goodell, "Slaves Can Possess Nothing," in *The American Slave Code in Theory and Practice: Its Distinctive Features Shown by Its Statutes, Judicial Decisions, and Illustrative Facts* (New York: American and Foreign Anti-Slavery Society, 1853), iv, 89–104.

30. Compare, e.g., Roger D. Congleton, "Information, Special Interests, and Single-Issue Voting," *Public Choice* 69, no. 1 (1991): 39–49, https://doi.org/10.1007/BF00123853; and Benoit Denizet-Lewis, "For Gay Conservatives, the Trump Era Is the Best and Worst of Times," *New York Times Magazine*, January 11, 2019, https://www.nytimes.com/2019/01/11/magazine/gay-conservative-trump-era.html.

31. Compare, e.g., Leonard Pitts Jr., "One Gets Used to Being Shoved Aside When Voting While Black," *Miami Herald*, January 13, 2018, https://www.miamiherald.com/opinion/opn-columns-blogs/leonard-pitts-jr/article194558454.html; Adam Howard, "The 'Magical Negro' Trope Makes a Comeback in Two New Movies," NBCNews.

com, September 1, 2016, https://www.nbcnews.com/news/nbcblk/magical-negro-meme-makes-comeback-two-new-movies-n641296; and Alan Rappeport, "Hillary Clinton's 'All Lives Matter' Remark Stirs Backlash," *New York Times*, June 24, 2015, https://www.nytimes.com/politics/first-draft/2015/06/24/hillary-clintons-all-lives-matter-remark-stirs-backlash/.

32. See, e.g., Jeffrey Haas, "The FBI's Clandestine Operations," in *The Assassination of Fred Hampton: How the FBI and the Chicago Police Murdered a Black Panther* (Chicago: Lawrence Hill Books, 2010), 171–230; and John Leland, "Who Really Killed Malcolm X?," *New York Times*, February 6, 2020, https://www.nytimes.com/2020/02/06/nyregion/malcolm-x-assassination-case-reopened.html.

33. Public Enemy, "Welcome to the Terrordome," *Fear of a Black Planet*, Def Jam and Columbia, 1990, 03:06–10. Compare Randall Kennedy, "The Idea of the Sellout in Black American History," in *Sellout: The Politics of Racial Betrayal* (New York: Pantheon, 2008), 32–57.

34. Compare, e.g., Amy Roeder, "America Is Failing Its Black Mothers," *Harvard Public Health*, Winter 2019, https://www.hsph.harvard.edu/magazine/magazine_article/america-is-failing-its-black-mothers/; John Hudak, "Biden Should End America's Longest War: The War on Drugs," Brookings Institution, September 24, 2021, https://www.brookings.edu/blog/how-we-rise/2021/09/24/biden-should-end-americas-longest-war-the-war-on-drugs/; and William A. Darity Jr. and A. Kirsten Mullen, "A Program of Black Reparations," in *From Here to Equality: Reparations for Black Americans in the Twenty-First Century* (Chapel Hill: University of North Carolina Press, 2020), 256–70.

35. See Erick Trickey, "Chief Justice, Not President, Was William Howard Taft's Dream Job," *Smithsonian*, December 5, 2016, https://www.smithsonianmag.com/history/chief-justice-not-president-was-william-howard-tafts-dream-job-180961279/. See also *Deep Divisions in Americans' Views of Nation's Racial History—and How to Address It*, Pew Research Center, August 12, 2021, https://www.pewresearch.org/politics/2021/08/12/deep-divisions-in-americans-views-of-nations-racial-history-and-how-to-address-it/. Compare, e.g., Derrick Bell, "Foreword: The Civil Rights Chronicles," *Harvard Law Review* 99, no. 1 (1985): 4–83, https://doi.org/10.2307/1341120.

36. See, e.g., William Howard Taft, "The South and the National Government," intro. Andrew Carnegie (speech, North Carolina Society, New York, December 7, 1908), 10–12 [9–16]. Compare Ellora Derenoncourt et al. [Chi Hyun Kim, Moritz Kuhn, and Moritz Schularick], "Wealth of Two Nations: The U.S. Racial Wealth Gap, 1860–2020," National Bureau of Economic Research, Working Paper 30101, June 2022, 2–4 [1–40], http://dx.doi.org/10.3386/w30101.

Section 9

1. Malcolm X (with Alex Haley), *The Autobiography of Malcolm X* (New York: Grove Press, 1965), 289–90.

2. Malcolm X, *Autobiography*, 382.

3. Compare, e.g., Ta-Nehisi Coates, "'Better Is Good': Obama on Reparations, Civil Rights, and the Art of the Possible," *Atlantic*, December 21, 2016, https://www.thea tlantic.com/politics/archive/2016/12/ta-nehisi-coates-obama-transcript-ii/511133/.

Forgiveness on Layaway: Reflections from South Africa

1. See Abraham Lincoln, "Fourth Joint Debate, at Charleston, September 18, 1858," in Abraham Lincoln and Stephen A. Douglas, *Political Debates between Hon. Abraham Lincoln and Hon. Stephen A. Douglas, in the Celebrated Campaign of 1858 in Illinois* (Columbus: Follett, Foster, 1860), 136. Compare, e.g., Dina Gerdeman, "The Clear Connection Between Slavery and American Capitalism," *Forbes*, May 3, 2017, https://www.forbes.com/sites/hbsworkingknowledge/2017/05/03/the-clear-connect ion-between-slavery-and-american-capitalism/; and Terri L. Snyder, *The Power to Die: Slavery and Suicide in British North America* (Chicago: University of Chicago Press, 2015).

2. See, e.g., Frederick Douglass, *Oration, Delivered in Corinthian Hall, Rochester, by Frederick Douglass, July 5th, 1852* (Rochester, NY: Lee, Mann, 1852), 17–19; and Toni Morrison, "A Humanist View" (speech, Portland State University, May 30, 1975), Portland State Library Special Collections, 31:48–36:45 [06:50–43:01], accessed February 14, 2021, https://soundcloud.com/portland-state-library/portland-state-black-studies-1.

3. See, e.g., Michael S. Rosenwald, "Slave-Owning Presidents Become Targets of Protesters," *Washington Post*, June 23, 2020, https://www.washingtonpost.com/hist ory/2020/06/23/slave-owning-presidents-become-targets-protesters/; Julie Zauzmer Weil, Adrian Blanco, and Leo Dominguez, "More than 1,700 Congressmen Once Enslaved Black People. This Is Who They Were, and How They Shaped the Nation.," *Washington Post*, January 10, 2022, https://www.washingtonpost.com/history/interact ive/2022/congress-slaveowners-names-list/; and Paul Finkelman, *Supreme Injustice: Slavery in the Nation's Highest Court* (Cambridge, MA: Harvard University Press, 2018).

4. Compare, e.g., Saidiya V. Hartman, *Scenes of Subjection: Terror, Slavery, and Self-Making in Nineteenth-Century America* (New York: Oxford University Press, 1997); and David M. Chalmers, *Hooded Americanism: The History of the Ku Klux Klan*, 3rd ed. (1965; Durham, NC: Duke University Press, 1981).

5. See, e.g., Joe Heim, "Teaching America's Truth," *Washington Post*, August 28, 2019, https://www.washingtonpost.com/education/2019/08/28/teaching-slavery-schools/ ; and Adam Sewer, "The Fight over the 1619 Project Is Not About the Facts," *Atlantic*, September 23, 2019, https://www.theatlantic.com/ideas/archive/2019/12/historians-clash-1619-project/604093/.

6. See, e.g., Adeel Hassan and Jack Healy, "America Has Tried Reparations Before. Here Is How It Went," *New York Times*, June 19, 2019, https://www.nytimes.com/2019/06/ 19/us/reparations-slavery.html; and John Tateishi, *Redress: The Inside Story of the Successful Campaign for Japanese American Reparations* (Berkeley: Heyday, 2020).

7. Rep. Steve Cohen, introducing *Apologizing for the Enslavement and Racial Segregation of African-Americans*, HR 194, 110th Cong., *Congressional Record* 154 (July 29, 2008): H7226 [H 7224–27]. Compare, e.g., "Partners in Apartheid: U.S. Policy on South Africa," *Africa Today* 11, no. 3 (1964): 2–17, http://www.jstor.org/stable/4187823.

8. See, e.g., Lee A. Harris, "'Reparations' as a Dirty Word: The Norm Against Slavery Reparations," *University of Memphis Law Review* 33, no. 2 (2003): 409–48, http://dx.doi.org/10.2139/ssrn.433020; and Tala Hadavi, "Support for a Program to Pay Reparations to Descendants of Slaves Is Gaining Momentum, but Could Come with a \$12 Trillion Price Tag," CNBC.com, August 12, 2020, https://www.cnbc.com/2020/08/12/slavery-reparations-cost-us-government-10-to-12-trillion.html. See also George Ingram, "What Every American Should Know About US Foreign Aid," Brookings Institution, October 15, 2019, https://www.brookings.edu/policy2020/votervital/what-every-american-should-know-about-us-foreign-aid/. Compare William A. Darity Jr. and A. Kirsten Mullen, *From Here to Equality: Reparations for Black Americans in the Twenty-First Century* (Chapel Hill: University of North Carolina Press, 2020).

9. Desmond Tutu, *No Future Without Forgiveness* (New York: Doubleday, 1999), 45.

10. Tutu, *No Future Without Forgiveness*, 49–51.

11. **Compare—on how "given vastly different starting conditions under slavery, racial wealth convergence would remain a distant scenario, even if wealth-accumulating conditions had been equal across the two groups [White versus Black Americans] since Emancipation"**—Ellora Derenoncourt et al. [Chi Hyun Kim, Moritz Kuhn, and Moritz Schularick], "Wealth of Two Nations: The U.S. Racial Wealth Gap, 1860–2020," National Bureau of Economic Research, Working Paper 30101, June 2022, 1 [1–40], http://dx.doi.org/10.3386/w30101.

12. Nick Smith, *I Was Wrong: The Meanings of Apologies* (New York: Cambridge University Press, 2008), 9–10.

13. Smith, *I Was Wrong*, 10.

14. **For an anti-retributivist account of justice, see** Erin I. Kelly, *The Limits of Blame: Rethinking Punishment and Responsibility* (Cambridge, MA: Harvard University Press, 2018).

15. Smith, *I Was* Wrong, 3.

16. See John Gramlich, "What the Data Says (and Doesn't Say) About Crime in the United States" Pew Research Center, November 20, 2020, https://www.pewresearch.org/fact-tank/2020/11/20/facts-about-crime-in-the-u-s/. See also Wendy Sawyer and Peter Wagner, *Mass Incarceration: The Whole Pie 2022*, Prison Policy Initiative, March 14, 2022, https://www.prisonpolicy.org/reports/pie2022.html; and Marc Mauer, "Long-Term Sentences: Time to Reconsider the Scale of Punishment," *UMKC Law Review* 87, no. 1 (2018): 113–31.

17. See Glenn Thrush, "Apologizing Isn't Easy for Congress," Politico, June 26, 2009, https://www.politico.com/story/2009/06/apologizing-isnt-easy-for-congress-024 233. Compare, e.g., William Howard Taft, "The South and the National Government," intro. Andrew Carnegie (speech, North Carolina Society, New York, December 7, 1908), 10–12 [9–16]; and Ta-Nehisi Coates, "Why Precisely Is Bernie Sanders Against

Reparations?," *Atlantic*, January 19, 2016, https://www.theatlantic.com/politics/arch
ive/2016/01/bernie-sanders-reparations/424602/.

18. Charles L. Griswold, *Forgiveness: A Philosophical Exploration* (New York: Cambridge
University Press, 2007), 149–50.

19. Griswold, *Forgiveness*, 174.

20. Griswold, *Forgiveness*, 69.

21. Griswold, *Forgiveness*, xvii.

22. Griswold, *Forgiveness*, 39.

23. See, e.g., Martha C. Nussbaum, *Anger and Forgiveness: Resentment, Generosity, Justice*
(New York: Oxford University Press, 2016); and Margaret Urban Walker, *Moral
Repair: Reconstructing Moral Relations after Wrongdoing* (New York: Cambridge
University Press, 2006).

24. Compare, e.g., William Park, "What Other Cultures Can Teach Us About Forgiveness,"
BBC.com, November 9, 2020, https://www.bbc.com/future/article/20201109-what-
other-cultures-can-teach-us-about-forgiveness.

25. See Margaret R. Holmgren, *Forgiveness and Retribution: Responding to Wrongdoing*
(New York: Cambridge University Press, 2012), 32. Compare, e.g., Thomas Brudholm,
Resentment's Virtue: Jean Améry and the Refusal to Forgive, foreword Jeffrie G.
Murphie (Philadelphia: Temple University Press, 2008).

26. Griswold, *Forgiveness*, 110.

27. See, e.g., Lisa Wong Macabasco, "Lynching Postcards: A Harrowing Documentary
About Confronting History," *Guardian*, March 8, 2022, https://www.theguard
ian.com/film/2022/mar/08/lynching-postcards-a-harrowing-documentary-
about-confronting-history; Paul Erasmus, *Confessions of a Stratcom Hitman*
(Johannesburg: Jacana Media, 2021); and Devlin Barrett, "FBI Pressured to Answer
for Domestic-Spying Program Tied to Black Panther Fred Hampton's Killing in 1969,"
Washington Post, May 4, 2021, https://www.washingtonpost.com/national-security/
fred-hampton-black-panthers-fbi-surveillance/2021/05/04/2b12f826-acd7-11eb-
b476-c3b287e52a01_story.html.

28. Griswold, *Forgiveness*, 110–11.

29. Tutu, *No Future Without Forgiveness*, 54–55.

30. See, e.g., Mahmood Mamdani, "Amnesty or Impunity? A Preliminary Critique of
the Report of the Truth and Reconciliation Commission of South Africa (TRC),"
Diacritics 32, no. 3–4 (2002), 56–58 [32–59], http://dx.doi.org/10.1353/dia.2005.0005.

31. Walker, *Moral Repair*, 214–15.

32. See James L. Gibson, *Overcoming Apartheid: Can Truth Reconcile a Divided Nation?*
(New York: Russell Sage Foundation, 2004), 132–33.

33. Walker, *Moral Repair*, 208.

34. Archbishop Desmond Tutu and Truth and Reconciliation Commission of South
Africa, *Truth and Reconciliation Commission of South Africa Report*, 5 vols.
(London: Macmillan, 1999).

35. See Gibson, *Overcoming Apartheid*, 133.

36. Pippa de Bruyn, "Truth + Guilt + Apology = Reconciliation?," in Pippa de Bruyn,
Frommer's South Africa, 6th ed. (Hoboken, NJ: Wiley, 2010), 12–13. **On the**

circumstances of European colonialism in South Africa before apartheid officially commenced in 1948, see Facing History and Ourselves, "Introduction: Before
Apartheid," August 3, 2018, https://www.facinghistory.org/resource-library/intro
duction-apartheid.

37. See de Bruyn, "Reconciliation?," 13.

38. See Antony Sguazzin, "S. Africa Wealth Gap Unchanged Since Apartheid, Group
Says," Bloomberg.com, August 4, 2021, https://www.bloomberg.com/news/artic
les/2021-08-04/apartheid-legacy-maintains-south-african-wealth-gap-group-
says#xj4y7vzkg. See also Katy Scott, "South Africa Is the World's Most Unequal
Country. 25 Years of Freedom Have Failed to Bridge the Divide," CNN.com, May 10,
2019, https://www.cnn.com/2019/05/07/africa/south-africa-elections-inequality-
intl/index.html.

39. Griswold, *Forgiveness*, 153.

40. Compare, e.g., Reuters, "Germany Apologises for Colonial-Era Genocide in
Namibia," May 28, 2021, https://www.reuters.com/world/africa/germany-officially-
calls-colonial-era-killings-namibia-genocide-2021-05-28/.

41. On the shortcomings of color-conscious "justice through subterfuge," see Bernard
R. Boxill, "The Surrender to Injustice," in *Blacks and Social Justice*, rev. ed. (Lanham,
MD: Rowman and Littlefield, 1992), 234–36 [226–70].

Index

For the benefit of digital users, indexed terms that span two pages (e.g., 52–53) may, on occasion, appear on only one of those pages.